Biological Basis and Therapy of Neuroses

Editor

Ashoka Jahnavi Prasad, D.Sc., M.B., MRC Psych.
Mental Health Research Institute of Victoria
Royal Park Hospital
Parkville, Victoria, Australia

CRC Press
Taylor & Francis Group
Boca Raton London New York

CRC Press is an imprint of the
Taylor & Francis Group, an **informa** business

First published 1988 by CRC Press
Taylor & Francis Group
6000 Broken Sound Parkway NW, Suite 300
Boca Raton, FL 33487-2742

Reissued 2018 by CRC Press

Library of Congress Cataloging-in-Publication Data

Biological basis and therapy of neuroses.

Includes bibliographies and index.
1. Neuroses--Treatment.   2. Biological psychiatry.
I. Prasad, Ashoka Jahnavi.   [DNLM: 1. Biological
Psychiatry.   2. Neurotic Disorders--therapy.
WM 170 B615]
RC530.B56   1988   616.85'2   87-27743
ISBN 0-8493-4899-4

ISBN 13: 978-1-315-89114-9 (hbk)
ISBN 13: 978-1-351-07024-9 (ebk)

Visit the Taylor & Francis Web site at http://www.taylorandfrancis.com and the
CRC Press Web site at http://www.crcpress.com

PREFACE

Like "hysteria", the term "neurosis" has aroused considerable controversy ever since its coinage by Cullen (1769) in Edinburgh. His definition was rather wide and included diverse entities, e.g., epilepsy, apoplexy, asthma, mania, anxiety, etc.

However, it would seem that the concept has not stood the test of time, and has had varied connotations over different periods. In colloquial usage, it connotes a rather heterogeneous group of seemingly unrelated disorders. The biggest textbook in psychiatry (viz. *Comprehensive Textbook of Psychiatry* by Kaplan, Freedman, and Sadock) defines it as "mental disorder characterised primarily by anxiety which may either be directly experienced and dominate the clinical picture or be unconsciously controlled and modified by various psychological mechanisms to produce other subjectively distressing and ego-alien symptoms. They are not accompanied by gross distortions of reality of severe personality disorganization; neurotic disorders *have no organic basis,* are relatively persistent and are treatable."

Paradoxically, it is the term "neurosis" more than perhaps any other which has the implication of organic basis. Possibly the popular but erroneous belief mentioned earlier impeded systematic biological research in this area. On careful scrutiny, it becomes obvious that there is considerable evidence for a biological basis to these hitherto ignored conditions.

To the best of my knowledge, this is the first endeavor to present this information in a systematic manner. I would consider our efforts successful if this book fulfills the function of generating further research in this interesting but relatively under-researched area.

Without the help of my colleagues who enthusiastically participated in this venture, this book would not have seen the light of day. To all of them, I owe a deep sense of gratitude. Prof. Merton Sandler, a past president of the British Association of Psychopharmacology and a leading authority in this field, provided constant encouragement. I am deeply grateful to him for writing the foreword. I should like to place on record my sincere appreciation to CRC Press; its President, Mr. B. J. Starkoff; and Ms. Marsha Baker for their support while this volume was under preparation. It has been a pleasure collaborating with them. And last, but certainly not least, I must thank Gillian Cochrane, Anna Stone, and Susan Schiltz for their secretarial assistance.

I would welcome comments from readers about the contents. If the exercise proves successful, it would be a tribute to all the contributors. However, I must bear sole responsibility and crave indulgence for any shortcomings which may have crept in inadvertently.

Ashoka Jahnavi Prasad

FOREWORD

Over the past 25 years, we have witnessed a quiet revolution in the theory and practice of psychiatry. The cult of personality — what we might call divination by introspection — fights a rear guard action. Slowly and hesitantly, the study of psychopathology has succumbed to the scientific method, and what is now termed biological psychiatry marches inexorably forward. But wait! Are there not pockets of resistance? Few would now deny the essential biological basis of the affective disorders, with their somatic features and biological markers; and although *direct* evidence is scanty, the schizophrenic syndromes too march triumphantly foward on a broad biological front, with the clinical response to neuroleptics as their battle standard. It is the neuroses which, until very recently, have been stuck in the psychoanalytical mud! And, as the reader will see in Dr. Prasad's excellent and imaginative compilation, even this last bastion is beginning to crumble.

Dr. Prasad has recruited a powerful and enthusiastic team of biological psychiatrists to describe for us this final onslaught. They do not underestimate the obstacles in their path, but neither do they find them insuperable. Continuous optimism is the watchword, and yet the story is powerful and compelling. Who can doubt that careful and painstaking application of the scientific method to the problem of the neuroses will lead to final therapeutic victory.

Prof. Merton Sandler M.D., F.R.C.P., F.R.C. Path., F.R.C.Psych.
Queen Charlotte's Hospital
London, England

THE EDITOR

Ashoka Jahnavi Prasad, M.D., is presently in charge of the Section on Psychopharmacology at the Mental Health Research Institute of Victoria in Melbourne, Australia. He also holds Consultant Psychiatrist positions at the Royal Park Hospital and the Prince Henry's Hospital within the same city, in addition to an academic position with the Department of Psychological Medicine, Monash University, Clayton, Victoria.

Dr. Prasad graduated in medicine in 1976 from the University of Kanpur, India and after holding a brief residency position in New Delhi, moved to the Republic of Ireland for a year before his move to the Royal Edinburgh Hospital in Edinburgh, Scotland, one of the leading centers for academic psychiatry, where he completed most of his residency in psychiatry. After obtaining his membership of the Royal College of Psychiatrists, he took up a research appointment with the University of Leeds in England where he was also briefly a Tutor in Psychiatry.

Very soon, he was offered the first Kate Stillman Fellowship with the University of London where he worked for his doctorate under Prof. Merton Sandler, a leader in the field of psychopharmacology. He then worked as a Consultant Psychiatrist at Whipps Cross and Claybury Hospitals in London before moving to his present position.

Among other distinctions, Dr. Prasad was the recipient of the Murphy award in 1986 and was the Catherine Waterman Visiting Professor for 1987. He has also published about 100 scientific papers and abstracts in different journals and sits on the International Editorial Board of the *Journal of Psychiatric Research,* a leading journal in the field, and is on the Board of Assessors of several other journals.

Dr. Prasad has taken an active interest in both postgraduate and undergraduate medical education, serving as an examiner to several universities, and has lectured widely all over the world. He has presented papers in over 25 international meetings, many of them as the Chairperson. He has also written a number of articles on health issues in national papers and appeared on television.

Apart from psychiatry, Dr. Prasad's other academic interests are medical anthropology and aviation medicine and he holds postgraduate qualifications in both subjects.

CONTRIBUTORS

Thomas R. E. Barnes, M.B., B.S., M.R.C. Psych.
Senior Lecturer in Psychiatry
Charing Cross and Westminster Medical
School
London, England

Gary Bell, B.A., M.B., B.S., M.R.C. Psych.
Lecturer in Psychiatry
University College and Middlesex School
of Medicine
London, England

Jonathan Davidson, M.D.
Director of Anxiety Disorders Program
Department of Psychiatry
Duke University Medical Center
Durham, North Carolina

Sanjay Dubé, M.B., B.S., M.D.
Director
Affective Disorders Unit
Lafayette Clinic
Detroit, Michigan

Nigel F. S. Hymas, M.A., M.R.C. Psych.
Senior Registrar
Goodmayes Hospital
Ilford, Essex, and
Department of Psychological Medicine
St. Bartholomew's Hospital
London, England

Cornelius Katona, M.A., M.B., B.Chir., M.R.C. Psych.
Senior Lecturer in Psychiatry
University College and Middlesex School
of Medicine
London, England

M. S. Keshavan, M.B., B.Sc., M.D., M.R.C. Psych., M.N.A.M.S.
Assistant Professor
Department of Psychiatry
Wayne State University
Detroit, Michigan

Peter F. Liddle, B.M., B.Ch., Ph.D., M.R.C. Psych.
Senior Lecturer in Psychiatry
Charing Cross and Westminster Medical
School
London, England

Rahul Manchanda, M.B.B.S., M.D., M.R.C. Psych., F.R.C.P.(C)
Assistant Professor
Department of Psychiatry
University of Western Ontario
London, Ontario, Canada

Harold Merskey, D.M.
Professor
Department of Psychiatry
University of Western Ontario
London, Ontario, Canada

Ashoka Prasad, D.Sc., M.B., M.R.C. Psych.
Director of Psychopharmacology
Mental Health Research Institute
Royal Park Hospital
Melbourne, Australia

Daniel Rogers, M.A., M.R.C., M.R.C. Psych.
Consultant Neuropsychiatrist
Burden Neurological Hospital
Bristol, Avon, England

Natraj Sitaram, M.D.
Professor
Department of Psychiatry
Wayne State University
Lafayette Clinic
Detroit, Michigan

**Janet Treasure, Ph.D., M.B.B.S.,
M.R.C.P., M.R.C. Psych.**
Lecturer in Psychiatry
Department of Psychiatry
Institute of Psychiatry
London, England

V. K. Yeragani, M.D.
Assistant Professor of Psychiatry
Lafayette Clinic
Wayne State University
Detroit, Michigan

TABLE OF CONTENTS

Chapter 1

NEUROBIOLOGY OF ANXIETY

M. S. Keshavan and V. K. Yeragani

TABLE OF CONTENTS

My limbs quail, my mouth goes dry,
my body shakes, and my hair stands on end.
The bow gandiva slips from my hand
and my skin is dry all over;
I am not able to stand steady
and my mind is reeling.

Bhagavadgita 1, 28-30, 600 B.C.[1]

Sigmund Freud referred to the problem of anxiety as a nodal point at which the most serious and important questions converge[2] and a riddle "whose solution must cast a floodlight on our whole mental existance." Concepts of anxiety have played a central role in all major theories of abnormal behavior developed over the past century. However, major discoveries in such diverse areas as genetics, psychopharmacology, and neurophysiology made in the last two decades have provided valuable insights into the origin, mechanisms, manifestations and treatment of anxiety disorders. In the light of these new data, concepts of anxiety states are being radically revised. This chapter is an attempt to review the current status of the neurobiological substrate of anxiety states.

I. NOSOLOGY OF ANXIETY STATES — A HISTORICAL OVERVIEW

Anxiety is a universal human experience, characterized by a fearful anticipation of an unpleasant event in the future. It is a central feature of most psychiatric illnesses. The experience of anxiety has been known for centuries. Perhaps the earliest clear description of anxiety may be seen in Arjuna's revelation of his reaction to the threat of battle in the Indian epic, Mahabharata (600 B.C.).[1] All through history, clinical descriptions of anxiety have been remarkably consistent. However, the concept of anxiety has undergone radical changes. A century ago, neurosis was thought to be a single disorder associated with a lack of moral fiber and due to degenerations of the nervous system. Since then, there has been a steady splitting of the condition into categories defined by the clinical manifestations. Hysteria was the first entity to be separated from the main body of neuroses. Freud and Hecker were responsible for the separation of anxiety neuroses as a distinct entity;[2,3] they delineated its somatic and psychological aspects, and gave the first clear description of panic attacks.

In the early 1960s Klein found that the pharmacological treatment of panic disorders (PD) differs from that of generalized anxiety disorder (GAD) in that the former responds specifically to tricyclic antidepressant drugs.[4] This, and the observation discussed in detail later, that PD and GAD may be genetically different, have strongly suggested that these two were different disorders.

During the early half of this century, most practicing psychiatrists accepted a simple psychoanalytically oriented diagnostic scheme for anxiety disorders, "anxiety neurosis", representing the simplest undisguised expression of neurotic disorder; and "anxiety hysteria", in which the anxiety was concealed due to the utilization of several defense mechanisms resulting in phobic states. The first edition of the Diagnostic and Statistical Manual of the American Psychiatric Association (DSM-I) adopted the terminology of "reaction", in keeping with the then-prevailing meyerian approach.[5] Thus, anxiety neurosis became anxiety reaction, and anxiety hysteria became phobic reaction. The second edition, (DSM-II), along with the International Classification of Diseases (9th revision), relied heavily on the concept of neurosis.[6] Causal assumptions were integral to the concept of neurosis. Anxiety was noted as the chief characteristic of neurosis, and could be expressed directly (anxiety neurosis) or indirectly by various psychological mechanisms (phobic, hysterical, etc.).

The third revision of the Diagnostic and Statistical Manual (DSM-III) represents a radical departure from the older taxonomy, as it bases itself on empirical clinical observations and not on causal assumptions.[7] What was termed a "reaction" in DSM-I and a "neurosis" in DSM-II or ICD-9 has become a "disorder" in DSM-III. The anxiety disorders are divided into phobic and anxiety states; the anxiety states comprise four categories: (1) Panic Disorder (PD), (2) Generalized Anxiety Disorder (GAD), (3) Obsessive Compulsive Disorder, and (4) Post-traumatic Stress Disorder.

Panic Disorder is defined as recurrent anxiety attacks manifested by sudden onset of intense apprehension, fear, or terror, often with feelings of impending doom. Common symptoms are shortness of breath, chest pain or discomfort, palpitations, choking, or smothering sensations, dizziness, vertigo, unsteadiness, feelings of unreality, paresthesias, hot/cold flashes, sweating, fainting, trembling, and fear of dying, or going crazy. The attacks usually last for minutes. More rarely they last hours. Varying degrees of nervousness and apprehension may develop between attacks.

Generalized Anxiety Disorder is characterized by a persistent generalized anxiety of at least a month's duration, without the specific symptoms associated with panic, phobic, or obsessive compulsive disorders. Generally there are signs of motor tension, autonomic hyperactivity, apprehensive expectation, and vigilance and scanning. Mild depressive symptoms are commonly associated with this disorder.

In this chapter, we have confined ourselves to a discussion of the neurobiological basis of panic and generalized anxiety disorders.

II. GENETIC FACTORS IN ANXIETY

A. Role of Genetic Factors in Anxiety Disorders

It has long been suspected that anxiety disorders are familial. Beard found that neurasthenia was largely a hereditary disorder.[8] A family history of nervous disorder was present in nearly half of patients with Da Costa's Syndrome, and in over two thirds of patients with neurocirculatory asthenia.[9,10] Earlier family studies showing increased morbid risk of anxiety neurosis in the first degree relatives of patient probands have been confirmed recently by a study using explicit diagnostic criteria.[11-13] Two subsequent family studies where the first degree relatives were personally interviewed have revealed discrepant concordance rates,[14,15] 41% and 8%, respectively. These differences may have been related to differing severities of the disorder in probands.[16] Harris et al. found a morbidity risk of 33% for anxiety in the first degree relatives of probands with panic disorder, compared to 15% in relatives of controls.[17] PD accounted for the majority of anxiety disorders in the relatives, and GAD was not more prevalent than among controls. Twin studies have shown higher concordance rates for anxiety neurosis between MZ twins than between DZ twins.[18,19] Torgerson showed that while 31% of MZ cotwins of PD probands had PD, none of the DZ cotwins did so.[20] As for GAD, they observed that the concordance rate was almost the same in MZ as in DZ pairs — 17% compared to 20%. Thus, hereditary factors are probably less important in the development of GAD, and PD and GAD may in fact be etiologically unrelated.

B. Are Genetic Factors Specific to Anxiety Disorders?

The question has been raised whether the genetic basis of anxiety disorders is specific. A genetic overlap with affective disorders has been suggested because of the beneficial effects of antidepressive drugs on panic disorder. Leckman et al.[21] have found the relatives of probands with both major depression and panic disorders have a higher frequency of major depression as well as PD than relatives of probands with only major depression. Torgerson[20] found a higher frequency of affective disorders among relatives, though not among cotwins of probands with different anxiety disorders. Cloninger et al.[15] did not find depression among relatives of probands with anxiety disorders to be more frequent than the general population. Thus, the genetic relationship between affective and anxiety disorders remains unsettled. There is some evidence of an excess of alcoholics among the relatives of probands with anxiety states,[13] but few studies have further explored this overlap. Although physical disorders such as mitral valve prolapse may occasionally mimic anxiety states, no genetic association with this disorder has been found.

C. How are Anxiety Disorders Transmitted?

Based on segregation analysis in 19 families with panic disorder, Crowe et al.[14] have suggested a dominant gene for this disorder. However, two findings suggest that single gene inheritance may be less likely. First, the prevalence of anxiety disorders is higher in the relatives of probands with more severe and chronic disorders than in the relatives of less severe and acutely ill patients.[14] Secondly, family studies show a risk for the siblings of probands of 8 to 15% if parents are unaffected, rising to 25% when one and 40% when two parents are affected.[22] These findings are compatible with a continuous polygenic or multifactorial liability to anxiety disorders.

III. NEUROPHARMACOLOGICAL AND NEUROCHEMICAL ASPECTS OF ANXIETY

Panic anxiety is a dysphoric experience and often is the cause of substantial functional impairment. The description of panic attacks is similar to the reaction of normal people confronted with a life-threatening situation. Thus, one could argue that the panic anxiety experienced by the patients is not qualitatively different from the anxiety experienced by normal people. And, there may not even be a quantitative difference in the intensity of the symptoms of panic anxiety in patients and anxiety associated with life-threatening situations in normal people. Thus, it is difficult to understand the panic anxiety which is not associated with life-threatening situations. Several questions remain unanswered. Is panic anxiety primarily a physiological or a psychological experience? How similar are the individual panic episodes in the same patient and among different patients? Is there a common diathesis or a biological substrate that is necessary for the development of panic anxiety?

Cannon and his colleagues established the first biologic model of fear based on their experiments between 1930 and 1935.[23] When young, physically healthy cats were held on their backs and confronted by a barking dog, the cats bared their teeth, laid back their ears, hissed, and made attempts to escape. Cannon showed that the tachycardia, piloerection, and pupillary dilatation accompanied by these intense behavioral changes were associated with the release of epinephrine from the adrenals and could be produced by i.v. administration of epinephrine. In 1935, Lindemann and Finesinger reported that adrenaline and methacholine, a cholinergic drug,[24] were able to induce anxiety attacks in 16 out of 20 psychoneurotic patients, and since then several lines of investigation have thrown new light on the neuropharmacological basis of panic attacks.

A. Lactate Model

Cohen and White and their associates were the first to show that patients with anxiety neurosis had higher plasma lactate levels than controls during moderate exercise.[25,26] This finding was confirmed by other investigators, who also attempted to control for the effects of prior physical condition of the subjects.[27-29]

In 1967, Pitts and McClure found that an infusion of 10 mℓ/kg of .5 M sodium lactate produced symptoms of panic anxiety in 13 of 14 anxiety neurotics, but only 2 of 16 normal controls.[30] Several other investigators have replicated the findings of Pitts and McClure using patients with similar diagnostic criteria.[31-39] (See Table 1.)

Table 1

**PERCENTAGES OF PATIENT AND CONTROL PANICKERS DURING
LACTATE AND PLACEBO INFUSIONS IN SOME STUDIES**

| | | | Panic attacks (%) | | | | |
| | | | Panic patients | | Controls | | |
Study	Patients (n)	Controls (n)	P	L	P	L	Mean time to panic (min)
Pitts and McClure (1967)[30]	14	10	0	93	0	20	1 — 2
Fink et al. (1971)	5	4	0	100	0	25	8 — 12
Kelly et al. (1971)[34]	20	10	5	80	0	10	12
Appleby et al. (1981)	25	15	16	64	0	0	15
Liebowitz et al. (1984)[36]	43	20	7	72	0	0	12
Pohl et al. (1985)[39]	52	29	23	87	0	24	—

Note: P, placebo; L, lactate.

Tricylic and monoamine oxidase inhibitors (MAOIs) are effective for the treatment of spontaneously occurring panic attacks.[40,41] These agents also blunt the panicogenic effects of lactate when successfully treated patients are reinfused with sodium lactate.[34,36,37] Carr et al.[42] found that treatment of panic disorder patients with alprazolam minimized the lactate-induced panic anxiety.

Fyer et al.[43] found that the lactate sensitivity reemerges in remitted patients who had been off medication for varying periods of time. There are studies suggesting that the provocation of panic anxiety is specific to panic attack patients. Cowley et al.[44] have found that the frequency of lactate-induced panic attacks was similar in patients with primary depression and secondary panic attacks and in patients with panic disorder or agoraphobia with panic attacks without depression. This may imply that lactate may provoke panic anxiety in patients who have panic attacks, whether or not they have a primary panic disorder or agoraphobia with panic attacks. The frequency of lactate-induced panic in patients with obsessive-compulsive disorder and social phobia is significantly lower than panic patients and similar to those of control subjects.[45,46]

Liebowitz et al.[47] reported that lactate-induced panic attacks were regularly accompanied by elevated heart rate and lowered PCO_2 and bicarbonate levels. There were less consistent elevations in plasma norepinephrine and cortisol levels. There were also no consistent changes in lactate, pyruvate or epinephrine levels, pH, phosphate level or systolic blood pressure at the point of lactate-induced panic. Rainey et al.[48] reported that plasma lactate levels increased at a faster rate during the infusions in the panic disorder patients than the control subjects before treatment and at the same rate as the controls after treatment with imipramine. Carr et al.[42] found no pituitary-adrenal activation during lactate-induced anxiety.

1. Possible Mechanisms of Lactate-Induced Panic Anxiety

Pitts and McClure hypothesized that lactate elevations lead to complexing of ionized calcium at the surface of excitable membranes.[30] They speculated that chronic overproduction of epinephrine, overactivity of the central nervous system, a defect in aerobic or anaerobic metabolism resulting in excess lactate production, a defect in calcium metabolism, or some combination of these may underlie the pathophysiology of panic anxiety. Grosz and Farmer argued,[49,50] however, that induction of metabolic alkalosis was the more likely precipitant of lactate-induced panic, claiming that the lactate doses administered by Pitts and McClure were insufficient to significantly depress ionized calcium. They also studied normal subjects and showed that bicarbonate infusion produced symptoms similar to those induced by lactate. However, the findings of Liebowitz et al.[47] do not support that alkalosis alone is panicogenic.

Pitts and Allen later found that infusion of ethylene diaminetetraacetic acid (EDTA),[51] a calcium chelating agent, produced symptoms of tetany in panic patients, but did not induce panic anxiety. They also argued that panic patients are hypersensitive to β-adrenergic agonists. This is supported by the findings of Rainey et al.[37] who were able to induce panic attacks using infusions of isoproterenol. However, Gorman et al.[52] found that pretreatment with propranolol did not prevent lactate-induced panic despite evidence of peripheral beta-adrenergic blockade.

Other postulated mechanisms of lactate-induced panic include central noradrenergic stimulation,[53] increased central carbon dioxide sensitivity, and chemoreceptor hypersensitivity.[47,54,55] However, Liebowitz et al.[56] argue that a peripheral catecholamine surge may not be the mechanism by which lactate causes panic, although elevated plasma epinephrine may play a predisposing role. They also suggest that lactate infusion may cause panic by alteration of the NAD+ to NADH ratio, shifting the ratio in favor of NADH. Rainey et al.[48] propose several possibilities for an abnormal aerobic metabolism during lactate-induced panic anxiety, as a result of decreased oxygen availability, alterations in the activity of enzymes of intermediary metabolism, or an increase in aerobic metabolism.

Ackerman and Sachar hypothesized that peripheral physiologic changes may generate anxiety via a conditioned emotional response or a learned perceptual association.[57] They argued that physiologic abnormalities were neither necessary nor sufficient for the production of an anxiety-state. Lactate infusions sometimes produce only bodily sensations of panic in patients.[34] Bonn et al.[58] have used repeated lactate infusions as a desensitization technique. Thus, it is also possible that some of the somatic symptoms during lactate infusions could trigger central anxiety in patients, a hypothesis that is consonant with the James-Lange theory of anxiety.[59]

2. Specificity of Lactate-Induced Panic — Methodological Issues

Maragraf et al.[60] have extensively reviewed the issue of specificity of lactate-induced panic in panic disorder patients. They raise the point that some of the studies have not used control groups and that most studies did not use a double-blind design. Their criticism about specific biochemical and physiological changes during lactate infusions is partly valid. They also raise the issue of the similarity of lactate-induced panic attacks to naturally occurring ones. In fact, there are some studies which illustrate the similarity of the lactate panic attacks to naturally occuring ones.[36,37]

Do subjects who panic with lactate differ from those who do not in regard to their baseline anxiety and autonomic measure? Liebowitz et al.[47] reported that the patients who panicked with lactate had significantly higher pre-infusion heart rate and systolic blood pressure compared to the nonpanickers, and that the lactate panickers also were anxious before the infusions.[36] However, Yeragani et al.[61] have found no significant differences in the preinfusion heart rates of first infusion panickers and nonpanickers in a double-blind placebo-controlled multiple infusion study designed to induce panic anxiety. This issue needs to be studied in more detail.

On the whole, the large amount of data on lactate infusions indicate that it is a valid experimental model to study panic anxiety. Future studies should aim at further delineating the neurochemical and metabolic concomitants of lactate-induced panic. It will also be important to determine more carefully whether lactate-induced panic is state or trait dependent. The follow-up of normal subjects who panic with lactate may also yield important information regarding the predictive value of the lactate test for the development of pathological anxiety.

B. Carbon Dioxide as an Anxiogenic Agent

Carbon dioxide (CO_2) has been reported to provoke symptoms of anxiety in patients with neurocirculatory asthenia,[26] irritable heart,[62] and neurotic anxiety states.[63] Many of these patients probably meet DSM III criteria for panic disorder or agoraphobia. Also, there is a similarity between symptoms of hyperventilation and panic attacks.[64-66]

Gorman et al.[54] reported that 7 of 12 patients with panic disorder or agoraphobia with panic attacks and none of four controls experienced panic attacks during 15 min of breathing 5% CO_2. Only three patients panicked on room air hyperventilation and this low rate of panic anxiety with hyperventilation occurred despite the development of metabolic alkalosis. Woods et al.[67] compared the ventilatory response to CO_2 in 14 medication-free patients with agoraphobia and panic attacks and 23 healthy subjects. Ventilatory response to CO_2 was similar in patients and controls, and they suggest that the patients do not have an abnormal chemoreceptor sensitivity. Anxiety ratings increased markedly during rebreathing, both in patient and controls, but the anxiety increases were significantly greater in patients. Alprazolam treatment in eight patients markedly attenuated anxiety increased during rebreathing. Woods et al.[67] suggest the differences in anxiogenic sensitivity to CO_2 between patients and controls may be due to differences in the regulation of noradrenergic or other neurotransmitter systems.

Single breath inhalation of 35% CO_2 results in production of many peripheral symptoms of panic in normal subjects.[68] Greiz and Van den Hout describe a panic disorder patient who had severe panic anxiety on initial administration of 35% CO_2, and frequent repeated exposure to this stimulus led to habituation of the response and also amelioration of clinical symptoms.[69]

These findings suggest that some panic disorder patients may be more sensitive than normal controls to CO_2. CO_2 may affect the central nervous system (CNS) directly, but the exact mechanism of its panicogenic effect is still to be established. Systematic studies using CO_2 inhalation in patients to determine its metabolic and neurochemical effects before and after treatment with different antipanic agents are needed.

C. Caffeine Model

Caffeine, a xanthine derivative, has been described as the most widely used psychotropic agent. Several studies suggest a significant association between caffeine intake and scores of self-rated anxiety and/or depression in both college studies and psychiatric inpatients.[70-73] Caffeine intake, in doses above 600 mg/day has been reported to induce symptoms of "caffeinism", characterized by anxiety, nervousness, sleep disturbances, and psychophysiologic symptoms that may be indistinguishable from the symptoms of anxiety neurosis.[74] In addition, there are case reports implicating caffeine as the cause of clinical states resembling panic and generalized anxiety disorders.[75,76] Uhde et al.[77] reported "unequivocal panic" in two normal subjects given a single 720 mg dose of caffeine (approximately equivalent to 10 cups of percolated coffee).

Boulenger et al.[78] have reported that patients with panic anxiety disorder, but not affectively ill patients or normal controls, had levels of self-rated anxiety and depression that correlated with their degree of caffeine consumption. In addition, this self report survey suggested that patients with panic disorder had an increased sensitivity to the effects of one cup of coffee. This apparent sensitivity to caffeine was also documented by observation that patients with panic disorder reported the discontinuation of coffee intake due to untoward side effects than controls. Thus, panic anxiety patients may be more sensitive to caffeine. Lee et al.[79] also found that patients with panic disorder were much more likely than a control group of medical inpatients to be low consumers of caffeine (0 to 249 mg/day). These patients also showed increased sensitivity to the anxiety-provoking effects of a single cup of coffee. These studies suggest that, like sodium lactate, caffeine may have panicogenic effects.

Charney et al.[80] studied the effects of oral administration of caffeine (10 mg/kg) on behavioral ratings, somatic symptoms, blood pressure, and plasma levels of MHPG and cortisol in 17 healthy subjects and 21 patients meeting DSM III criteria for agoraphobia with panic attacks or panic disorder. Caffeine produced significantly greater increases in subject-rated anxiety, nervousness, fear, nausea, palpitations, restlessness, and tremors in the patients compared to healthy subjects. In the patients, but not the healthy subjects, these symptoms were significantly correlated with plasma caffeine levels. Of these patients, 71% reported that the behavioral effects of caffeine were similar to those experienced during panic attacks. Caffeine did not alter plasma MHPG levels in either the healthy subjects or the patients. Caffeine increased plasma cortisol levels equally in both groups.

There are certain methodological problems in studies of the caffeine model similar to those involved in the lactate studies. Charney et al.[80] used a single-blind strategy with placebo given on the first experimental day and caffeine on the second. The explanation for this was that the experience of an increase in anxiety would be expected to "unblind" the second test day. The expectation effects for patients and controls might also have been a factor in this study. Also, the panic ratings were based on the patients' subjective reports.

Several mechanisms have been hypothesized to be responsible for caffeine's anxiogenic effects and include inhibition of phosphodiesterase,[81] increased brain catecholamine activity,[82] antagonism of benzodiazepine and adenosine receptor function,[83,84] and a dysfunction in calcium-mediated neuronal excitability.[80]

D. Isoproterenol Model

There is evidence to suggest β-adrenergic hypersensitivity in some patients with panic anxiety. The racing and pounding heart beats, shortness of breath, rapid breathing, and chest pain suggest β-adrenergic activation. Isoproterenol hydrochloride, a β-adrenergic agonist, causes normal subjects to experience many somatic symptoms of panic, including shortness of breath and rapid pounding heart beats.[85] Patients with hyper-dynamic β-adrenergic state develop these symptoms when given low doses of isoproterenol.[86] Propranolol, a β-adrenergic blocker, is effective in the treatment of somatic symptoms of anxiety.[87]

Rainey et al.[37] reported that 29 out of 39 panic disorder patients developed symptoms of panic anxiety during isoproterenol infusions, while one out of 18 normal controls did. This was part of a double-blind placebo-controlled study designed to induce panic anxiety using isoproterenol and sodium lactate as the active pharmacological agents. The isoproterenol panic attacks were reported to be less severe compared to lactate attacks. Further, the majority of lactate and isoproterenol attacks were reported to be similar to spontaneously occurring panic attacks. In this study, Rainey et al. also report their findings on reinfusions after treatment. Five out of seven panic disorder patients had panic attacks during isoproterenol infusions before treatment with imipramine, and none out of seven after treatment. This further validates the use of isoproterenol model in studying panic anxiety in the laboratory. Since isoproterenol does not cross the blood-brain barrier well, the question of precisely how isoproterenol-induces panic anxiety remains unclear.

Nesse et al.,[88] in their study, compared the β-adrenergic receptor sensitivity in panic disorder patients with that of controls, using a series of bolus i.v. doses of isoproterenol hydrochloride. This study did not aim to induce panic anxiety. Patients had higher resting heart rates but showed a decreased heart rate response to isoproterenol compared to controls. This study does not support the hypothesis that panic disorder patients have β-adrenergic hypersensitivity and, in fact, this may be a subsensitivity of the β-adrenergic receptors in the patients.

Isoproterenol does seem to induce panic anxiety at least in a subgroup of panic disorder patients, and this may be due to mechanisms other than cardiac β-adrenergic sensitivity. One possibility is that patients may be more sensitive than controls to detect peripheral physiological changes and may associate the somatic symptoms due to β-adrenergic stimulation with panic anxiety.[89]

E. Yohimbine Model

Yohimbine, an alpha-2 adrenoceptor antagonist, can produce anxiety in psychiatric patients and normal volunteers.[90] In contrast, clonidine which decreases noradrenergic activity by activating alpha-2 adrenergic autoreceptors, seems to have some antianxiety action in anxious subjects.[91,92] Abrupt discontinuation of opiates, antidepressant drugs and clonidine have been associated with anxiety states and an increase in noradrenergic turnover as reflected by changes in plasma MHPG.[53,93,94]

A 20-mg oral dose of yohimbine causes robust increases in plasma free MHPG and small increases in systolic blood pressure, anxiety, and autonomic symptoms in normal subjects.[95] Charney et al.[96] have also reported that yohimbine produces significantly greater anxiety, palpitations, hot and cold flashes, restlessness, tremors, and piloerection in panic disorder patients than in normal controls. There was also a significant drug effect on plasma MHPG both in patients and controls. However, there was a significant patient control difference in MHPG response when patients with frequent panic attacks were compared to controls. MHPG correlated with anxiety ratings in low- but not high-frequency panickers. Charney et al.[96] suggest that this may be evidence of an increased noradrenergic sensitivity in anxiety states associated with panic and that this may be related to impaired presynaptic noradrenergic neuronal regulation in these patients.

Charney and Heninger[97] report that alprazolam produced a small but significant decrease in baseline MHPG and blunting of the yohimbine-stimulated increase. Imipramine induced a significant fall in baseline MHPG but had no effect on the MHPG response to yohimbine stimulation.[98] Charney et al.[99] also reported that both diazepam and clonidine significantly antagonized yohimbine-induced anxiety, but only clonidine significantly attenuated the yohimbine-induced increases in plasma MHPG, blood pressure and autonomic symptoms. This implies that diazepam reverses yohimbine-induced anxiety in a nonspecific way without affecting several physiological and biochemical indicators of noradrenergic activity in humans. Uhde et al.[100] have confirmed the findings of Charney et al. and have found that six of seven panic disorder patients experienced profound anxiety following oral administration of 20 mg of yohimbine. Clonidine effectively blocked these yohimbine responses.

Stimulation of Locus Coeruleus (LC), a brain stem nucleus that contains the majority of norepinephrine-containing neurons in the CNS, produces anxiety-like behavior in the monkey.[101] Several anxiolytic drugs decrease LC unit activity, and yohimbine which activates the LC, produces anxiety in humans.[90,95]

In summary, yohimbine appears to be a specific pharmacological agent with panicogenic properties. This effect can be observed at low doses and also in normal controls. Even though there is an increase in MHPG associated with an increase in subjective anxiety, the evidence for a differential increase of plasma MHPG levels between patients and controls is less convincing. An increase in plasma MHPG levels during anxiety states may be attributable to enhanced peripheral norepinephrine release, which may reflect central sympathetic outflow. Yohimbine-induced panic anxiety in panic disorder patients with frequent panic attacks is accompanied by a greater increase in plasma MHPG than in normal controls.[96] However, the baseline levels of plasma MHPG in panic disorder patients were not higher than in controls in several studies.[96,102] It is also important to note that lactate-induced panic and caffeine-induced panic or anxiety are not accompanied by increases in plasma-free MHPG.[42,100,80] Kaitin et al.[103] report LC stimulation made their patient "more comfortable" without any other subjective effects, in their studies on the effects of electrical stimulation of LC in a patient with cerebral palsy who had stimulation electrodes implanted in the vicinity of LC for the symptoms of spasticity. This is in sharp contrast to the findings in monkeys reported by Redmond et al.[101]

Thus, the evidence for an increase in plasma MHPG associated with panic anxiety is not convincing. This may be due to several factors. Plasma MHPG may be an unreliable index of CNS noradrenergic turnover because the major fraction of circulating plasma MHPG is derived from peripheral sources.[104] Factors such as diet, exercise, circadian rhythm, and psychological stress may have a substantial influence on noradrenergic turnover and MHPG levels.

F. The Benzodiazepine and GABA Receptor Model of Anxiety Disorders

The established superiority of benzodiazepines (BZs) as antianxiety drugs has generated considerable interest in their use as molecular probes for studying the biochemical and neurophysiological basis of anxiety disorders.[105] In 1977, two groups of investigators discovered the presence of high affinity saturable and stereospecific binding sites for BZs in the CNS[106,107] of a variety of species including man. Several groups have subsequently demonstrated the presence of BZ receptors using in vivo labeling techniques and have shown that the potencies of various BZs for displacing the specific binding of (3H) Diazepam and (3H) Nitrazepam in vivo is also highly correlated with their behavioral and clinical potencies.[108]

Even before the discovery of the BZ receptors, there was electro-physiological evidence that BZs potentiate the inhibitory actions of Gamma Aminobutyric Acid (GABA).[109] In 1978, Tallman and his colleagues showed that GABA increased the affinity of BZ for its receptor.[110] Subsequently, Costa et al.[111] reported that small permeable anions such as chloride also enhanced the affinity of BZ receptors. Conversely, BZs potentiate the effects of GABA.[112] It is now generally known that BZ receptors are structurally coupled with the GABA receptor and, along with an associated chloride channel, form a supramolecular receptor complex.[113] It has been suggested that the mechanisms of action of benzodiazepines and GABA may involve the unmasking of hidden binding sites or the induction of conformational changes in a portion of binding sites that was previously a nonfunctional structure.[91] Opening of the chloride channels directly limited to GABA receptors results in influx of chloride ions into the cell, and thus hyperpolarizes the neurons.[114]

Several drugs may produce their pharmacological effects through an action on the GABA-benzodiazepine-chloride ionophore complex. Barbiturates, for example, seem to interact directly with a chloride channel coupled to BZ receptors, resulting in an increase in the time of channel opening.[115] Ethanol, which also has anxiolytic effects, has been reported to increase BZ binding and thereby increase the action of GABA.[113]

1. Where in the CNS do Benzodiazepines Act?

Using an autoradiographic technique to actually visualize the distribution of BZ receptors in the brain,[116] it is possible to identify brain regions mediating the pharmacological effects of BZs. Studied in this way, BZ receptors are rather diffusely distributed. It has been suggested that the frontal, cortical, and hippocampal regions which have high densities of BZ receptors may mediate the anxiolytic effects of BZs.[117] Likewise, the BZ receptors in the amygdala may be the site of BZ neuroendocrine effects, and cerebellar BZ receptors probably mediate the ataxia produced by BZ in high doses; the effect of BZ on fear-induced immobility may be mediated by receptors in the substantia nigra,[118] and the BZ receptors in the superior colliculus may be involved in the control of visual and auditory attention.[119] The relatively small magnitude of differences between the different brain regions makes the above anatomical speculations at best tentative. However, this approach may prove useful in delineating the neuroanatomical pathways mediating BZ effects, and perhaps help to define the neuroanatomical correlates of anxiety disorders.

2. Benzodiazepine Antagonists

The observations of specific BZ receptors in the brain stimulated the search for brain's own anxiolytic or anxiogenic substances (i.e., naturally occurring ligands for the BZ receptor). In 1980, Braestrup et al.[120] identified a novel high-affinity BZ receptor ligand from human urine that was subsequently identified as beta-carboline-3-carboxylic acid ethylester (beta CCE). The initial impressions that this is an endogenous ligand have not been substantiated. However, the extremely high affinity of beta CCE for the BZ receptor has made it a very valuable tool for studying the BZ receptor. Several other beta carboline derivatives, as well as novel BZs (e.g., RO-15-1788) and chemically unrelated compounds (e.g., CGS-8216) have been discovered that antagonize all of the pharmacological actions of BZs.[121,122] Beta carboline derivates possess intrinsic pharmacological properties of their own, aside from BZ receptor blocking effects. Newer BZ receptor ligands like RO-15-1788 do not possess these properties, and hence appear to be relatively selective antagonists.

Beta-CCE produces dramatic behavioral, endocrine, and physiological effects in subhuman primates very reminiscent of the features of acute anxiety in humans.[123] These effects are prevented by pretreatment with BZs,[124] suggesting that this syndrome may represent a reliable pharmacological animal model of anxiety. Dorrow et al.[125] have recently found that a beta carboline derivative (FG 7/72) produced, in normal volunteers, symptoms identical to severe anxiety, with "inner tension, excitation, and sensations of physical disturbance." Benzodiazepines reversed these symptoms, supporting the validity of the "BZ-GABA receptor model" of anxiety.

3. Benzodiazepine Endogenous Ligands

Several naturally occurring compounds that bind to BZ receptors have been identified and proposed as the endogenous effectors. These include the purines, inosine and hypoxanthine, which have been isolated from brain tissue;[126] melatonin;[127] nicotinamide;[128] nepenthine;[129] and thromboxane A_2.[129] Some investigators have questioned even the requirement of an endogenous ligand on the basis that a protein that can affect the BZ binding site could itself interact during a BZ-induced conformational change in the receptor.[130] However, the search for a potential natural ligand for BZ receptors continues, in the hope that such knowledge will undoubtedly open up new insights into the biological basis of anxiety.

G. Opioids and Panic Anxiety

Spontaneous panic attacks may be associated with separation anxiety.[131] Single doses of morphine, in nonsedating doses, significantly reduced distress vocalizations in infant guinea-pigs and puppies separated from their normal social environments.[132] Naloxone, an opioid antagonist, reliably increased distress vocalizations in both chicks and guinea-pigs.[132] Theodora et al.[133] reported that naloxone attenuates both the anxiolytic and hypotensive actions of diazepam in man, suggesting an involvement of CNS endorphinergic systems in responses to anxiety-provoking situations. However, Liebowitz et al.[134] reported that naloxone did not produce panic attacks or alter responses to sodium lactate infusions, when 12 patients with panic attacks were challenged with i.v. naloxone alone or combined with sodium lactate.

IV. PSYCHOPHYSIOLOGICAL ASPECTS OF ANXIETY

There is no direct or single physiological measure of anxiety, which is a complex, largely subjective experience. What can be measured are physiological measures related directly to the somatic elements and indirectly to the subjective experience of anxiety. A physiological concept that underlies all affective states, and particularly such emotions as anger and panic, is arousal. Arousal is defined as an individual's level of behavioral activity ranging from relaxed sleep to emotional excitement.[135] A number of psychophysiological measures of arousal differentiate between anxious patients and normal controls. Anxious patients have higher mean forearm EMG levels, pulse rates, skin conductance and increased forearm blood flow.[135] Increasing anxiety is also associated with decreases in alpha activity in EEG, and there are differences in dominant alpha frequencies between anxious and normal subjects.[135] Spontaneous fluctuations in skin conductance are more frequent in anxious patients than controls.[136] The psychophysiological measures can be quite specifically related to the symptoms of anxiety. For example, Sainsbury and Gibson[137] have shown that patients with aches and pains in the limbs have high forearm EMGs and contrasted with patients with headaches who had high frontalis EMG levels.

Sleep is a state in which the physiological parameters of the individual can be measured without the potential confound of subjective situational variables. Ackner[138] has shown that anxious patients show greater drops in pulse rates with the onset of sleep than do controls. In a recent study, we have seen no differences in sleep heart rates between patients with primary anxiety and normal controls.[139] Interestingly, however, patients with coexisting panic attacks and depression had higher sleep heart rates than those with either disorder alone, or normal controls. Anxious patients also have poorer sleep efficiency and more frequent awakenings, and longer (more normal) Rapid Eye Movement (REM) latencies and REM percentage compared to depressives.[140,141]

Anxious patients show some evidence of impaired homeostatic processes in relation to external stimuli. Normal subjects soon lose a response of pupillary dilation after a pain stimulus has been discontinued, whereas in anxious patients the pupils remain dilated for some time. Anxious patients also habituate more slowly to discrete repeated stimuli than normals.[135] In states of severe anxiety, responses may even increase with repetition rather than decrease, a form of "recruitment", a possible mechanism of maintaining anxiety, once generated.

Kelly et al.[142] found that forearm blood flow and heart rate were significantly different between anxious patients and normal controls and that resting forearm blood flow correlated significantly with clinical and subjective ratings of anxiety. However, the patients had smaller stress-induced increases in forearm blood flow and heart rate. Roth et al.[143] have also recently found that patients with panic attacks had only weak and inconsistent differences in autonomic reactivity to, or recovery from, stimuli with diverse qualities of novelty, startlingness, intensity, or phobicity, even though the tonic levels of heart rate and skin conductance were higher among patients compared to controls. Yeragani et al.[61] found that the panic attack patients had an increase in their heart rates from baseline during a psychologically stressful situation, but this arousal was not different between those patients who went on to panic and those who did not during an infusion procedure designed to induce panic anxiety. Thus, the tonic levels of autonomic measures appear to be higher in anxious subjects, while the physiological reactivity to different stimuli may not be higher.

A. A Psychophysiological Model of Anxiety

How can we integrate the known facts about the factors involved in the generation of anxiety, and the subjective physiological and behavioral changes accompanying anxiety? A general model encompassing various observations and theories has been proposed by Lader (Figure 1).[135] According to this model, childhood experiences determine the extent to which a person scans the environment for threat and appraises potentially noxious stimuli. If threat is discerned, arousal ensues, heightening vigilance and preparing the body for appropriate action (fight or flight). Fear is experienced with concomitant physiological changes considered adaptive if the stimulus is indeed threatening. If there is no real threat, the response is maladaptive. Genetic predisposition and past experience may in some way influence the pattern of these responses.

V. NEUROPHYSIOLOGY OF ANXIETY

The tremendous recent increase in our knowledge of the neuroanatomical pathways me-

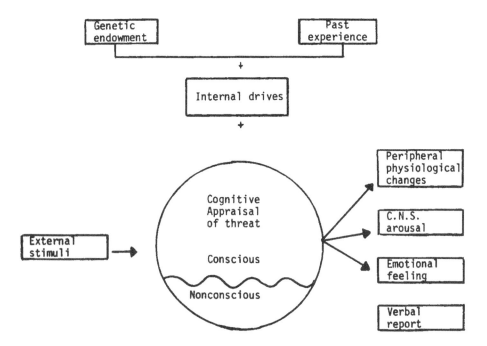

FIGURE 1. A general model of anxiety. (From Lader, M., *J. Clin. Psychiatry*, 44, 11, 7, 1983. With permission.)

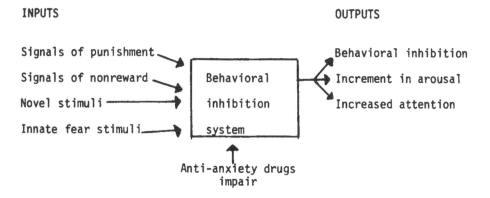

FIGURE 2. Model of the behavioral inhibition system. (From Gray, J. A., *Neuropsychology of Anxiety*, Oxford University Press, Oxford 1982. With permission.).

diating behavior has derived from a variety of anatomic, behavioral and pharmacological approaches in animals, and from clinical observations, notably from psychosurgical studies. Two major approaches to understanding the neurophysiological mechanisms in anxiety are those of Gray and Redmond.[144,53] Gray postulates a behavioral inhibitory system (Figure 2) comprising inputs provoking anxiety (stimuli that warn of punishment, non-reward, novel stimuli, and innate fear stimuli) and outputs of anxiety behavior (inhibition of ongoing motor behavior, increased level of arousal, and increased attention to the novel aspects of the environment). Gray[144] attempts to neuroanatomically localize these inputs and outputs. According to him, the ascending norepinephrine (NE) projection to the septohippocampal system may select whether certain incoming information is "important." The ascending 5-HT projection may add information on whether the system is associated with "punishment";

the ascending cholinergic system facilitates stimulus analysis, and the ascending NE system primes the effector, autonomic, and motor systems for the action required. Gray and Marks have speculated that escape may be mediated by periaqueductal gray matter,[144,145] some hypothalamic nuclei, and amygdala. The cingulate cortex may mediate active avoidance, and the septo hippocampal system may be involved in passive avoidance behavior. Our knowledge of this elaborate system is incomplete, but these hypotheses are of considerable heuristic value.

Redmond has argued that the norepinephrine pathways emerging from the locus coeruleus is essential to the mechanism of anxiety.[53] The locus coeruleus is a small brain stem nucleus, supplying about 70% of the NE in the primate brain — cerebral cortex, hippocampal, cingulate gyrus, and amygdala — a major portion of the limbic cortex. According to Redmond,[53] the locus coeruleus is activated by sensory or cortical pathways conveying information with noxious potential to the individual. The locus coeruleus efferents are postulated to be associated with motivation, learning, and memory, as well as the physiological aspects of anxiety via the hypothalamus. There may be a reciprocal feedback from visceral organs to the locus coeruleus, forming a physiological basis for the James-Lange hypothesis.[59]

Evidence for the involvement of the septohippocampal system in the neurobiology of panic disorder has recently been obtained from a number of neurophysiological studies. Gloor et al.[147] reported that stimulation of the hippocampus, parahippocampal gyrus, and amygdala most commonly produced a sensation of fear in awake human subjects. Reiman and his colleagues, in a remarkable PET Scan study of panic patients vulnerable to lactate induced panic, have observed an asymmetry in blood flows, blood volume, and oxygen metabolism suggestive of abnormal increases in the right parahippocampal region.[148] These asymmetries might represent an increase in neuronal activity, an asymmetry in the cellular processes, or an increase in permeability of the blood-brain barrier in that area. These authors hypothesize that high NE input from the locus coeruleus could lead to the observed increases in blood flow and oxygen metabolism, and make this area overly responsive to sensory stimuli, resulting in a panic attack.[149] This work provides a cogent model, consistent with both the Gray and Redmond systems, and capable of generating a series of testable predictions concerning pathways in a CNS "anatomy of anxiety."

VI. NEUROENDOCRINE ASPECTS OF ANXIETY DISORDERS

Affective disorders are consistently associated with alterations in neuroendocrine function. In view of the considerable clinical overlap between anxiety and affective disorders, it has been questioned whether similar neuroendocrine changes occur in anxiety disorders as well.

Several studies have found that a small and variable proportion of patients with panic disorders have an abnormal Dexamethasone Suppression Test (DST).[150] The frequency of abnormal DST in these patients seem to fall between the values usually reported for melancholic patients and normal controls.[151] Hamlin and Potash[152] have reported that patients with panic disorder have a lower mean maximal thyroid stimulating hormone (TSH) to thyrotropin releasing hormone (TRH) than normal controls. Byrne and colleagues have recently confirmed this observation, and also found that as a group, panic disorder patients have lower mean TSH and prolactin responses to TRH.[153] It has been suggested that this blunted TSH response to TRH may be related to noradrenergic hyperactivity seen in some patients with panic disorders.[153] However, such suggestions must be considered at best tentative, in view of the complex neurotransmitter regulation of hypothalamo-pituitary-thyroid axis.

VII. NEUROBIOLOGY OF ANXIETY — AN INTEGRATION

From the above discussion it is apparent that numerous circuits and neuroregulators are involved in the acquisition and expression of the many different forms of behavior subsumed under the term anxiety. The immense complexity of this problem has defied attempts to integrate this vast knowledge into a meaningful model. The task, however, is no longer hopeless.

Similar basic strategies of defense behavior, withdrawal, aggressive defense, and deflection of attack, analogous to features of human anxiety disorders are found in both invertebrates and vertebrates.[145] Likewise, basic forms of learning such as habituation, sensitization, and conditioning are also similar across phyla. Recent research provides strong reasons to believe that common physiological mechanisms may underlie these shared defense strategies. Serotoninergic systems are related to defense behavior involving cessation of ongoing activity in a variety of species.[154] However, there are no data showing that serotonergic manipulation affects any form of human anxiety. In mammals, noradrenergic systems can play a parallel activating role more for alarm with increased motor and autonomic activity than immobility and freezing.[53] These two defense activating systems (noradrenergic and serotonergic) have their nuclei of origin in closely related brainstem areas and are activated by sensory inputs.

There might be additional defense activating systems. One could be dopaminergic, from the ventral tegmentum to the frontal cortex.[145] Dopaminergic mechanisms might be involved in the action of the anxiolytic, buspirone.[155] The diffuse inhibitory GABAergic circuits seem to act by modulating the ascending activating systems (serotonergic, noradrenergic, and probably dopaminergic) rather than by directly inhibiting the deep structures that express fear. Anxiolytic drugs such as benzodiazepines seem to act in this indirect fashion by potentiating the GABA-BZ receptor complex. Peptides, such as endogenous opiates,[156] may attenuate defensive behavior, perhaps by short circuits present diffusely throughout the nervous system. There is a good deal of evidence for complex interactions between these neurochemical systems in the genesis and alteration of anxiety. The anxiogenic effect of yohimbine, an alpha$_2$ receptor antagonist, is blocked by diazepam. Conversely, clonidine, an alpha$_2$ agonist, partially blocked the behavioral activation caused by B-CCE in nonhuman primates and the anxiety symptoms during benzodiazepine withdrawal.[157,158] Clonidine also effectively blocks opiate withdrawal symptoms.[159] Caffeine, an anxiogenic drug, counteracts the cognitive, psychomotor and muscle relaxant effects of the BZs.[160]

Thus, pathological anxiety states may serve as end-products of malfunction at any level in a complex defense system involving several inter-linked neuronal systems. It is possible that clinically different aspects of anxiety may be mediated by dysfunctions in different defense reaction circuits. For example, the noradrenergic and the BZ/GABA receptor "models" of anxiety may describe somewhat different, but overlapping phenomena; thus, in panic disorder, the locus coeruleus may provide the triggering stimulus that sets off the panic attack, but the associated anticipatory and residual anxiety may, in part, be mediated through BZ-GABA receptor containing neuronal systems.[161] Likewise, different treatment approaches may operate by affecting different aspects of malfunction.

VIII. CONCLUSIONS AND NEW DIRECTIONS

Whereas half a century ago our times were termed the "age of anxiety", today we may be facing the "age of anxiety research".[162] The anxiety researcher of today has two important tasks: (1) the available knowledge of the biological correlates of anxiety states needs to be deployed in improved diagnosis, treatment, and prediction of treatment response in anxiety; (2) the vastly expanding awareness of the neurobiology of behavior needs to be applied to resolve concrete clinical problems of anxiety disorders. We can anticipate that exciting times are ahead.

ACKNOWLEDGMENTS

We gratefully acknowledge Mrs. M. A. Schemm and Ms. C. M. Cussigh for secretarial assistance. Drs. N. Sitaram and S. Gershon provided valuable support.

REFERENCES

1. **Kunham, R. C.**, A survey of Sanskrit literature, *Bharatiya vidya bhavan*, Bombay, 1962, 65.
2. **Freud, S.**, The justification for detaching from neurasthenia a particular syndrome: The anxiety neurosis, in *Collected Papers*, (transl. J. Stratchey), London, Hogarth Press, 1924, 76.
3. **Hecker, E.**, Ueber larvierte und Abortiv Angstzustande bei neurasthenie, *Z. Z. Nervenheil K.*, 16, 565, 1893.
4. **Klein, D. F.**, Delineation of two drug responsive syndromes, *Psychopharmacology*, 5, 397, 1964.
5. American Psychiatric Association, *Diagnostic and Statistical Manual of Mental Disorders*, Washington, D.C., American Psychiatric Association, 1952.
6. American Psychiatric Association, *Diagnostic and Statistical Manual of Mental Disorders*, 2nd ed., Washington, D.C., American Psychiatric Association, 1968.
7. American Psychiatric Association, *Diagnostic and Statistical Manual of Mental Disorders*, Washington, D.C., American Psychiatric Association, 1980.
8. **Beard, M.**, Neurasthenia or nervous exhaustion, *Boston Med. Surg. J.*, 3, 217, 1869.
9. **Dacosta, J. M.**, An irritable heart: A clinical form of functional cardiac disorder and its consequences, *Am. J. Med. Sci.*, 61, 17, 1871.
10. **Oppenheimer, B. S. and Rothschild, M. A.**, The psychoneurotic factor as the irritable heart of soldiers, JAMA, 70, 1919, 1958.
11. **McInness, R. G.**, Observations on heredity in neurosis, *Proc. R. Soc. Med.*, 30, 895, 1937.
12. **Brown, F. N.**, Heredity in the psychoneuroses, *Proc. R. Soc. Med.*, 35, 785, 1942.
13. **Noyes, R., Clancy, J., Crowe, R. R., Hoenk, P. R., and Slymen, D. J.**, The familial prevalence of anxiety neurosis, *Arch Gen. Psychiatry*, 35, 1057, 1978.
14. **Crowe, R. R., Noyes, R., Pauls, D. L., and Slymen, D.**, A family study of panic disorder, *Arch. Gen. Psychiatry*, 40, 1065, 1983.
15. **Cloninger, C. R., Lewis, C., Rice, J., and Reich, T.**, Strategies for resolution of biological and cultural inheritance, in *Genetic Strategies in Psychobiology and Psychiatry*, Gershon, E. S., Matthysee, S., Breakfield, S. D., and Ciaranello, R. D., Eds., Boxwood Press, California, 1981.
16. **McGuffin, P.**, Genetic influences on personality neurosis and psychosis, in *Scientific Principles of Psychopathology*, McGuffin, P., Shanks, M. F., and Hodgson, R. J., Eds., Grune and Stratton, 1984, 191.
17. **Harris, E. L., Noyes, R., and Crowe, R. R.**, A family study of agoraphobia, *Arch. Gen. Psychiatry*, 40, 1061, 1983.
18. **Slater, E. and Shields, J.**, Genetical aspects of anxiety, in *Studies of Anxiety*, Lader, M. H., Ed., *British Journal of Psychiatry Special Publication No. 3*, Headley, Ashford, England, 1969.
19. **Torgerson, S.**, The contribution of twin studies to Psychiatric nosology, in *Twin Research, Part A: Psychology and Methodology*, Nance, W. E., Ed., Alan R. Liss, New York, 1978, 125.
20. **Torgerson, S.**, Genetic factors in anxiety disorders, *Arch. Gen. Psychiatry*, 40, 1085, 1983.
21. **Leckman, J. F., Weissmann, M. M., and Merikangas, K. R.**, Panic disorder and major depression, *Arch. Gen. Psychiatry*, 40, 1055, 1983.
22. **Carey, G. and Gottesman, I. I.**, Twin and family studies of anxiety, phobic and obsessive disorders, in *Anxiety: New Research and Changing Concepts*, Klein, D. F., and Rabkin, J., Eds., Raven Press, New York, 1981, 117.
23. **Cannon, W.**, *Bodily Changes in Pain, Hunger, Fear and Rage*, Appleton Century Croft, New York, 1929.
24. **Lindemann, E. and Finesinger, J.**, The effect of adrenalin and mecholyl in states of anxiety in psychoneurotic patients, *Am. J. Psychiatry*, 95, 353, 1935.
25. **Cohen, M. and White, P.**, Life situations and emotions and neurocirculatory asthenia, anxiety neurosis, neurasthenia, effort syndrome, *Proc. Assoc. Res. Nerv. Ment. Dis.*, 29, 832, 1950.
26. **Cohen, M. E., Consolazio, F. C., and Johnson, R. E.**, Blood lactate response during moderate exercise in neurocirculatory asthenia, anxiety neurosis or effort syndrome, *J. Clin. Invest.*, 26, 339, 1947.
27. **Jones, M. and Mellersh, V.**, Comparison of exercise response in anxiety states and normal controls, *Psychosom. Med.*, 8, 180, 1946.

28. **Linko, E.,** Lactic acid response to muscular exercise in neurocirculatory asthenia, *Ann. Med. Int. Fenniae,* 39, 161, 1950.

29. **Holmgren, A. and Strom, G.,** Blood lactate concentration in relation to absolute and relative work load in normal men and in mitral stenosis, atrial septal defect and vasoregulatory asthenia, *Acta Med. Scand.,* 163, 185, 1959.

30. **Pitts, F. N. and McClure, J.,** Lactate metabolism in anxiety neurosis, *N. Engl. J. Med.,* 277, 1329, 1967.

31. **Bonn, J., Harrison, J., and Rees, L.,** Lactate induced anxiety — therapeutic applications, *Br. J. Psychiatry,* 119, 468, 1971.

32. **Fink, M., Taylor, M., and Volavka, J.,** Anxiety precipitated by lactate, *N. Engl. J. Med.,* 281, 1429, 1969.

33. **Haslam, M.,** The relationship between the effect of lactate infusion on anxiety, and its amelioration by carbon dioxide inhalation, *Br. J. Psychiatry,* 119, 129, 1974.

34. **Kelly, D., Mitchell-Heggs, N., and Sherman, D.,** Anxiety and the effect of sodium lactate assessed clinically and physiologically, *Br. J. Psychiatry,* 119, 129, 1971.

35. **Lapierre, Y., Knott, V., and Gray, R.,** Psychological correlates of sodium lactate, *Psychopharmacol. Bull.,* 20(1), 50, 1984.

36. **Liebowitz, M. R., Fyer, A. J., Gorman, J. M., Dillon, D., Appleby, I. L., Levy, G., Anderson, S., Levitt, M., Palij, M., Davies, S. O., and Klein, D. F.,** Lactate provocation of panic attacks: I. Clinical and Behavioral findings. *Arch. Gen. Psychiatry,* 41, 764, 1984.

37. **Rainey, J., Ettedgui, E., Pohl, R., Balon, R., Weinberg, P., Yelonek, S., and Berchou, R.,** The beta receptorisoproterenol anxiety states, *Psychopathology,* 17(3), 40, 1984.

38. **Sheehan, D. V., Coleman, J. H., Greenblatt, D. J., Jones, K. J., Levine, P. H., Orsulak, P. J., Peterson, M., Schildkraut, J. J., Uzogara, E., and Watkins, D.,** Some biochemical correlates of panic attacks with agoraphobia and their response t a new treatment, *J. Clin. Psychopharmacol.,* 4, 66, 1984.

39. **Pohl, R., Rainey, J., Ortiz, A., Balon, R., , ˈngh, H., and Berchou, R.,** Isoproterenol-induced anxiety states, *Psychopharmacol. Bull.,* 21(3), 424, 1℃ ℃5.

40. **Zitrin, C. M., Klein, D. F., and Woerner, , ℃. G.,** Treatment of agoraphobia with group exposure in vivo and imipramine, *Arch. Gen. Psychiatry,* 3℃, 63, 1980.

41. **Sheehan, D. V., Ballenger, J., and Jacobsen, G.** Treatment of endogenous anxiety with phobic, hysterical and hypochondriacal symptoms, *Arch. Gen. Psychiatry,* 37, 51, 1980.

42. **Carr, D. B., Sheehan, D. V., Susman, O. S., Coleman, J. H., Greenblatt, D. J., Heninger, G. R., Jones, K. J., Levine, P. H., and Watkins W. D.,** Neuroendocrine correlates of lactate-induced anxiety and their response to chronic alprazolam therapy, *Am. J. Psychiatry,* 143, 483, 1986.

43. **Fyer, A. J., Liebowitz, M. R., Gorman, J. M., Davies, S. O., and Klein, D. F.,** Lactate vulnerability of remitted panic patients, *Psychiatry Res.,* 14, 143, 1985.

44. **Cowley, D. S., Dager, S. R., and Dunner, D. L.,** Lactate-induced panic in primary affective disorder, 143, 646, 1986.

45. **Gorman, J. M., Liebowitz, M. R., Fyer, A. J., Dillon, D., Davies, S. O., Stein, J., and Klein, D. F.,** Lactate infusions in compulsive disorder, *Am. J. Psychiatry,* 142, 864, 1985.

46. **Liebowitz, M. R., Fyer, A. J., Gorman, J. M., Dillon, D., Davies, S., Stein, J. M., Cohen, B. S., and Klein, D. F.,** Specificity of lactate infusions in social phobia vs. panic disorder, *Am. J. Psychiatry,* 142, 947, 1985.

47. **Liebowitz, M. R., Gorman, J. M., Fyer, A. J., Levitt, M., Dillion, D., Levy, G., Appleby, I. L., Anderson, S., Palij, M., Davies, S. O., and Klein, D. F.,** Lactate provocation of panic attacks: II. Biochemical findings, *Arch. Gen. Psychiatry,* 42, 709, 1985.

48. **Rainey, J. M., Frohman, C. E., Warner, K., Bates, S., Pohl, R. B., and Yeragani, V.,** Panic anxiety and lactate metabolism, *Psychopharmacol. Bull.,* 21, 434, 1985.

49. **Grosz, H. J. and Farmer, B. B.,** Blood lactate in the development of anxiety symptoms, *Arch. Gen. Psychiatry,* 21, 611, 1969.

50. **Grosz, H. J. and Farmer, B. B.,** Pitts and McClure's lactate anxiety study revisited, *Br. J. Psychiatry,* 120, 415, 1972.

51. **Pitts, F. N. and Allen, R. E.,** Biochemical induction of anxiety, in *Phenomenology and Treatment of Anxiety,* Fann, W. E., Karacan, I., Pokorny, A. D., and Williams, R. L., Eds., Spectrum, New York, 1979, 125.

52. **Gorman, J. M., Levy, G. F., Liebowitz, M. R., McGrath, P., Appleby, I. L., Dillon, D. J., Davies, S. O., and Klein, D. F.,** Effect of acute beta-adrenergic blockade on lactate-induced panic, *Arch. Gen. Psychiatry,* 40, 1079, 1983.

53. **Redmond, D. E.,** New and old evidence for the involvement of a brain norepinephrine system in anxiety, in *Phenomenology and Treatment of Anxiety,* Fann, W. E., Karacan, I., Pokorny, A. D., and Williams, R. L., Eds., Spectrum, New York, 1979, 153.

54. **Gorman, J. M., Askanazi, J., Liebowitz, M. R., Fyer, A. J., Stein, J., Kinney, J. M., and Klein, D. F.,** Response to hyperventilation in a group of patients with panic disorder, *Am. J. Psychiatry,* 141, 857, 1984.
55. **Carr, D. and Sheehan, D.,** Panic anxiety: A new biological model, *J. Clin. Psychiatry,* 45, 323, 1984.
56. **Liebowitz, M. R., Gorman, J. M., Fyer, A., Dillon, D., Levitt, M., and Klein, D. F.,** Possible mechanisms for lactate's induction of panic, *Am. J. Psychiatry,* 143, 495, 1986.
57. **Ackerman, S. H. and Sachar, E. J.,** The lactate theory of anxiety: A review and re-evaluation, *Psychosom. Med.,* 36, 69, 1974.
58. **Bonn, J. A., Harrison, J., and Rees, L.,** Lactate infusion in the treatment of free floating anxiety, *Can. J. Psychiatry,* 18, 41, 1973.
59. **James, W.,** *The Principles of Psychology,* Dover, New York, 1950.
60. **Maragraf, J., Ehlers, A., and Roth, W. T.,** Sodium lactate infusions and panic attacks: A review and critique, *Psychosom. Med.,* 48, 23, 1986.
61. **Yeragani, V. K., Pohl, R., Rainey, J. M., Balon, R, Ortiz, A., Lycaki, H., and Gershon, S.,** Pre-infusion heart rates and laboratory-induced panic anxiety, *Acta Psychiatric Scand.,* in press.
62. **Drury, A. N.,** The percentage of carbon dioxide in the alveolar air and the tolerance to accumulating of carbon dioxide in cases of so-called "irritable heart" of soldiers, *Heart,* 7, 165, 1919.
63. **Singh, B. S.,** Ventilatory response to carbon dioxide II. Studies in neurotic psychiatric patients and practitioners of transcendental meditation, *Psychosom. Med.,* 46, 347, 1984.
64. **Hill, O.,** The hyperventilation syndrome, *Br. J. Psychiatry,* 135, 367, 1979.
65. **Lum, L. C.,** Hyperventilation: The time and the iceberg, *J. Psychosom. Res.,* 19, 375, 1975.
66. **Ley, R.,** Agoraphobia, the panic attack and the hyperventilation syndrome, *Behav. Res. Ther.,* 23, 79, 1985.
67. **Woods, S. W., Charney, D. S., Loke, J., Goodman, W. K., Redmond, D. E., and Heninger, G. R.,** Carbon dioxide sensitivity in panic anxiety, *Arch. Gen. Psychiatry,* 43, 900, 1986.
68. **Van den Hout, M. A. and Griez, E.,** Panic symptoms after inhalation of carbon dioxide, *Br. J. Psychiatry,* 144, 503, 1984.
69. **Griez, E. and Van den Hout, M. A.,** Treatment of phobophobia by exposure of CO_2-induced anxiety, *J. Nerv. Ment. Dis.,* 171, 506, 1983.
70. **Gilliland, K. and Andress, D.,** Ad lib caffeine consumption symptoms of caffeinism and academic performance, *Am. J. Psychiatry,* 138, 512, 1981.
71. **Primavera, L. H., Simon, W., and Lamiza, J.,** An investigation of personality and caffeine use, *Br. J. Addict.,* 70, 213, 1975.
72. **Winstead, D. K.,** Coffee consumption among psychiatric inpatients, *Am. J. Psychiatry,* 133, 1447, 1976.
73. **Greden, J. P., Fontaine, P., Lubetsky, M., and Chamberlin, K.,** Anxiety and depression associated with caffeinism among psychiatric inpatients, *Am. J. Psychiatry,* 135, 963, 1978.
74. **Greden, J. F.,** Anxiety or Caffeinism, A diagnostic dilemma, *Am. J. Psychiatry,* 131, 1089, 1974.
75. **Lutz, E. G.,** Restless legs, anxiety and caffeinism, *J. Clin. Psychiatry,* 39, 693, 1978.
76. **MacCallum, W. A. G.,** Excess coffee and anxiety states, *Int. J. Soc. Psychiatry,* 25, 209, 1979.
77. **Uhde, T. W., Boulenger, J. P., Jimerson, D. C., and Post, R. M.,** Caffeine: Relationship to human anxiety, plasma MHPG and cortisol, *Psychopharmacol. Bull.,* 20, 426, 1984.
78. **Boulenger, J., Uhde, T. W., Wolff, E. A., and Post, R. M.,** Increased sensitivity to caffeine in patients with panic disorders, *Arch. Gen. Psychiatry,* 41, 1067, 1984.
79. **Lee, M. A., Cameron, O. C., and Greden, J. F.,** Anxiety and caffeine consumption in people with anxiety disorders, *Psychiatry Res.,* 15, 211, 1985.
80. **Charney, D. S., Heninger, G. R., and Jatlow, P. I.,** Increased anxiogenic effects of caffeine in panic disorders, *Arch. Gen. Psychiatry,* 42, 233, 1985.
81. **Butcher, R. W. and Sutherland, E. W.,** Adenosine 3', 5'-phosphate in biological materials, *J. Biol. Chem.,* 237, 1244, 1962.
82. **Berkowitz, B. A., Tarver, J. H., and Spector, S.,** Release of norepinephrine in the central nervous system by theophylline and caffeine, *Eur. J. Pharmacol.,* 10, 64, 1970.
83. **Marangos, P. J., Paul, S. M., Parma, A. M., Goodwin, F. K., Syapin, P., and Skolnic, P.,** Purinergic inhibition of diazepam binding to rat brain (in vitro), *Life Sci.,* 24, 851, 1979.
84. **Snyder, S. H. and Sklar, P.,** Behavior and molecular actions of caffeine: Focus on adenosine, *J. Psychiatr. Res.,* 18, 91, 1984.
85. **Weiner, N.,** Norepinephrine, epinephrine and the sympathomimetic amines, in *The Pharmacological Basis of Therapeutics,* Gilman, A. C., Goodman, R. S., and Gilman, A., Eds., MacMillan, New York, 1980, 138.
86. **Frolich, E. D., Tarazi, R. C., and Dustan, H. P.,** Hyperdynamic beta-adrenergic circulatory state: increased beta receptor responsiveness, *Arch. Int. Med.,* 123, 1, 1969.
87. **Kathol, R. C., Noyes, R., Slymen, P. J., Crowe, R. R., Clancy, J., and Kerber, R. E.,** Propranolol in chronic anxiety disorders, *Arch. Gen. Psychiatry,* 37, 1361, 1980.

88. **Nesse, R. M., Cameron, O. G., Curtis, G. C., McCann, D. S., and Huber-Smith, M. J.,** Adrenergic function in patients with panic anxiety, *Arch. Gen. Psychiatry,* 41, 771, 1984.

89. **Schandry, R.,** Heart beat perception and emotional experience, *Psychophysiology,* 18, 483, 1981.

90. **Holmberg, C. and Gershon, S.,** Autonomic and psychiatric effects of yohimbine hydrochloride, *Psychopharmacologia,* 2, 93, 1961.

91. **Hoehn-Saric, R., Merchant, A. F., Kayser, M. L., and Smith, V. K.,** Clonidine and anxiety disorders, *Arch. Gen. Psychiatry,* 38, 1278, 1981.

92. **Liebowitz, M. R., Fyer, A. J., McGratter, P., and Klein, D. F.,** Clonidine treatment of panic disorder, *Psychopharmacol. Bull.,* 17, 122, 1981.

93. **Charney, D. S., Heninger, G. R., Sternberg, D. E., and Landis, H.,** Abrupt discontinuation of tricyclic antidepressant drugs: Evidence of noradrenergic hyperactivity, *Br. J. Psychiatry,* 141, 377, 1982.

94. **Engberg, G., Elam, M., and Svensson, T. H.,** Clonidine withdrawal: Activation of brain noradrenergic neurons with specifically reduced alpha-2 receptor sensitivity, *Life Sci.,* 30, 235, 1982.

95. **Charney, D. S., Heninger, G. R., and Steinberg, D. E.,** Assessment of alpha$_2$-adrenergic autoreceptor function in humans: Effects of oral yohimbine, *Life Sci.,* 30, 2033, 1982.

96. **Charney, D. S., Heninger, G. R., and Breier, A.,** Noradrenergic function in panic anxiety, *Arch. Gen. Psychiatry,* 41, 751, 1984.

97. **Charney, D. S. and Heninger, G. R.,** Noradrenergic function and the mechanism of action of antianxiety treatment. I. The effect of long-term alprazolam treatment, *Arch. Gen. Psychiatry,* 42, 450, 1985.

98. **Charney, D. S. and Heninger, G. R,** Noradrenergic function and the mechanism of action of antianxiety treatment II: Effect of long-term imipramine treatment, *Arch. Gen. Psychiatry,* 42, 473, 1985.

99. **Charney, D. S., Heninger, C. R., and Redmond, D. E. Jr.,** Yohimbine-induced anxiety and increased noradrenergic function in humans: Effects of diazepam and clonidine, *Life Sci.,* 33, 19, 1983.

100. **Uhde, T. W., Boulenger, J. P., and Vittone, B. J.,** Historical and modern concepts anxiety: A focus on adrenergic function, in *Biology of Agoraphobia,* Ballenger, J. C, Ed., American Psychiatric Press, Washington, D.C., 1984, 1.

101. **Redmond, D. E. Jr., Huang, Y. H., Snyder, D. R., and Mass, J. W.,** Behavioral effects of stimulation of the locus Ceruleus in the Stumptail Monkey (Macace arctoides), *Brain Res.,* 116, 502, 1976.

102. **Hamlin, C. L., Lydiard, R. B., Martin, D., Dackis, C. A., Pottash, A. C., Sweeney, D., and Cold, M. S.,** Urinary excretion of noradrenaline metabolite decreased in panic disorder, *Lancet,* 2, 740, 1983.

103. **Kaitin, K. I., Bliwise, D. L., Gleason, C., Nino-Murcia, G., Dement, W. C., and Libet, B.,** Sleep disturbance produced by electrical stimulation of the locus ceruleus in a human subject, *Biol. Psychiatry,* 21, 710, 1986.

104. **Kopin, I. J.,** Avenues of investigation for the role of catecholamines in anxiety, *Psychopathology,* 17(1) 83, 1984.

105. **Greenblatt, D. J., Shader, R. I., and Abernathy, D. R.,** Current status of benzodiazepines, *N. Engl. J. Med.,* 309, 354, 1983.

106. **Mohler, H. and Okada, T.,** Benzodiazepine receptor: demonstration in the central nervous system, *Science,* 198, 849, 1977.

107. **Squires, R. F. and Braestrup, C.,** Benzodiazepine receptors in rat brain, *Nature,* 266, 732, 1977.

108. **Chang, R. S. L. and Snyder, S. H.,** Benzodiazepine receptors: labeling in intact animals with 3H-flunitrazepam, *Eur. J. Pharmacol.,* 48, 213, 1978.

109. **Haefely, W., Kulesar, R., Mohler, H., Pieri, L., Polc, P., and Schaffner, R.,** Possible involvement of GABA in the central actions of benzodiazepines, *Adv. Biochem. Psychopharmacol.,* 41, 131, 1975.

110. **Tallman, J., Thomas, J., and Gallager, D.,** GABAergic modulation of benzodiazepine site sensitivity, *Nature,* 174, 383, 1978.

111. **Costa, E. and Gindotti, A.,** Molecular mechanisms in the receptor action of benzodiazepines, *Ann. Rev. Pharmacol.,* 19, 531, 1979.

112. **Tallman, J. F., Pusl, S. M., Skolnick, P., and Gallagher, D. W.,** Receptors for the age of anxiety: pharmacology of the benzodiazepines, *Science,* 207, 274, 1980.

113. **Skolnick, P. and Paul, S. M.,** Molecular pharmacology of the benzodiazepines, *Int. Rev. Neurobiol.,* 23, 103, 1982.

114. **Koella, W. P.,** GABA systems and behavior, in *Amino Acid Neurotransmitters,* DeFeudis, F. V. and Mandel, P., Eds., Raven Press, New York, 1981, 11.

115. **Olsen, R. W.,** GABA-benzodiazepine-barbiturate receptor interactions, *J. Neurochem.,* 37, 1, 1981.

116. **Young, W. S. and Kuhar, M. J.,** Radiohistochemical localization of benzodiazepine receptors in rat brain, *J. Pharmacol. Exp. Ther.,* 212, 337, 1980.

117. **Ghoneim, M. M., Hinrichs, J. V., and Menaldt, S. P.,** Dose-response analysis of the behavioral effects of diazepam: I. Learning and memory, *Psychopharmacology,* 82, 291, 1984.

118. **Ross, R., Waszczak, B., Lee, E., and Walters, J.,** Effects of benzodiazepines on single unit activity in the substantia nigra pars reticulata, *Life Sci.,* 31, 1025, 1982.

119. **Wurtz, R. H. and Goldberg, M. E.,** Visual-motor function of the primate superior colliculus, *Ann. Rev. Neurosci.*, 3, 189, 1980.

120. **Braestrup, C., Nielson, M., and Olsen, C. F.,** Urinary and brain beta-carboline-3-carboxylates as potent inhibitors of brain benzodiazepine receptors, *Proc. Natl. Acad. Sci. (USA)*, 77, 2288, 1980.

121. **Hunkeler, W., Mohler, H., Pieri, L., Polc, P., Bonetti, E. P., Cumin, R., Schaffner, R., and Haefely, W.,** Selective antagonists of benzodiazepines, *Nature*, 290, 514, 1981.

122. **Czernik, A. J., Petrack, B., Kalinsky, H. J., Psychoyos, S., Cash, W. D., Tsai, C., Rinehart, R. K., Granat, F. R., Loell, R. A., Brundish, D. E., and Wade, R.,** CGS 8216: Receptor binding characteristics of a potent benzodiazepine antagonist, *Life Sci.*, 30, 363, 1982.

123. **Costa, E. and Biggio, G.,** The action of stress on beta-carbolines, diazepam and Ro-15-1788 on GABA receptors in the rat brain, in *Advances in Biochemical Psychopharmacology*, Vol. 38, Biggio, G. and Costa, E., Eds., Raven Press, New York, 1983, 87.

124. **Ninan, P. T., Insel, T. M., Cohen, R. M., Cook, J. M., Skolnick, P., and Paul, S. M.,** Benzodiazepine receptor-mediated experimental "anxiety" in primates, *Science*, 218, 1332, 1982.

125. **Dorrow, R., Horowski, R., Pashelke, G., Amin, M., and Braestrup, C.,** Severe anxiety induced by FG-7142, a beta-carboline ligand for benzodiazepine receptors, *Lancet*, 2, 98, 1983.

126. **Skolnick, P., Paul, S. M., and Marangos, P. J.,** Purines as endogenous ligands of the benzodiazepine receptor, *Fed. Proc.*, 39, 3050, 1980.

127. **Marangos, P. J., Patel, J., Hirata, F., Sondhein, D., Paul, S. M., Skolnick, P., and Goodwin, F. K.,** Inhibition of diazepam binding by tryptophan derivatives including melatonin and its brain metabolite n-acetyl-5-methoxy kynurenamine, *Life Sci.*, 29, 259, 1981.

128. **Mohler, H., Polc, P., Cumin, R., Pieri, L., and Kettler, R.,** Nicotinamide is a brain constituent with benzodiazepine-like actions, *Nature*, 278, 563, 1979.

129. **Woolf, J. H. and Nixon, J. C.,** Endogenous effector of the benzodiazepine binding site: Purification and characterization, *Biochemistry*, 20, 4263, 1981.

130. **Braestrup, C. and Nielson, M.,** Anxiety, *Lancet*, 2, 1030, 1982.

131. **Klein, D. F.,** Anxiety reconceptualized, in *Anxiety: New Research and Changing Concepts*, Klein, D. F. and Rabkin, J. C., Eds., Raven Press, New York, 1981.

132. **Panksepp, J., Herman, B., and Vilberg, T.,** Endogenous opioids and social behavior, *Neurosci. Behav. Res.*, 4, 473, 1978.

133. **Theodora, D., Millan, M. J., Ulsamer, B., and Doenicke, A.,** Naloxone attenuates the anxiolytic action of diazepam in man, *Life Sci.*, 31, 1833, 1982.

134. **Liebowitz, M. R., Gorman, J. M., Fyer, A. J., Dillon, D. J., and Klein, D. F.,** Effects of naloxone on patients with panic attacks, *Aus. J. Psychiatry*, 141, 995, 1984.

135. **Lader, M.,** *The psychophysiology of Mental Illness*, Routledge and Kegan Paul, London, 1975.

136. **Lader, M. and Wing, J. K.,** *Physiological Measures, Sedative Drugs and Morbid Anxiety*, Oxford University Press, Oxford, 1966.

137. **Sainsbury, P. and Gibson, J. G.,** Symptoms of anxiety and tension and accompanying physiological changes in the muscular system, *J. Neurol. NeuroSurg. Psychiatr.*, 17, 216, 1954.

138. **Ackner, B.,** Emotions and the Peripheral Vasomotor System. A review of previous work, *J. Psychosom. Res.*, 1, 3, 1956.

139. **Abramson, L. B., Keshavan, M. S., Brown, A., and Sitaram, N.,** Increased arecoline-induced heart-rates in anxious depressives, *Am. J. Psychiatry*, (submitted).

140. **Akiskal, H. S., Cowan, R., Simmons, R., Reimao, R., and Lemmi, H.,** Sleep EEG differentiation of anxiety and depressive disorders, in *Biological Psychiatry*, Shagass, C., Josi assen, R. C., Bridger, W. H., Weiss, K. J., Stoff, D., and Simpson, G. M., Eds., Elsevier, New York, 1985, 466.

141. **Dube, S., Kumar, N., Ettedgui, E., Pohl, R., Jones, D., and Sitaram, N.,** Cholinergic REM induction response: Separation of anxiety and depression, *Biol. Psychiatry*, 20, 408, 1985.

142. **Kelly, D., Brown, C. C., and Schaffer, J. W.,** A comparison of physiological and psychological measurements in anxious patients and normal controls, *Psychophysiology*, 6, 429, 1970.

143. **Roth, W. T., Telch, M. J., Taylor, C. B., Sachitano, J. A., Gallen, C. C., Kopell, M. L., McClenahan, K. L., Agras, W. S., and Pfefferbaum, A.,** Autonomic characteristics of agoraphobia with panic attacks, *Biol. Psychiatry*, 21, 1133, 1986.

144. **Gray, J. A.,** *The Neuropsychology of Anxiety: An Enquiry into the Functions of the Septo-Hippocampal System*, Oxford University Press, New York, 1982.

145. **Marks, I. and Tobena, A.,** What do neurosciences tell us about anxiety disorders? A comment, *Psychol. Med.*, 16, (1)9, 1986.

146. **Lader, M.,** Behavior and anxiety: Physiological mechanisms, *J. Clin. Psychiatry*, 44(11)5, 1983.

147. **Gloor, P., Olivier, A., Quesney, L. F., Anderman, F., and Horowitz, S.,** The role of the limbic stem in experimental phenomena of temporal lobe epilepsy, *Ann. Neurol.*, 12, 129, 1982.

148. **Reiman, E. M, Raichle, M. E., Butler, F. K., Herskovitch, P., and Robins, E.,** A focal brain abnormality panic disorder, a severe form of anxiety, *Nature*, 310, 683, 1984.

149. **Reiman, E. M., Raichle, M. E., Robins, E., Butler, K., Hesscovitch, P., Fox, P., and Perlmutter, J.,** The application of positron emission tomography to the study of panic disorder, *Am. J. Psychiatry,* 143, 469, 1986.

150. **Curtis, G. C., Cameron, O. G., and Nesse, R. M.,** The dexamethasone suppression test in panic disorder and agoraphobia, *Am. J. Psychiatry,* 139, 1043.

151. **Bridges, M., Yeragani, V. K., Rainey, J. M., and Pohl, R.,** Dexamethasone suppression test in patients with panic attacks, *Biol. Psychiatry,* 21, 849, 1981.

152. **Hamlin, C. L. and Pottash, A. L. C.,** Evaluation of anxiety disorders, in *Diagnostic and Laboratory Testing in Psychiatry,* Wood, S., Ed., New York, Plenum (in press).

153. **Roy Byrne, P. P., Uhde, T. W., Rubinow, D. R., and Post, R. M.,** Reduced TSH and prolactin responses to TRH in patients with panic disorder, *Am. J. Psychiatry,* 143, 503, 1986.

154. **Kandel, E. R.,** From metapsychology to molecular biology: Explorations into the nature of anxiety, *Am. J. Psychiatry,* 140, 1277, 1983.

155. **Taylor, D. P., Ribet, L. A., and Stanton, H. C.,** Dopamine and anxiolytics, in *Anxiolytics Neurochemical Behavioral and Clinical Perspectives,* Malic, K. B., Enna, S. J., and Yamamura, H. J., Eds., Raven Press, New York, 1983.

156. **Post, R. M., Pickar, D., Ballenger, J. C., Norber, D., and Rubinow, D. R.,** Endogenous opiates in cerebrospinal fluid: relationship to mood and anxiety, in Neurobiology of Mood Disorder, Post, R. M. and Ballenger, J. C. Eds., Williams & Wilkins, Baltimore, 1984, 356.

157. **Insel, T. R., Ninan, P. T., Aloi, J., Jimerson, D. C., Skolnick, P., and Paul, S. M.,** A benzodiazepine receptor mediated model of anxiety: Studies in nonhuman primates and clinical implications, *Arch. Gen. Psychiatry,* 41, 741, 1984.

158. **Keshavan, M. S. and Crammer, J. L.,** Clonidine and benzodiazepine withdrawal, *Lancet,* 1, 325, 1985.

159. **Gold, M. S., Redmond, D. E., and Kleber, H. E.,** Clonidine blocks acute opiate withdrawal symptoms, *Lancet,* 2, 599, 1978.

160. **File, S. E., Deakin, J. F. W., Longden, A., and Crow, T. J.,** Investigation of the role of the locus coeruleus in anxiety and antagonistic behavior, *Brain Res.,* 169, 411, 1979.

161. **Hommer, D. W. and Paul, S. M.,** Benzodiazepines, neurotransmitters and the neurobiology of anxiety, in *The Biological Foundations of Clinical Psychiatry,* Giananni, A. J., Ed., Medical Books, New York, 1986, 235.

162. **Freedman, D. X.,** New dimensions in anxiety: Physiologic and psychiatric perspectives, *J. Clin. Psychiatry,* 11,(2)3, 1983.

Chapter 2

THE BIOLOGICAL STATUS OF NEUROTIC DEPRESSION

Sanjay Dubé and Natraj Sitaram

TABLE OF CONTENTS

I. INTRODUCTION

The affective disorders include a number of clinical conditions whose common and essential feature is a disturbance of mood, accompanied by cognitive, psychomotor, psychological, and interpersonal difficulties. Grouping the affective disorders according to patients' predominant symptoms is not an ideal system for a nosology. A system based on the genetic, psychodynamic, and biological causes of the disorders would be ideal, and investigations are underway to establish their precise roles. In view of our limited clinical knowledge, however, classification by type of psychological impairment has had great heuristic value, e.g., impaired intelligence (mental retardation), thinking and cognition (dementias, schizophrenias, etc.), and social behavior (character and personality disorders).

II. NOSOLOGICAL CONTROVERSIES

Since the original classification of affective illness, i.e., manic-depressive insanity by Kraepelin[1] and its subsequent modification[2] several new concepts have proliferated. The currently accepted and widely used DSM-III Category of Affective Disorders groups all the affective disorders together, regardless of presence or absence of psychotic features or precipitating life experiences. Within that group, the subcategory Bipolar Disorder includes Mixed, Manic, or Depressed forms, and the subcategory Major Depression includes Single Episode or Recurrent forms. They include two additional subcategories: Other Specific Affective Disorders (including Cyclothymic Disorder and Dysthymic Disorder), and Atypical Affective Disorders. "Depressive neurosis" used in earlier classifications is now classified as either Major Depression, Single Episode or Recurrent, without Melancholia as well as Dysthymic Disorder, or as Adjustment Disorder with Depressed Mood. The term "neurosis" has been dropped in entirety from the DSM-III, due to the "etiological" connotation accompanying the term, while lacking empirical evidence to substantiate the etiological role played in the causation of illness. Despite the clearer descriptive phenomenological characteristics provided by the DSM-III, a large body of clinicians, and particularily those subscribing to the psycho-analytic school of thought, accuse the DSM-III of reducing psychiatry to a behavioral and purely empirical field.

To focus on "neurotic" depression, a term always surrounded by ambiguities, the issue becomes more complex. Sometimes, neurotic depressions have been referred to as nonpsychotic forms of depression. At other times "neurotic" has been used synonymously with "reactive", although little evidence exists that precipitating events are highly correlated with nonpsychotic symptom states. "Neurotic" has been used to refer to those depressions that arise in patients with long-standing character traits or maladaptive personalities. Attempts to separate psychotic and neurotic patients into distinct groups have so far been largely unsuccessful. Current evidence indicates that psychotic-neurotic distinctions, like those within an endogenous-reactive dichotomy, occur on a continuum that would be better called psychotic-nonpsychotic than psychotic-neurotic. The psychotic-neurotic dichotomy has been confused by etiologic assumptions. "Biological" depressions have been used synonymously with psychotic depressions. "Neurotic", on the other hand, has been ascribed to social and psychological causes that lead to impairments in personality function. Evidence for these etiological assumptions is minimal at best. In clinical practice, moreover, the endogenous-reactive dichotomy has unfortunately been used interchangeably with the psychotic-neurotic distinction. Since psychotic states may follow reactions to severe stressful life events, i.e., losses, this usage appears invalid. Similarily, "endogenous" depressions may present without any psychotic features such as delusions or hallucinations.

Most older classifications and textbooks treat neurotic depression separately from affective psychoses and manic depressive illness, the latter being grouped with the psychoses. Neurotic depressions were grouped along with the other neuroses. The newer classifications such as the DSM-III combine all forms of affective disturbance, independent of whether or not they have been previously categorized as neurotic or psychotic with regard to symptoms, severity, or impairment of functioning.

III. THE ENDOGENOUS-REACTIVE CONTINUUM

A decade ago investigators at Washington University, St. Louis, introduced criteria for diagnosis of depression and mania[3,4] which were recently modified[5] and referred to as "Primary Affective Disorder". These were incorporated with little change into the Research Diagnostic Criteria (RDC)[6] and the DSM-III. This system gives special consideration to exclusion criteria, with special emphasis on primary/secondary distinction. Thus, the presence of a previous psychiatric illness precludes the diagnosis of a primary affective disorder and requires a secondary designation. However, the symptoms of primary and secondary affective disorder are similar in cross section.[3] As a result, a history of a pre-existing nonaffective illness continued to receive emphasis in this system, while symptomatic criteria were considered less specific and used broadly to define a depressive syndrome. Though the term "primary affective disorder" would suitably describe many patients with endogenous depression, primary affective disorder is not synonymous with endogenous depression. Yet, as indicated by Klerman,[7] endogenous depression has come to imply more than just a lack of external cause and suggests the presence of certain state characteristics. Several investigators have emphasized the "autonomous" quality of the depressive state.[8-12] Endogenous depression is characterized by a distinct quality of mood[13,14] and certain signs and symptoms described in several studies that attempt to differentiate endogenous from reactive depression with factor analysis,[10,15-27] cluster analysis,[28-31] discriminant function analysis,[20,32-34] or symptom checklist.[35-37] Various descriptions referring to this state as autonomous,[13,10] primary,[8] vital,[14] melancholic,[12] or endogenomorphic[11] are similar in suggesting that the essence of diagnosis is the description of the characteristics, rather than a mere absence of precipitating stress. The status of "neurotic depression", however, is not so clear. The term "neurotic-reactive depression" has been used by several investigators to differentiate it morphologically from the "endogenous" depressives. Recent studies based on follow-up, outcome, prognostic, treatment, and family history data suggest that "neurotic reactive depression" may not be a diagnosis of exclusion alone, but exist as a separate entity.

On the other hand, there is an opposite view held by those who believe that the "milder" depressive states represent a residual category diagnosed by excluding the positive features of "major" forms of depressive disorders.[20,49] Kiloh et al.[26] believe that the support provided by factor analytic techniques for the existence of neurotic depression as a distinct syndrome is an artifact.

Whereas the more severe conditions are well-established entities, the "milder" disorders are subject to considerable controversy. It is uncertain whether these conditions are in continuum with the unipolar disorders or represent fundamentally different forms of psychopathology.

Early reports about the status of neurotic depression described brief and mild attacks of manic-depressive depression and suggested that they were the more common forms of depressions.[50,51] Winokur and Pitts in their earlier work were unable to differentiate the neurotic from psychotic depressions based on family history alone, a view which has since been rejected.[52,53] Beck[54] concluded that the two could not be distinguished on symptomatological grounds. Despite the increased interest in identifying the neurotic forms of depressive disorders, our current psychiatric repertoire is not sophisticated enough with respect to these conditions to permit differentiation among (1) self-limited depressions of a situational nature, (2) mild depressive states representing early abortive attacks of the major affective disorders (unipolar and bipolar), (3) atypical depressions constituting the prodromal manifestations of certain nonaffective psychiatric syndromes such as schizophrenia, senile dementia, and finally (4) those depressive clinical pictures occurring in the context of character disorders.[38] In their follow-up study of 100 patients with the "mild" depressive states variously referred to as "situational", "reactive", or "neurotic", Akiskal et al.[38] found the neurotic depressives characterized by a mild illness with nonendogenous and non-psychotic features, occurring in the context of psychologically understandable reaction of depression in an individual with varying degrees of characterological vulnerability. Char-acterologically, they were described as pathologically dependent, highly manipulative, im-pulsive, and unstable. Hostile depressive outbursts, occasioned by impending or actual separation and manifested by dramatic behaviors or impulsive suicidal attempts with ma-nipulative attempts was the most common theme. A 3 to 4 year follow-up revealed that 18% had developed a bipolar illness, and 22% developed unipolar disorders with predominantly favorable outcome. Irrespective of the diagnostic subtype, a "characterological" component occurring in 24% of the sample appeared to predict an unfavorable outcome. This study did not validate the existence of neurotic depression as a distinct nosological entity and argues that most studies which have attempted to validate the two-type thesis of depressive disorders[39] have rarely listed the symptomatic stability of their study populations by prospective designs. They also criticize the findings of the New Castle group,[22] which tested the stability of the neurotic syndrome in a 5 to 7 year follow-up study, as being retrospectively biased or due to the effect of a selection bias.

Various groups have failed to demonstrate a difference between the subtypes of depression based on psycho-social precipitants in cross sectional design.[40-42] An interesting concept suggests that "endogenicity" and "reactivity" are orthogonal dimensions rather than mu-tually exclusive categories.[43]

It seems, therefore, that diagnostically and prognostically heterogenous groups of de-pressed patients are being lumped under the rubric of "neurotic depression". Unless psy-chiatric classification and nomenclature achieve the sophistication of being able to provide an etiological classification of such conditions, a biological understanding would at best be presumptuous. Validation and comparison of studies dealing with biological markers, etc. in the "neurotic depressions" are difficult to conduct, due to the lack of a unified concept and heterogenous viewpoints. Recently, however, familial and genetic data suggest that a subgroup of depressions, most likely resembling the "neurotic depressives" of the earlier classifi-cations, may justify their existence as a separate entity.

IV. FAMILY HISTORY AND GENETIC DATA

At the present time, the primary-secondary dichotomy is perhaps the most useful clas-sificatory scheme in the dualistic approach to depressive conditions. Winokur[44,45] suggested that neurotic depression is synonymous with secondary depression. Similar observations were made by other investigators.[38,46,47] In a family study of unipolar and neurotic-reactive depressives,[48] the morbidity risk (MR%) of bipolar illness among neurotic-reactive depres-sives was very low (1% among parents and 0.7% among siblings), whereas the overall MR% for affective disorders was high (12.1% among parents and 6.7% among siblings) The MR% for alcoholism was higher in the neurotic-reactive depressives (3.7% among parents and 1.2% among siblings), as compared to the unipolar probands (0% among parents and 1.1% among siblings). Although the morbidity for affective disorders is significantly higher among relatives of unipolar than neurotic-reactive probands, the high MR% found among relatives of the latter suggests that a hereditary factor is important in the occurrence of a neurotic-depressive reaction.

D'Elia et al.[55] point out the term "neurotic-reactive depression" can be defined purely on the basis of phenomenological factors and corresponds to the neurotic-depressive reaction in the WHO manual. Although the term has multiple meanings, if it is limited to a significant depressive syndrome in a person who has a life-long history of a stormy life-style and personality problems, the reliability should improve.[53] The diagnosis of neurotic-reactive depression should be based on positive criteria and not solely by exclusion of endogenous features. Such a diagnosis should be validated by clearly independent factors unrelated to personality or life events. Of the familiar subgroups of their 288 female neurotic depressive probands, 85 to 95% fit Feighner's criteria for primary unipolar depression.[5] There was a highly significant increase in alcoholism in families of the neurotic-reactive depression compared to the endogenous subtype (55 vs. 31%, $p < 0.005$). Those with a family history of alcoholism were characterized by personality traits suggestive of emotional instability, i.e., demanding behavior, life-long nervousness, complaining and irritability, and fear. Based on these data, this group of patients has been classified as the depression spectrum diseases (DSD).[56] This is described as an ordinary depression occurring in a person with a family history of alcoholism or antisocial personality. Depression may exist in the family, though the marker for the illness is a family history of alcoholism (mostly) and antisocial personality (less likely). DSD shares commonality with neurotic-reactive distinction.[57] Like neurotic-depressive reaction, DSD occurs in individuals who have a stormy life history of multiple social and personal problems, and life-long irritability. Also, unlike the familial pure depressive disease (FPDD) which manifests as recurrent episodes, DSD is less likely to show this course. Furthermore, the validity of this concept was demonstrated by Perris et al.[48] In their sample, 43 patients with FPDD, 16 (32%) had a cross-diagnosis of reactive-neurotic depression, whereas 12 out of 15 patients (80%) with a diagnosis of DSD had a cross-diagnosis of reactive-neurotic depression and the differences were statistically significant.

Supportive data to this concept[58] indicate an increased alcoholism in families of primary depressives with a co-existing personality disorder, a greater likelihood of having a diagnosis of neurotic-reactive depression and lesser likelihood of being nonsuppressors on the dexamethasone suppression test (DST). Zimmerman et al.[59] compared the psychosocial, demographic, and clinical correlates of familial subtypes of primary unipolar depression, i.e., within Winokur's trichotomy in a prospective fashion. Their results are consistent with the view that DSD overlaps neurotic depression and FPDD overlaps endogenous depression. In comparison with patients having FPDD, those with DSD had more frequent marital separations or divorces, experienced more life events, had poorer social support, and more frequently made nonserious suicide attempts. There was, however, only a trend in the predicted direction of more personality disorder in patients with DSD. On the basis of the above, studies which have used clinical diagnoses or personality factors indicate that a family history of alcoholism is related to the diagnosis of a neurotic-reactive depression. This provides a strong support for the validity of the concept of the neurotic-reactive depression.

V. BIOLOGICAL STATUS OF THE PSYCHOTIC-NEUROTIC DISTINCTION: FURTHER EVIDENCE

Shagass[60] reviewed the various studies providing evidence for the psychotic-neurotic distinction in depression. With the increasing use of electroconvulsive therapies (ECT), it was observed that clinically distinguishable subgroups of depression responded quite differently to these treatments, i.e., major depressive disorders experienced a high rate of remission, whereas the minor depressive disorders did not respond well. Bipolar depressives have shown a reduced salivary secretion as compared to the other depressives.[61] Mecholyl and epinephrine challenge tests to study the autonomic balance revealed that patients with an elevated blood pressure (140 mm systolic or higher) prior to the challenge, and who showed a greater and prolonged decrease with mecholyl were those who responded well to ECTs, irrespective of diagnosis.[62-64] Sedation thresholds tested by electrophysiological monitoring during amytal injections were found elevated in neurotic depressives and found unrelated to the level of anxiety.[60] Similar observations[65,66] were made using galvanic skin response (GSR) as a measure of sedation threshold. They formulated a hypothesis that the hypothalamus of neurotic depressives was in a state of hyperexcitability. The results were substantiated by demonstrating an increase in the 17-hydroxy corticosterone, and urine levels of norepinephrine, metanephrine, and 3-methoxy-4 hydroxy vanillylmandelic acid (VMA) in the neurotic depressives, as compared to psychotic depressives and normals. Stimulation thresholds measured by changes in pulse and blood pressure following methamphetamine or methylphenydate were found low in neurotic depressives and high in the psychotic depressives (7.1 vs. 17 mg/kg of methamphetamine required to produce a 25% change in pulse or 15% change in blood pressure respectively).[67]

The DST has been extensively used to differentiate between the subgroups of depression.[68] Although 70% of patients suffering from Major Depressive Disorder have an abnormal DST,[69] with figures being higher for endogenous depression using the New Castle Diagnostic Score (81%), about 49% of nonendogenous depressives also present with an abnormal response. The latter figure is similar for ICD diagnosed neurotic depressive outpatients. The common occurrence of an abnormal DST in depressives as well as other neurotics, suggests a commonality in the underlying biological changes. Although the sensitivity of the DST is high in endogenous depression, its relative lack of specificity limits its diagnostic value. Lack of diagnostic specificity of the DST has been extensively reported in studies of endogenous vs. neurotic depressives,[70-73] psychotic vs. nonpsychotic depressives,[74] primary vs. secondary depressives,[75] Winokur's subtypes,[76] and in comparison of endogenous vs. neurotic and minor depressives.[77]

Rinieris et al.[78] compared the thyroid function tests in psychotic and neurotic depressives. Comparison revealed a significantly higher score for depression and lower values for serum thyroxine and free thyroxine index in the group of psychotic depressives. The thyroid function tests were generally normal in the group of patients with neurotic depression. These results may be interpreted in light of recent findings implicating catecholamines in the release of hypothalmic hormones[79] and the neurochemical mechanism of affective disorders.[80,81]

A reduced growth hormone (GH) response to amphetamine in endogenous as compared to neurotic depressives has been reported.[82] Matussek et al.[83-84] demonstrated a diminished GH response to 0.15 mg clonidine in endogenous depression as compared to neurotic depressives and controls. Evidence toward the involvement of hypothalmo-pituitary-axis (HPA) in depressed patients[85] has been usually interpreted as evidence of psychological stress[86] or of a limbic system disturbance in this illness.[87] Elevation of total plasma cortisol levels, of cortisol metabolite excretion, of urinary free cortisol and glucocorticoid production production rates have been reviewed and studied extensively.[88-91] Carroll et al. examined the csf cortisol levels of Unipolar (UP) and Bipolar (BP) depressives and compared them to those suffering from neurotic depressives and nondepressives.[85] Patients with a UP and BP depression had values twice as high as the neurotic depressives and the nondepressives. Furthermore, psychotic UP and BP depressives had values four times as high as the neurotic depressives.

Recent findings have demonstrated a relationship between the levels of 24 hr urinary excretion of cyclic AMP and affective illness. This nucleotide has been implicated as a second messenger substance for many neurotransmitter and hormone induced responses.[92] Cyclic AMP is functionally closely related and possibly fundamental to the action of catecholamines and serotonin, both of which have been implicated in the amine hypothesis of depression.[93] An increase in the level of urinary Cyclic AMP has been observed after improvement from depression,[94,95] after ECT,[96] and during manic illness following a switch from depression to mania.[97,98] Furthermore, severely depressed patients had a significant decrease in the 24 hr cyclic AMP levels. Sinanan et al.[99] found no difference in the 24 hr urinary cyclic AMP levels between the endogenous and neurotic depressives, although the levels were reduced in both the disorders. They suggested that the lower levels of Cyclic AMP in psychotic depression as compared to neurotic depressives[97] is due to the introduction of an experimental bias as a result of a 2-week washout period which has caused a spontaneous remission in the neurotic depressives. The findings of this study supports the unitary theory of depression of Sir Aubrey Lewis.[215]

Whereas considerable evidence to suggest that UP depressives have a decrease in brain monoamine levels during depression,[80,100-102] the evidence for neurotic depressives is not very supportive. Furthermore, the platelet serotonin levels in neurotic and characterological depression have been found no different from normal controls.[103] Kaiya et al.[104] investigated the plasma glutamate decarboxylase (GAD) activity as a probe to evaluate the GABA (gamma-amino-butyric-acid) functions and found low GAD activity in neurotic patients. However, the description of "neurotic" patients was not provided, and, therefore, the results attributable to neurotic depression are not available. In addition to reduced GAD activity in neurotic patients, reduced GAD levels were also found in depressives. These observations were ascribed to a related clinical symptom common to both disorders, i.e., anxiety. Controversy exists over the reduced serum Dopamine β Hydroxylase activity in depressive disorders.[105,106] In patients with affective disorders, contradictory results have been reported as concerns the activity of platelet mono-amine-oxidase (MAO). Platelet MAO activity has been studied as a possible biological marker for vulnerability since it appears to be under genetic control,[107-111] a significantly higher morbidity risk for neurotic-reactive depression as well as for alcoholism in the first-degree relatives of probands with affective illness and low platelet MAO activity. These results seem to be in line with Winokur's[113] description of depressive spectrum disease (DSD). Similar observations[114,115] have been made by other investigators. However, Puchall et al.[111] demonstrated an increased incidence of "high MAO related disorders" (major or minor depression, depressive personality) in relatives of probands with high MAO activity.

Serum glutamate levels were found elevated in a group of endogenous and neurotic depressives.[116] In an earlier study[117] free tryptophan levels were found reduced in both endogenous and neurotic depressives. No inter-group differences were observed. In contrast to reduced levels of free and total tryptophan in the plasma of five endogenous depressive patients, neither total nor free tryptophan was reduced in the plasma of five neurotic depressives.[118] Possibly the reduced levels of free and total tryptophan in the endogenous and neurotic depressives may not have been a cause but a consequence of the depression. However, the validity of their diagnoses is questionable. A decrease in the 5 HT uptake into platelets of endogenous and neurotic depressives[119] reflects the decreased central 5 HT function in depressed patients.[120] The absence of difference of pretreatment monoamine uptake between the endogenous and the neurotic/reactive unipolar depressives suggest that the effect on the platelets is nonspecific to the type of depression and that platelet amine uptake measurements alone are not sufficient to distinguish biochemically the two different types of depression.

VI. NEUROPSYCHOLOGICAL STUDIES

Neuropsychological evaluation of 60 depressed patients revealed selective impairment in the processing of visio-spatial material and retention of both nonverbal and verbal information by both neurotic and psychotic depressives, with a greater impairment shown by the latter group, permitting a clear separation between the groups. Posttreatment improvement did not permit a separation between them.[121] The results suggest a difference in the severity and not in the pattern of the neuropsychological deficits between the groups, with involvement of the nondominant hemisphere. These findings are consistent with previous research suggesting an association between the nondominant hemisphere dysfunction and depression.[122-127] Remission of asymmetrical cognitive deficits with attenuation of clinical symptoms suggests that functional alterations, as opposed to structural abnormalities, may underlie the neuropsychological deficits. The investigation of the neuropsychological deficits in depressed patients before and after antidepressant therapy provides evidence for a single continuum for neurotic/psychotic depressive subtypes, with greater impairment in the latter group.

A. The Relationship of "Mixed" Syndromes and Electrophysiological Studies

Considerable overlap between the symptoms of depression and anxiety has led to the study of co-morbid syndromes of depression and anxious states.[128-131] These "mixed" syndromes are associated with chronicity and poor psychosocial outcome.[132-134] Silberman and Post[135] report a poorer premorbid social adjustment in their atypical depressives. An overlap exists between the neurotic, anxious, phobic, hostile or characterological depressives and those classified as atypical depressives; and this is on a continuum which overlaps the typical syndrome.[136-138] Weissman and associates have shown a strong association between comorbidity of major depression and anxiety disorder or secondary alcoholism and early age of onset with presence of major depression in the first-degree relatives.[139] A striking similarity exists between the "co-morbid" syndrome of Wiessman's group and Winokur's subclassification of the Depressive Spectrum Disorders. Significant overlap between syndromes earlier believed to be mutually exclusive has been determined since the development of structured diagnostic assessments.[140,141]

A short first REM latency has been reported in primary depressives,[135,142,143] "characterological depressives",[144] and atypical depressives.[147] Although delta sleep is found reduced in primary depression,[142,145,146] a two-fold increase is seen in atypical depression. The other REM parameters, lie between endogenous depressives and normals.[135] These patients may, in fact, have been "neurotic" or "characterological" depressives. The fact that they have a shortened REM latency of the first REM period suggests a biological diathesis and that they share some pathophysiological features with other groups of depressives. In our own sample[134] the "mixed" group (i.e., MDD plus panic) occupied an intermediate position between the pure MDD group and the panic patients, suggesting that it consisted of a heterogenous population.

A functional cholinergic hypersensitivity in endogenously depressed patients has been observed by some investigators.[148-151] In one study[152] the shortening of REM latency by cholinergic infusion did not differentiate between endogenous, neurotic, and other depressed patients. However, arousal responses during the infusion procedure underscores the impact of their findings. Shortened REM latency was not found in a group of patients with "character spectrum disorder".[153] Based on their observations of family history, REM latency, and pharmacological responsiveness, they have proposed a nosological framework for the understanding of Dysthymic disorder.[153]

VII. TREATMENT

The term neurotic depression obfuscates several useful distinctions and prompts unjustified implications with regard to etiology and treatment. Furthermore, empirical evidence suggests that the term does not help in differentiating patients with mild, endogenous depressions, who are presumably drug sensitive, from the larger heterogenous group of other mild depressions.

A. Mono-Amine Oxidase Inhibitors (MAOIs)

Evidence clearly established the value of phenelzine as an antidepressant for patients suffering from atypical or neurotic depressions. The heterogenous nature of such diagnostic subtypes accounts for some of the inconsistent results with phenelzine. Reports of positive results for phenelzine in controlled trials comes from several studies.[154-161] On the other hand, several investigators[162-167] have reported negative results. In a double-blind controlled investigation of phenelzine, amitriptyline, and placebo in outpatients of depression or mixed anxiety and depression,[168] both drugs had significant antidepressant effects, without evidence of a preferential response in any subgroup. Kay et al.[166] reported similar results. Klein et al.[169] have reviewed 9 studies involving a phenelzine/placebo contrast in neurotic or atypical depressives. Of these, seven studies showed a significant superiority of phenelzine over placebo. Superiority of Moclobemide, a new MAOI, was seen over placebo in a group of psychotic and neurotic depressed patients.[170] The utility of tranylcypromine is strongly suggested, but is not as well supported as phenelzine, due to a paucity of controlled studies. There is evidence that, as is the case for phenelzine, tranylcypromine is more useful in neurotic or reactive depressives. Furthermore, it is now well accepted that patients who appear phenomenologically similar may have depressive illness mediated by diverse biochemical mechanisms. This may account for the endogenous depressives responding to MAOIs rather than tricyclic antidepressants (TCAs). MAOIs have also been recommended in atypical depressives.[169]

B. Tricyclic Antidepressants (TCAs)

Several studies have documented the positive effects of TCAs in neurotic outpatient depressives.[171-173] The relative efficacy of TCAs compared to MAOIs seems an outdated question. The assumption that TCAs have a broader spectrum of effectiveness results from earlier studies conducted on endogenously depressed inpatients. Current psychiatric literature provides no conclusive evidence with regard to the superiority of treatment with any particular TCA.[174,177] A poor response to imipramine[175] was correlated to severity of psychic anxiety, and mild depression accompanied by anxiety, somatization, and hypochondriasis.[176] A combination of amitriptyline and chlordiazepoxide has been observed to be most effective in the high depression/high anxiety group, chlordiazepoxide alone in the high anxiety/low depression group, and amitriptyline alone in the high depression/low anxiety group.[178] While imipramine and doxepin were both found effective in a group of psychoneurotic patients, analysis of covariance found imipramine to be superior.[179] In other studies, imipramine and doxepine have been found to be equally effective.[180-184,186] Though studies on drug specificity[175,185] have indicated that amitriptyline and imipramine are each effective in controlling certain symptoms, more recent studies[177] contradict these findings. Clomipramine was favored over doxepine in another study of 99 out-patients with neurotic depression.[187] Maprotiline was found superior to imipramine in terms of slightly better response and faster onset of action.[188] Chronic treatment with nomifensine, (a potent inhibitor of noradrenaline and dopamine uptake) in a group of neurotic depressives[189] caused a significant reduction of morning elevations of cortisol levels and an increase in catecholamine levels by the sixth week of treatment (dopamine: 106%; noradrenaline: 14%; adrenaline: 10%). The enhanced catecholamine transmission and lowering of plasma cortisol was suggested as a possible cause of improvement.

C. Anxiolytics

Several studies have recommended the use of anti-anxiety medication in the treatment of neurotic depression, some having found them superior to antidepressants.[90-192,194-200] Others have cautioned against the use of combinations of anti-anxiety and antidepressants due to lack of advantage over single drug treatments.[193,201-203]

In summary, various pharmacotherapeutic agents have been tried and recommended in neurotic conditions. The role of MAOIs and TCAs is well accepted, without any clear evidence of one drug over the other. However, the role of anxiolytics and benzodiazepines is not well established. With the exception of some of the newer anxiolytics, such as alprazolam, it is doubtful if the benzodiazepines have any antidepressant effects. Trials with newer agents such as L-tryptophan, L-5 hydroxytryptophan, L-dopa, thyroid hormones, pyridoxine (B_6), combinations of MAOIs and TCAs, and the newer class of antidepressant drugs, i.e., Maprotiline and Mianserine, are warranted to establish their role as possible therapeutic agents in neurotic depression.

VIII. CONCLUSION

The term "neurotic depression" has been a subject of major controversy in the last decade. There has been no empirical basis for assuming the universal presence of an intrapsychic conflict resulting in symptom formation in those disorders. The term "neurosis" has even lost its earlier specificity, as contemporary psychoanalytic theory has shifted its focus from "symptom neurosis" to the "character neurosis" (personality disorders). These are some of the reasons that have led to the dropping of the term "neurosis" in entirety from the DSM-III.[204] Since psychiatry today does not know with certainty the causes of several maladies it attempts to treat, it is recommended that classification should be based on shared phenomenological characteristics.[205]

A critical task facing investigators today is the valid interpretation of research results. Several studies which, in effect, may have actually studied the same patient population, have variously referred to their sample populations as: neurotic, nonendogenous, reactive, situational, character spectrum disorder, neurotic-reactive, characterological, or "mixed" anxiety depression. Hence, interpretation and integration of the available information is crucial to understanding the relationship of these conditions to each other. Angst and Dobler-Mikola[206] have reported data consistent with the Winokur's subgroup of neurotic-reactive depression. Their term "conflictual depression" is characterized by recurrent brief depressions in individuals with chronic personality problems, who are in conflict with parents and others. They also experience intimate relationships as being distressing. On the other hand, the "character spectrum disorders" of Akiskal[207,208] show an increased family history of alcoholism, along with a normal REM latency, different from the reduced REM latency of endogenous depressives. Terms such as "chronic characterological depression", and "character spectrum disorders" share a commonality with neurotic reactive depression, as does depression spectrum disease. The diagnosis of neurotic reactive depression can be made by exclusion of endogenous depression, but it may also be made based on positive criteria, i.e., an illness with a chronic, stormy, disabling lift-style and a lack of repeated hospitalizations for major depressive episodes. There may be an associated pattern of life-long irritability and a tendency to complain.[209] Mendels[210] reported that one group of depressives showed "reactive items", such as neurotic traits in childhood and adulthood, precipitating factors, initial insomnia, inadequate personality, and emotional lability, which in essence are similar to the neurotic-reactive depressives. Winokur[53] provides an extensive review and discussion providing validity for the concept of neurotic-reactive-depression. This theme espouses a philosophy which is at variance with that of the DSM-III. Neurotic-reactive depressives may or may not be in absolute congruence with the concept of Dysthymia, since the depression may or may not be severe or chronic. What seems to be of special interest is that the validity of the concept is provided by a unifying theme of a pre-existent familial factor, i.e., familial alcoholism, in addition to phenomenological criteria. A significant overlap is seen between the concept of neurotic-reactive depression and depression spectrum disease. The former is preferable, as many patients with neurotic reactive depression do not have a family history of alcoholism. Even though such a family history is a validating factor for the concept, it may not be present in the individual patient.

At the present time, the only set of criteria including neurotic reactive traits and endogenous symptoms is the Newcastle scale.[211] Further studies using the Newcastle scale and Winokur's schema may help elucidate the differences between endogenous and the neurotic reactive depressives. Hirschfeld[212] reported on a group of "situational" major depressions which are defined by an apparent precipitant for the illness episode, but are quite similar to a group of patients without an apparent precipitant. The question of presence or absence of precipitants in classification of depression needs further exploration.

Winokur's hypothesis raises the question of whether individuals with similar personality traits but without any history of major depressive episode would have an equally high family history of alcoholism. The finding that two to three times increased alcoholism has been seen in relatives of probands with borderline personality disorder than in other patient groups lends support to this concept.[213] It appears, therefore, that if an apparently reactive depression occurs in the context of verifiable pre-existing personality traits, this may serve as the basis for a valid subclassification of depression, i.e., neurotic-reactive-depression.

Important evidence regarding the neurotic reactive condition is derived from studies of Weissman and her colleagues. The "mixed" syndromes of co-morbid anxiety-depression manifest a remarkable similarity to the description of the neurotic-reactive-depressives, i.e., early onset of depression, chronicity of course, greater impairment in functioning, poor treatment response to antidepressants, increased alcoholism in family members, and polysomnographic characteristics,[134] unlike the classical endogenous depressives.[214] Further studies regarding this group's similarity to Winokur's trichotomy are on the way and appear promising. In order to present a coherent framework of the neurotic-reactive-depression, a careful selection of this patient group and its validation with studies of biological markers, biochemical substrates, and treatment outcome are required.

REFERENCES

1. **Kraepelin, E.,** *Manic Depressive Insanity and Paranoia,* E and S Livingstone, Edinburgh, 1921.
2. **Bleuler, E.,** *Textbook of Psychiatry,* Dover, New York, 1951.
3. **Woodruff, R. A., Murphy, E. E., and Herjanic, M.,** The natural history of affective disorders: I. Symptoms of 72 patients at the time of index hospital admission, *J. Psychiatr. Res.,* 5, 255, 1967.
4. **Robins, E. and Guze, S. B.,** Classification of affective disorders: The primary-secondary, the endogenous-reactive and the neurotic-psychotic concepts, in Williams, T. A., Katz, M. M., and Shield, J. A., Eds., Recent Advances in the Psychobiology of the Depressive Illnesses. DHEW Publication No. (HSM) 79-9053, U.S. Government Printing Office, 1972.
5. **Fieghner, J. P., Robins, E., Guze, S. B., et al.,** Diagnostic criteria for use in psychiatric research, *Arch. Gen. Psychiatry,* 26, 57, 1972.
6. **Spitzer, R. L., Endicott, J., and Robins, E.,** Research Diagnostic Criteria, New York, New York State Department of Mental Hygiene, New York State Institute, Biometrics Research, 1975.
7. **Klerman, G. L.,** Clinical phenomenology of depression: Implications for research strategy in the psychobiology of the affective disorders, in Williams T. A., Katz, M. M., and Shield, J. A., Eds., Recent Advances in the Psychobiology of the Depressive Illnesses. DHEW Publication No. (HSM) 70-9053, U.S. Government Printing Office, 1972.
8. **Kline, N. S.,** Depression: Diagnosis and treatment, *Med. Clin. North Am.,* 45, 1041, 1961.
9. **Roth, M.,** Depressive state and their borderlands: Classification, diagnosis and treatment, *Compr. Psychiatry,* 4, 135, 1960.
10. **Rosenthal, S. H. and Gudeman, J. E.,** The endogenous depressive pattern. *Arch. Gen. Psychiatry,* 16, 241, 1967.
11. **Klein, D. F.,** Endogenomorphic depression. *Arch. Gen. Psychiatry,* 31, 447, 1974.
12. **Akiskal, S. H. and McKinney, W. T.,** Depressive disorders: Toward a unified hypothesis, *Science,* 182, 20, 1973.

13. **Gillespie, R. D.,** The clinical differentiation of types of depression, *Guy's Hosp. Rep.,* 79, 306, 1929.
14. **Van Praag, H. M., Uleman, A. M., and Spitz, J. C.,** The vital syndrome interview, *Psychiatr. Neurol. Neurochir.,* 68, 329, 1965.
15. **Hamilton, M. and White, J. M.,** Clinical syndromes in depressive states, *J. Ment. Sci.,* 105, 985, 1959.
16. **Kiloh, L. G. and Garside, R. F.,** The independence of neurotic depression and endogenous depression, *Br. J. Psychiatry,* 109, 451, 1963.
17. **Hordern, A., Burt, C. G., Holt, N. F., et al.,** *Depressive States: A Pharmacotherapeutic Study,* Charles C Thomas, Springfield, Ill., 1965.
18. **Carner, M. W. P., Roth, M., and Garside, R. F.,** The diagnosis of depressive syndromes and the prediction of E.C.T. response, *Br. J. Psychiatry,* 3, 659, 1965.
19. **Rosenthal, S. H. and Klerman, G. L.,** Content and consistency in the endogenous depressive pattern, *Br. J. Psychiatry,* 112, 471, 1966.
20. **Mendels, J. and Cochrane, C.,** The nosology of depression: The endogenous-reactive concept, *Am. J. Psychiatry,* 124, 1, 1968.
21. **Kendell, R. E.,** *The Classification of Depressive Illness,* Oxford University Press, London, 1968.
22. **Kay, W. K., Garside, R. F., Beamish, P., et al.,** Endogenous and neurotic syndromes of depression: A factor analytic study of 104 cases. Clinical features, *Br. J. Psychiatry,* 115, 377, 1969.
23. **Fahy, T. J., Brandon, S., and Garside, R. F.,** Classification of depressive illness, *Proc. R. Soc. Med.,* 62, 331, 1969.
24. **Garside, R. F., Kay, D. W. K., Wilson, I. C., et al.,** Depressive syndromes and the classification of patients, *Psychol. Med.,* 1, 333, 1971.
25. **Paykel, E. S., Prusoff, B., and Klerman, G. L.,** The endogenous-neurotic continuum in depression: Rater independence and factor distributions, *J. Psy. Res.,* 8, 73, 1971.
26. **Kiloh, L. G., Andrews, G., Neilson, M., et al.,** The relationship of the syndromes called endogenous and neurotic depression, *Br. J. Psychiatry,* 212, 183, 1972.
27. **Lewinsohn, P. M., Zeiss, A. M., Robert, M. A., et al.,** Endogeneity and reactivity as orthogonal dimensions in depression, *J. Nerv. Ment. Dis.,* 164, 327, 1977.
28. **Overall, J. E., Hollister, L. E., Johnson, M., et al.,** Nosology of depression and differential response to drugs, *JAMA,* 195, 946, 1966.
29. **Pilowsky, I., Levine, S., and Boulton, D. M.,** The classification of depression by numerical taxonomy, *Br. J. Psychiatry,* 115, 937, 1969.
30. **Everitt, B. S., Goulay, A. J., and Kendell, R. E.,** An attempt at validation of traditional psychiatric syndromes by cluster analysis, *Br. J. Psychiatry,* 119, 399, 1971.
31. **Paykel, E. S.,** Classification of depressed patients. A cluster analysis derived grouping, *Br. J. Psychiatry,* 118, 275, 1971.
32. **Weckowicz, T. E., Yonge, K. A., Cropley, A. J., et al.,** Objective therapy publications in depression: A multivariate approach, *J. Clin. Psychol.,* 27, 3, 1971.
33. **Fleiss, J. L.,** Classification of the depressive disorders by numerical typology, *J. Psychiatr. Res.,* 9, 141, 1972.
34. **Gurney, C., Roth, M., Garside, R. F., et al.,** Studies in the classification of affective disorders, *Br. J. Psychiatry,* 121, 162, 1972.
35. **Cassidy, W. L., Flanagan, N. B., Spellman, B. A., et al.,** Clinical observations in manic-depressive disease, *JAMA,* 164, 1535, 1957.
36. **Foulds, M. A.,** Psychotic depression and age, *J. Ment. Sci.,* 106, 1394, 1960.
37. **Sandifer, M. Y. G., Wilson, I. C., and Green, L.,** The two-type thesis of depressive disorders, *Am. J. Psychiatry,* 123, 93, 1966.
38. **Akiskal, H. S., Bitar, A. H., Puzantian, V. R., Rosenthal, T. L., and Walker, P. W.,** The nosological status of neurotic depression. A prospective three-to four-year follow-up examination in light of the primary secondary and unipolar-bipolar dichotomies, *Arch. Gen. Psychiatry,* 35, 756, 1978.
39. **Klerman, G.,** Clinical research in depression, *Arch. Gen. Psychiatry,* 24, 305, 1971.
40. **Lewis, A.,** Melancholia: A clinical survey of depressive states, *J. Ment. Sci.,* 80, 277, 1934.
41. **Leff, M. J., Roatch, J. F., and Bunney, W. E.,** Environmental factors preceding the onset of severe depressions, *Psychiatry,* 33, 293, 1970.
42. **Paykel, E., Myers, J., Dienelt, M., et al.,** Life events and depression: A controlled study, *Arch. Gen. Psychiatry,* 21, 753, 1970.
43. **Lewinsohn, P. M., Zeiss, A. M., Zeiss, R. A., et al.,** Endogeneity and reactivity as orthogonal dimension in depression, *J. Nerv. Ment. Dis.,* 164, 327, 1977.
44. **Winokur, G.,** Family history studies: VIII, Secondary depression is alive and well and . . . , *Dis. Nerv. Syst.,* 33, 94, 1972.
45. **Winokur, G.,** Diagnostic and genetic aspects of affective illness, *Psychiatry Ann.,* 3, 6, 1973.
46. **Guze, S. B., Woodruff, R. A., and Clayton, P. J.,** Preliminary communication-secondary affective disorder: A study of 95 cases, *Psychol. Med.,* 1, 426, 1971.

47. **Wood, D., Othmer, S., Reich, T., et al.,** Primary and secondary affective disorder: I. Past history and current episodes in 92 depressed inpatients, *Comp. Psychiatry,* 18, 201, 1977.
48. **Perris, C., Perris, H., Ericsson, U., and vonKnorring, L.,** The Genetics of Depression, a family study of unipolar and neurotic-reactive depressed patients, *Arch. Psychiatr. Nervenkr.,* 232, 137, 1982.
49. **Foulds, G. A.,** The relationship between the depressive illness, *Br. J. Psychiatry,* 123, 531, 1975.
50. **Paskind, H. A.,** Brief attacks of manic-depressive depression, *Arch. Neurol. Psychiatry,* 22, 123, 1929.
51. **Ascher, E.,** A criticism of the concept of neurotic depression, *Am. J. Psychiatry,* 108, 901, 1952.
52. **Winokur, G. and Pitts, F. N.,** Affective disorder. Is reactive depression an entity? *J. Nerv. Ment. Dis.,* 138, 541, 1964.
53. **Winokur, G.,** The validity of neurotic-reactive depression, new data and reappraisal, *Arch. Gen. Psychiatry,* 42, 116, 1985.
54. **Beck, A.,** *Depression: Clinical, Experimental and Theoretical Aspects,* Harper & Row, New York, 1967.
55. **D'Elia, G., vonKnorring, L., and Perris, C.,** Non-psychotic depressive disorders: A ten year follow-up, *Acta Psychiatr. Scand.,* suppl. 255, 173, 1974.
56. **Winokur, G., Behar, D., VanValkenburg, C., and Lowry, M.,** Is a familial definition of depression both feasible and valid? *J. Nerv. Ment. Dis.,* 166, 764, 1978.
57. **Winokur, G.,** Controversies in depression or do clinicians know something after all? in *Treatment of Depression: Old Controversies and New Approaches,* Clayton, P. and Barret, J., Eds., New York, Raven Press, New York, 1983, 153.
58. **Pfohl, B., Stangl, D., and Zimmerman, M.,** The implications of DSM-III personality disorders for patients with major depression, *J. Affective Disord.,* 7, 309, 1984.
59. **Zimmerman, M., Coryell, W., and Pfohl, B.,** Validity of familial subtypes of primary unipolar depression, clinical, demographic and psychosocial correlates, *Arch. Gen. Psychiatry,* 43, 1090, 1986.
60. **Shagass, C.,** Neurophysiological evidence for different types of depression. Great Britain, *J. Behav. Ther. Exp. Psychiatr.,* 12(2), 99, 1981.
61. **Strongin, E. L. and Hinsie, L. E.,** A method to differentiate manic-depressive depressions from other depressions by means of parotid secretions, *Psychiat. Quart.,* 13, 697, 1939.
62. **Funkenstein, D. H., Greenblatt, M., and Solomon, H. C.,** Norepinephrine-like substances in psychotic and psychoneurotic patients, *Am. J. Psychiat.,* 108, 652, 1952.
63. **Funkenstein, D. H., Greenblatt, M., and Solomon, H. C.,** An autonomic nervous system test for prognostic significance in relation to electroshock treatment, *Psychosom. Med.,* 14, 347, 1952.
64. **Funkenstein, D., King, S., and Drolette, M.,** The direction of anger during a laboratory stress-induced situation, *Psychosom. Med.,* 16, 404, 1954.
65. **Perez-Reyes, M.,** Differences in sedative susceptibility between types of depression: Clinical and neuro-physiological significance, *Recent Advances in the Psychobiology of the Depressive Illnesses,* 1972, 119.
66. **Perez-Reyes, M.,** Differences in the capacity of the sympathetic and endocrine systems of depressed patients to react to a physiological stress, *Recent Advances in the Psychobiology of the Depressive Illness,* 1972, 131.
67. **Gilberti, F. and Rosse, R.,** Proposal of a psychopharmacological test ("Stimulation threshold") for differentiating neurotic from psychotic depressions, *Psychopharmacologia,* 3, 128, 1962.
68. **Carroll, B. J., Martin, F. I. R., and Davies, B. M.,** Resistance to suppression by dexamethasone of plasma 11-OCHS levels in severe depressive illness, *Br. Med. J.,* 3, 285, 1968.
69. **Coppen, A., Abou-Saleh, M., Milln, P., Metcalfe, M., Harwood, J., and Bailey, J.,** Dexamethasone suppression test in depression and other psychiatric illness, *Br. J. Psychiatry,* 142, 498, 1983.
70. **Klein, H. E., Bender, W., Mayr, H., Niederschweiberer, A., and Schmauss, M.,** The DST and its relationship to psychiatric diagnosis, symptoms and treatment outcome, *Br. J. Psychiatry,* 145, 591, 1984.
71. **Stokes, P. E., Stoll, P. M., Mattson, M. R., and Sollod, R. N.,** Diagnosis and psychopathology in psychiatric patients resistant to dexamethasone, *Hormones, Behavior and Psychopathology,* Sachar, E. J., Ed., *American Psychopathological Series, Raven Press.* 1976.
71. **Stokes, P. E., Stoll, P. M., Mattson, M. R., and Sollod, R. N.,** Diagnosis and psychopathology in psychiatric patients resistant to dexamethasone, *Hormones, Behavior and Psychopathology,* Sachar, E. J., Ed., *American Psychopathological Series, Raven Press.* 1976.
72. **Berger, M., Doerr, P., Lund, R., Bronisch, T., and vonZerssen, D.,** Neuroendocrinological and neurophysiological studies in major depressive disorders: Are there biological markers for the endogenous sub-type? *Biol. Psychiatry,* 17, 1217, 1982.
73. **Berger, M., Pirke, K. M., Doerr, P., Krieg, J. C., and vonZerssen, D.,** The limited utility of the dexamethasone suppression test for the differential-diagnostic process in psychiatry, *Br. J. Psychiatry,* 145, 1984.
74. **Caroff, S., Winokur, A., Rieger, W., Schweizer, E., and Amsterdam, J.,** Response to dexamethasone in psychotic depression, *Psychiatry Res.,* 8, 59, 1983.
75. **Reus, V. J., Joseph, M. S., and Dallmann, M. F.,** ACTH levels after the dexamethasone suppression test in depression, *New Engl. J. Med.,* 360(4), 238, 1982.

76. **Swartz, C. M.**, Biologically derived depression and the dexamethasone suppression test, *Comp. Psychiatry*, 23(4), 339, 1982.

77. **Kasper, S. and Beckmann, H.**, Dexamethasone suppression test in a pluridiagnostic approach: its relationship to psychopathological and clinical variables, *Acta Psychiatr. Scand.*, 68, 31, 1983.

78. **Rinieris, P. M., Christodoulou, G. N., Souvatzoglou, A. M., Koutras, D. A., and Stefanis, C. N.**, Free-thyroxine index in psychotic and neurotic depression, *Acta Psychiatr. Scand.*, 58, 56, 1978.

79. **Tuomisto, J., Ranta, T., Saarinen, A., and Leppaluoto, J.**, Neurotransmission and secretion of thyroid-stimulating hormone, *Lancet*, ii, 510, 1973.

80. **van Praag, H. M. and Korf, J.**, Retarded depression and the dopamine metabolism, *Psychopharmacologia* (Berlin) 19, 199, 1971.

81. **Goodwin, F. K., Post, R. M., Dunner, D. L., and Gordon, E. K.**, Cerebrospinal fluid amine metabolites in affective illness: The probenecid technique, *Am. J. Psychiatr.*, 130, 73, 1973.

82. **Langer, G., Heinze, G., Reim, B., and Matussek, N.**, Reduced growth hormone responses to amphetamine in endogenous depressive patients, *Arch. Gen. Psychiatr.*, 33, 1471, 1976.

83. **Matussek, N., Ackenheil, M., Hippius, H., Schroder, T., Schultes, H., and Wasilewski, B.**, Effect of clonidine on HGH release in psychiatric patients and controls. VI World Congress of Psychiatry, Hawaii, 1977.

84. **Matussek, N., Ackenheil, M., Hippius, H., Muller, F., Schroder, H. T., Schultes, H., and Wasilewski, B.**, Effect of clonidine on growth hormones release in psychiatric patients and controls, *Psychiatry Res.*, 2(1), 25, 1980.

85. **Carroll, B. J., Curtis, G. C., and Mendels, J.**, Cerebrospinal fluid and plasma free cortisol concentrations in depression, *Psychol. Med.*, 6, 235, 1976.

86. **Sachar, E. J., Hellman, L., Fukushima, D. K., and Gallagher, T. F.**, Cortisol production in depressive illness: a clinical and biochemical clarification, *Arch. Gen. Psychiatry*, 23, 289, 1970.

87. **Carroll, B. J.**, Limbic system-adrenal cortex regulation in depression and schizophrenia, *Psychosom. Med.*, 1976.

88. **Michael, R. P. and Gibbons, J. L.**, Interrelationships between the endocrine system and neuropsychiatry, *International Review of Neurobiology*, Vol. 5, Academic Press, New York, 1963.

89. **Carpenter, W. T. Jr. and Bunney, W. E. Jr.**, Adrenal cortical activity in depressive illness, *Am. J. Psychiatry*, 128, 31, 1971.

90. **Carroll, B. J.**, Psychoendocrine relationships in affective disorders, *Modern Trends in Psychosomatic Medicine*, Hill, O. W., Ed., Butterworth, London, 1976.

91. **Carroll, B. J. and Mendels, J.**, Neuroendocrine regulation in affective disorders. *Hormones, Behavior and Psychopathology*, Sachar, E. J, Ed., Raven Press, New York, 1976.

92. **Sutherland, E. W. and Rall, T. W.**, Fractination and characterization of a cycle adenine ribonucleotide formed by tissue particles, *J. Biol. Chem.*, 232, 1079, 1958.

93. **Granville-Grossman, K.**, *Recent Advances in Clinical Psychiatry*, J & A Churchill, London, 1971, 27.

94. **Abdulla, Y. H. and Hamadah, K.**, 3'5' Cyclic adenosine monophosphate in depression and mania, *Lancet*, i, 378, 1970.

95. **Paul, M. I., Gamor, H., Bunney, W. E., and Goodwin, F. K.**, Urinary cyclic AMP excretion in depression and mania, effects of levodopa and lithium carbonate, *Arch. Gen. Psychiatry*, 24, 327, 1971.

96. **Hamadah, K., Holmes, H., Barker, G. B., Hartman, G. C., and Parke, D. V. W.**, Effect of electric convulsion therapy on urinary excretion of 3'5' cyclic adenosine monophosphate, *Br. Med. J.*, iii, 439, 1972.

97. **Paul, M. I., Ditzion, B. R., Paul, G. L., and Janowsky, D. S.**, Urinary adenosine 3'5'-monophosphate excretion in affective disorders, *Am. J. Psychiatry*, 126(10), 1493, 1970.

98. **Paul, M. I., Cramer, H., and Bunney, W. E.**, Urinary adenosine 3'5'-monophosphate in the switch process from depression to mania, *Science*, 171, 300, 1971.

99. **Sinanan, K., Keatinge, A. M. B., Beckett, P. G. S., and Love, W. C.**, Urinary cyclic AMP in "Endogenous" and "Neurotic" depression, *Br. J. Psychiatry*, 126, 49, 1975.

100. **Mendels, J. and Stinnett, J.**, Biogenic amine metabolism, depression and mania. *Biological Psychiatry*, Mendels, J., Ed., Interscience — John Wiley & Sons, New York, 1973, 99.

101. **Schildkraut, J. J.**, The catecholamine hypothesis of affective disorders: A review of supporting evidence, *Am. J. Psychiatry*, 122, 509, 1965.

102. **Schildkraut, J. J.**, Neuropharmacology of the affective disorders, *Ann. Rev. Pharmacol.*, 13, 427, 1973.

103. **Takahashi, S.**, Reduction of Blood Platelet Serotonin Levels in Manic and Depressed Patients, *Folia Psychiatrica Neurologica Japonica*, 30(4), 475, 1976.

104. **Kaiya, H., Namba, M., Yoshida, H., and Nakamura, S.**, Plasma glutamate decarboxylase activity in neuropsychiatry, *Psychiatry Res.*, 6, 355, 1982.

105. **Meltzer, H. Y., Cho, H. W., Carroll, B. J., and Russo, P.**, Serum Dopamine-B-Hydroxylase activity in the affective psychoses and schizophrenia, *Arch. Gen. Psychiatry*, 33, 585, 1976.

106. **Eisemann, M., Ericsson, U., VonKnorring, L., Perris, C., Perris, H., and Ross, S.,** Serum Dopamine-Beta-Hydroxylase in Diagnostic subgroups of Depressed Patients and in Relation to their personality characteristics, *Neuropsychobiology,* 9, 193, 1983.

107. **Murphy, D. L.,** Clinical genetic hormonal and drug influences in the activity of human platelet monoamine oxidase, in *Monoamine Oxidase and its Inhibition,* Lundstenholme, G. E. W., and Knight, J., Eds., Elsevier/Exception Medical/North Holland, Amsterdam, 1976, 341.

108. **Winter, H., Herschel, M., Propping, P., Friedl, W., and Vogel, F.,** A twin study on three enzymes (DBH, COMT, MAO) of catecholamine metabolism: Correlations with MMPI, *Psychopharmacology,* 57, 63, 1978.

109. **Pandey, G. N., Dorus, E., Shaughnessy, R., and David, J. M.,** Genetic control of platelet monoamine oxidase activity: Studies of normal families, *Life Sciences,* 25, 1173, 19079.

110. **Rice, J., McGuffin, P., Goldin, L. R., Shaskan, E. G., and Gershon, E. S.,** Platelet monoamine oxidase activity: Evidence for a single major locus, *Psychiatry Res.,* 7, 325, 1982.

111. **Puchall, L. B., Coursey, R. D., Buchsbaum, M. S., and Murphy, D. L.,** Parents of high-risk subjects defined by levels of monoamine oxidase activity, *Schizophrenia Bull.,* 6, 338, 1980.

112. **vonKnorring, L., Perris, C., Oreland, L., Eisemann, M., Holmgren, S., and Perris, H.,** Morbidity risk for psychiatric disorders in families of probands with affective disorders divided according to levels of platelet MAO activity, *Psychiatry Res.,* 15, 271, 1985.

113. **Winokur, G.,** Familial (genetic) subtypes of pure depressive disease. *Am. J. Psychiatry,* 136, 911, 1979.

114. **Buchsbaum, M. S., Coursey, R. D., and Murphy, D. L.,** the biological high-risk paradigm: Behavioral and familial correlates of low platelet monoamine oxidase activity, *Science,* 194, 339, 1976.

115. **Coursey, R. D., Buchsbaum, M. S., and Murphy, D. L.,** Two-year followup of subjects and their families defined as at risk for psychopathology on the basis of platelet MAO activities, *Neuropsychobiology,* 8, 51, 1982.

116. **Kim, J. S., Schmid-burgk, W., Claus, D., and Kornhuber, H. H.,** Increased serum glutamate in depressed, *Arch. Psychiatr. Nervenkr.,* 232, 299, 1982.

117. **Schmid-Burgk, W., Kim, J. S., Lischewski, R., and Rabmann, W.,** Levels of total and free tryptophan in the plasma of endogenous and neurotic depressives, *Arch. Psychiatr. Nervenkr.,* 231, 35, 1980.

118. **Fiore, C. F., Malatino, I. S., and Petrone, G.,** Differences between plasma tryptophan patterns in endogenous and neurotic depression, *IRCS Med. Sci.,* 7, 525, 1979.

119. **Ehsanullah, R. S. B.,** Uptake of 5-hydroxtryptamine and dopamine into platelets from depressed patients and normal subjects-influence of clomipramine, desmethylclomipramine and maprotiline, *Postgrad. Med. J.,* 56(Suppl 1), 31, 1980.

120. **Murphy, D. L., Campbell, I. C., and Costa, J. L.,** The brain serotonergic system in the affective disorders, *Prog. Neuropsychopharmacol.,* 2, 1, 1978.

121. **Fromm, D. and Schopflocher, D.,** Neuropsychological test performance in depressed patients before and after drug therapy, *Biol. Psychiatry,* 19(1), 1984.

122. **Brumback, R. A., Staton, R. D., and Wilson, H.,** Neuropsychological study of children during and after remission of endogenous depressive episodes, *Perceptual Motor Skills,* 50, 1163, 1980.

123. **Flor-Heny, P., Fromm-Auch, D., Tapper, M., and Schopflocher, D.,** A neuropsychological study of the stable syndrome of hysteria, *Biol. Psychiat.,* 16, 601, 1981.

124. **Stanton, R. D., Wilson, H., and Brumback, R. A.,** Cognitive improvement associated with tricyclic antidepressant treatment of childhood major depressive illness, *Perceptual Motor Skills,* 53, 219, 1981.

125. **Yucker, D. M.,** Lateral brain function, emotion and conceptualization, *Psychol. Bull.,* 89, 19, 1981.

126. **Tucker, D. M., Craig, E., Stenslie, R. S., Roth, S., and Shearer, L.,** Right frontal lobe activation and right hemisphere performance, *Arch. Gen. Psychiatry,* 38, 160, 1981.

127. **Yeudall, L. T. and Fromm-Auch, D.,** Neuropsychological impairments in various psychopathological populations, in *Hemisphere Asymmetries of Function in Psychopathology,* Gruzelier, J. and Flor-Henry, P., Eds., Elsevier, New York, 1979.

128. **Hamilton, M.,** Symptoms and assessment of depression, *Handbook of Affective Disorders,* Paykel, E. S., Ed., Guilford Press, New York, 1982.

129. **Fawcett, J. and Kravitz, H. M.,** Anxiety syndromes and their relationship to depressive illness, *J. Clin. Psychiatry,* 44, 8, 1983.

130. **Leckman, J. F., Weissman, M. M., Merikangas, K. R., Pauls, D. L., and Prusoff, B. A.,** Panic disorder and major depression: Increased risk of depression, alcoholism, panic and phobic disorders in families of depressed probands with panic disorder, *Arch. Gen. Psychiatry,* 40, 1055, 1983.

131. **Breir, A., Charney, D. S., and Heninger, G. R.,** Major depression in patients with agoraphobia and panic disorders. *Arch. Gen. Psychiatry,* 41, 1129, 1984.

132. **Schapira, K., Roth, M., Kerr, T. A., and Gurney, C.,** The prognosis of affective disorders: The differentiation of anxiety states from depressive illnesses, *Br. J. Psychiatry,* 121, 175, 1972.

133. **Van Valkenburg, C., Akiskal, H. S., Puzantian, V., and Rosenthal, T.**, Anxious depressions: Clinical, family history and naturalistic outcome—Comparison with panic and major depressive disorders, *J. Affective Disorders*, 6, 67, 1984.

134. **Dubé, S., Jones, D. A., Bell, J., Davies, A., Ross, E, and Sitaram, N.**, Interface of Panic and Depression: Clinical and sleep EEG correlates, *Psychiatry Res.*, 19, 119, 1986.

135. **Silberman, E. K. and Post, R. M.**, Atypicality in primary depressive illness: A preliminary Survey, *Biol. Psychiatry*, 17(3) 285, 1982.

136. **Kay, D. W. K., Garside, R. F., Beamish, P., and Roy, J. R.**, Endogenous and neurotic syndromes of depression: A factor analytic study of 104 cases. Clinical features, *Br. J. Psychiatry*, 115, 377, 1969.

137. **Kiloh, L. G., Andrews, G., Neilson, M., and Bianchi, G. N.**, The relationship of the syndromes called endogenous and neurotic depression, *Br. J. Psychiatry*, 121, 183, 1972.

138. **Everitt, S. B., Gourlay, A. J., and Kendall, R. E.**, An attempt at validation of traditional psychiatric syndromes by cluster analysis, *Br. J. Psychiatry*, 119, 399, 1971.

139. **Weissman, M. M., Merikangas, K. R., Wickramaratne, P., Kidd, K. K., Prusoff, B. A., Leckman, J. F., and Pauls, D. L.**, Understanding the Clinical heterogeneity of major depression using family data, *Arch. Gen. Psychiatry*, 43, 430, 1986.

140. **Hirschfeld, R. M. A.**, Situational depression: Validity of the concept, *Br. J. Psychiatry*, 139, 297, 1981.

141. **Spitzer, R. H., Endwitt, J., and Robins, E.**, *Research Diagnostic Criteria of a Selected Group of Functional Disorders*, 3rd ed., New York State Psychiatric Institute, New York, 1978.

142. **Kupfer, D. J.**, REM latency: A psychobiologic marker for primary depressive disease, *Biol. Psychiatry*, 11, 159, 1976.

143. **Gillin, J. C., Duncan, W., Pettigrew, K. D., Frankel, B. L., and Snyder, F.**, Successful separation of depressed, normal and insomniac subjects by EEG sleep data, *Arch. Gen. Psychiatry*, 36, 8590, 1979.

144. **Akiskal, H. S., Rosenthal, T. L., Haykal, R. F., Rosenthal, L. H., and Scott-Strauss, A.**, Characterological depression. Clinical and EEG findings separating "subaffective dysthymias" from "character spectrum disorders:, *Arch. Gen. Psychiatry*, 37, 777, 1980.

145. **Mendels, J. and Hawkins, D. R.**, Sleep and depression: A controlled EEG study. *Arch. Gen. Psychiatry*, 16, 344, 1967.

146. **Lowy, F. H., Cleghorn, J. M., and McClure, D. J.**, Sleep patterns in depression. Longitudinal study of six patients and brief review of the literature, *J. Nerv. Ment. Dis.*, 153, 10, 1971.

147. **Quitkin, F. M., Rabkin, J. G., Stewart, J. W., McGrath, P. J., Harrison, W., Davies, M., Goetz, R., and Puig-Antich, J.**, Sleep of atypical depressives, *J. Affect. Disorders*, 8, 61, 1985.

148. **Sitaram, N., Wyatt, R. J., Dawson, S., and Gillin, J. C.**, REM sleep induction by physostigmine infusion during sleep, *Science*, 191, 1281, 1976.

149. **Kupfer, D. J.**, Application of EEG sleep for the differential diagnosis and treatment of affective disorders, *Pharmacopsychiatria*, 11, 17, 1978.

150. **Sitaram, N. and Gillin, J. C.**, Development and use of pharmacological probes of the CNS in man. Evidence of cholinergic abnormality in primary affective illness, *Biol. Psychiatry*, 15, 925, 1980.

151. **Jones, D., Kelwala, S., Bell, J., Dube, S., Jackson, E., and Sitaram, N.**, Cholinergic REM sleep induction response correlation with endogenous major depressive subtype, *Psychiatry Res.*, 14, 1985.

152. **Berger, M., Lund, R., Bronisch, T., and vonZerssen, D.**, REM Latency in neurotic and endogenous depression and the cholinergic REM induction test, *Psychiatry Res.*, 10, 113, 1983.

153. **Akiskal, H. S.**, Dysthymic disorder: Psychopathology of proposed chronic depressive subtypes, *Am. J. Psychiatry*, 140, 1, 1983.

154. **Rees, L. and Davies, B. A.**, Controlled trial of phenelzine (Nardil) in the treatment of severe depressive illness, *J. Ment. Sci.*, 197, 407, 1973.

155. **Martin, M. E.**, A comparative trial of imipramine and phenelzine in the treatment of depression, *Br. J. Psychiatry*, 109, 279, 1963.

156. **Imlah, N. W., Fahy, P. T. P., and Harrington, J. A.**, A comparison of two antidepressant drugs, *Psychopharmacologia*, 6, 472, 1964.

157. **Johnstone, E. C. and Marsh, W.**, Acetylator status and response to phenelzine in depressed patients, *Lancet*, i, 567, 1973.

158. **Robinson, D. S., Nies, A., Ravaris, C. L., and Lamborn, K. R.**, The mono-amine oxidase inhibitor, phenelzine, in the treatment of depressive anxiety states, *Arch. Gen. Psychiatry*, 29, 407, 1973.

159. **Solyom, L., Heseltine, G. F. D., McClure, D. J., Soltom, C., Ledwidge, B., and Steinberg, G.**, Behavior therapy versus drug therapy in the treatment of phobic neurosis, *Can. Psychiatric Assoc., J.*, 18, 25, 1973.

160. **Tyrer, P., Candy, J., and Kelly, D.** Phenelzine in phobic anxiety: a controlled trial, *Psychol. Med.*, 3, 120, 1973.

161. **Ravaris, C. L., Nies, A., Robinson, D. S., Ives, J. O., Lamborn, K. R., and Korson, L.**, A multiple dose controlled study of phenelzine in depression and anxiety states, *Arch. Gen. Psychiatry*, 33, 347, 1976.

162. **Hare, E. H., Dominian, J., and Sharpe, L.** Phenelzine and dexamphetamine in depressive illness, *Br. Med. J.*, 1, 9, 1962.

163. **Glick, B. S.**, Double blind study of tranylcypromine and phenelzine in depression, *Dis. Nerv. Syst.*, 25, 617, 1964.

164. **Greenblatt, M., Grosser, G. H., and Weschler, H.**, Differential response of depressed patients to somatic therapy, *Am. J. Psychiatry*, 120, 935, 1964.

165. Medical Council Research Reports, Clinical trial of the treatment of depressive illness, *Br. Med. J.*, i, 881, 1965.

166. **Kay, D. W. K., Garside, R. F., and Fahy, T. J.**, A double blind trial of phenelzine and amitriptyline in depressed outpatients: a possible differential effect of the drugs on symptoms, *Br. J. Psychiatry*, 123, 63, 1973.

167. **Raskin, A., Schulterbrant, J. G., Reatig, N., Crook, T. H., and Odle, D.** Depression subtypes and response to phenelzine, diazepam and placebo, *Arch. Gen. Psychiatry*, 30, 66, 1974.

168. **Rowan, P. R., Payke, E. S., and Parker, R. R.**, Phenelzine and Amitriptyline: Effects on the Symptoms of Neurotic Depression, *Br. J. Psychiatry*, 140, 475, 1982.

169. **Klein, D. F., Gittelman, R., Quitkin, F., and Rifkin, A.**, Review of literature on mood stabilizing drugs, in *Diagnosis and Drug Treatment of Psychiatric Disorders: Adults and Children*, 2nd ed., Williams & Wilkins, Baltimore, 1980, 268.

170. **Casacchia, M., Carolei, A., Barba, C., Frontoni, M., Rossi, A., Meco, G., and Zylberman, M. R.**, A placebo controlled study of the antidepressant activity of Moclobemide, A new MAO-A inhibitor, *Pharmacopsychiatry*, 17, 122, 1984.

171. **Bielski, R. J. and Friedel, R. O.**, Prediction of tricyclic antidepressant response, *Arch. Gen. Psychiatry*, 33, 1479, 1976.

172. **Morris, J. B. and Beck, A. T.**, The efficacy of antidepressant drugs. A review of research (1958-1972), *Arch. Gen. Psychiatry*, 30, 667, 1974.

173. **Paykel, E. S.**, Predictors of treatment response, in Psycho-pharmacology of Affective Disorders, Paykel, E. S. and Coppen, A., Eds., University Press, Oxford.

174. **Wittenborn, J. R., Kiremitci, N., and Weber, E.**, The choice of alternative antidepressants, *J. Nerv. Ment. Dis.*, 156, 97, 1973.

175. **Hordern, A., Holt, N. F., Burt, C. G., et al.**, Amitriptyline in depressive states, *Br. J. Psychiatry*, 109, 815, 1963.

176. **Rickels, K., Ward, C. H., and Schut, L.**, Different population, different drug responses, *Am. J. Med. Sci.*, 247, 328, 1962.

177. **Goldberg, H. L. and Finnerty, R. J.**, Which tricyclic for depressed outpatients, Imipramine pamoate or Amitriptyline, *Dis. Nerv. Syst.*, 785, 1977.

178. **Rickels, K., Hesbacher, P., and Downing, R. W.**, Differential drug effects in neurotic depression, *Dis. Nerv. Syst.*, 468, 1970.

179. **Finnerty, R. J., Goldberg, H. L., and Rickels, K.**, Doxepin versus imipramine in psychoneurotic depressed patients with sleep disturbance: A double blind study, *J. Clin. Psychiatry*, 32, 852, 1978.

180. **Hackett, E. and Kline, N. S.**, Antidepressant activity of doxepin and amitriptyline: A double blind evaluation, *Psycho-somatics*, 10(Sect. II) 21, 1969.

181. **Solis, H. G., Molina, G. B., and Pineyrol, A.**, Clinical evaluation of doxepin and amitriptyline in depressed patients, *Curr. Ther. Res.*, 12, 524, 1970.

182. **Bianchi, G. N., Borr, R. F., and Kilah, L. G.**, A comparative trial of doxepin and amitriptyline in depressive illness, *Med. J. Austria*, 1, 843, 1971.

183. **Wittenborn, J. R., Kiremitci, N., and Weber, E.**, The choice of alternative antidepressants, *J. Nerv. Ment. Dis.*, 156, 97, 1973.

184. **Burt, C. G., Gordon, W. F., and Holt, N. F.**, Amitriptyline in depressive states: A control trial, *J. Ment. Sci.*, 108, 711, 1962.

185. **Kiloh, L. G., Ball, J. R. B., and Garside, R. F.**, Prognostic factors in the treatment of depressive states with imipramine, *Br. Med. J.*, 1, 1225, 1962.

186. **Kimura, M., Nakagowa, T., Suchiro, K., et al.**, A double blind comparison of clinical efficacy of doxepin and imipramine in depressive patients. Presentation at International Congress of Psychosomatic Disease, Italy, August, 1975.

187. **Kornhaber, A. and Horwitz, I. M.**, A comparison of clomipramine and doxepine in neurotic depression, *J. Clin. Psychiatry*, 45,(8) 337, 1984.

188. **Van der Velde, C.**, Maprotiline versus imipramine and placebo in neurotic depression, *J. Clin. Psychiatry*, 42(4) 138, 1981.

189. **Rettori, V., Rubio, M., Seilicovich, A., Malik, A., and De los Santos, A.**, Effect of nomifensine on cortisol, prolactin and biogenic amines in depressed patients, *Neuropsychobiology*, 10, 2098, 1983.

190. **Kellner, R. and Sheffield, B. F.**, A cross-over trial of a combination of chlordiazepoxide with amitriptyline, *Br. J. Clin. Pract.*, 11, 459, 1969.

191. **Rickels, K., Raab, E., DeSilverio, R., and Etemad, B.,** Drug treatment of depression: Antidepressant or tranquilizer? *JAMA,* 201, 105, 1967.

192. **Rickels, K., Gordon, P. E., Jenkins, B. W., Perloff, M., Sachs, T., and Stepansky, W.,** Drug treatment in depressive illness: Amitriptyline and chloridazepoxide in two neurotic populations, *Dis. Nerv. Syst.,* 31, 30, 1970.

193. **Freeman, H.,** The therapeutic value of combinations of psychotropic drugs: a review, *Psychopharmacol. Bull.,* 4, 1, 1967.

194. **Keller, R.,** Chlordiazepoxide in depressive neurosis: an open label cross-study with amitriptyline, *Can. Psychiatr. Assoc. J.,* 18, 393, 1973.

195. **Henry, B. W., Overall, J. E., and Markette, J. R.,** Comparison of major drug therapies for alleviation of anxiety and depression, *Dis. Nerv. Syst.,* 32, 655, 1971.

196. **Stonehill, E., Lee, H., and Bant, A.,** A comparative study with benzodiazepines in chronic psychotic patients, *Dis. Nerv. Syst.,* 27, 411, 1966.

197. **Kerry, R. J., Jenner, F. A., and Pearson, I. B.,** A double blind cross over comparison of RO 5-4450, bromazepam, diazepam (Valium) and chlordiazepoxide (Lithium) in the treatment of neurotic anxiety, *Psychosomatics,* 13, 122, 1972.

198. **Shammas, E.,** Controlled comparison of bromazepam, amitriptyline and placebo in anxiety-depressive neurosis, *Dis. Nerv. Syst.,* 201, 1977.

199. **Imlah, N. W.,** An evaluation of alprazolam in the treatment of reactive or neurotic (secondary) depression, *Br. J. Psychiatry,* 146, 515, 1985.

200. **Fabre, L. F.,** Pilot open label study with alprazolam in out-patients with neurotic depression, *Curr. Ther. Res.,* 19, 661, 1976.

201. **Hyde, C. and Goldberg, D.,** Labelling and drug effects in the treatment of neurotic affective disorders: an experimental investigation, *Psychol. Med.,* 13, 399, 1983.

202. **Chesrow, E. J., Kapfitz, S. E., and Breme, J. T.,** Nortriptyline for the treatment of chronically ill and geriatric and geriatric patients, *J. Am. Ger. Soc.* 12, 271, 1964.

203. **Freeman, H. L., Bourne, M. S., and Schiff, A. A.,** The treatment of anxious/depressive states, *J. Intl. Med. Res.,* 2, 197, 1974.

204. **Bayer, R. and Spitzer, R. L.,** Neurosis, Psychodynamics and DSM-III, *Arch. Gen. Psych.,* 42, 187, 1985.

205. American Psychiatric Association, Task force on nomenclature: Progress report on preparation of DSM-III, II, 1976.

206. **Angst, J. and Dobler-Mikola, A.,** The Zurich study, a perspective epidemiological study: IV. Recurrent and non recurrent brief depression, *Eur. Arch. Psychiatry Neurol. Sci.,* in press.

207. **Akiskal, H.,** Characterologic manifestations of affective disorders: Toward a new conceptualization, *Integrative Psychiatry,* 83, 1984.

208. **Yeravian, B. and Akiskal, H.,** "Neurotic," characterological and dysthymic depressions, in *The Psychiatric Clinics of North America,* Vol 2, No. 3, Akiskal, H., Ed., W. B. Saunders, Philadelphia, 595, 1979.

209. **Van Valkenberg, G., Lowry, M., Winokur, G., and Cadoret, R.,** Depression spectrum disease versus pure depressive disease, *J. Nerv. Ment. Dis.,* 165, 341, 1977.

210. **Mendels, J.,** Depression: Distinctions between syndromes and symptom, *Br. J. Psych.,* 114, 1549, 1968.

211. **Davidson, J., Strickland, R., Turnbull, C., Belyea, M., and Miller, R.,** The new castle endogenous depression diagnostic index: Validity and reliability, *Acta Psych. Scand.,* 69, 220, 1984.

212. **Hirschfeld, R. M. A., Klerman, G. L., Andreasen, N. C., Clayton, P. J., and Keller, M. B.,** Situational major depressive disorder, *Arch. Gen. Psych.,* 42, 1109, 1985.

213. **Lomanger, A. W. and Tulis, E. H.,** Family history of alcoholism in borderline personality disorder, *Arch. Gen. Psychiatry,* 42, 153, 1985.

214. **Kupfer, D. J.,** REM latency: A psychobiologic marker for primary depressive disease, *Biol. Psychiatry,* 11, 159, 1976.

215. **Lewis, A. J.,** Psychological Medicine, in *A Textbook of the Practice of Medicine,* Price, F. W., Ed., 8th ed., H.M.S.O., London, 1950, 225.

Chapter 3

OBSESSIVE-COMPULSIVE DISORDER: ITS CLINICAL NEUROLOGY AND RECENT PHARMACOLOGICAL STUDIES

N. F. S. Hymas and Ashoka Jahnavi Prasad

TABLE OF CONTENTS

PART A: CLINICAL NEUROLOGY — A HISTORICAL REVIEW

I. INTRODUCTION

Nearly a century and a half ago, Esquirol[1] described the case of a woman who was plagued in her daily life by a number of absurd but irresistible ideas, by hesitation, indecision, and slowness, and by a compulsion to perform a number of bizarre rituals before being able to retire to bed at night. It was under the general rubric of "The Monomanias" that Esquirol classified his patient's mental illness — an illness which would now almost certainly be regarded as a severe, but nonetheless typical, example of an obsessive-compulsive disorder. Esquirol thereby provides one of the earliest descriptions of the syndrome.

Subsequent authors, notably Janet,[2] attempted to describe and classify the psychopathology of similar patients, and such an endeavor continues to be one of the preoccupations of more recent literature.[3-7] This continuing concern with phenomenological description and classification must partly reflect the fact that the etiology of this group of disorders remains an enigma. In 1936, in his paper on "Obsessional Illness", Lewis[8] bemoaned the fact that "about these phenomena . . . we still know too little"; some two decades later, he commented that there had been "no marked increase in our knowledge of it in recent years".[9] Salzman and Thaler[10] concluded more recently, in their review of the literature from 1953 to 1978 on the subject of obsessive-compulsive disorder, that "few significant changes in our views . . . of this disorder have been made".

Patients with obsessive-compulsive disorder have been studied from widely differing perspectives, and, since earliest descriptions, one perspective has been that the syndrome reflects a dysfunction of the brain. Such a view has been molded by two kinds of observations: first, the observation that in certain specific diseases of the brain (e.g., postencephalitic parkinsonism or psychomotor epilepsy) obsessive-compulsive phenomena may be seen; second, that in patients suffering from obsessive-compulsive disorder, certain specific signs and symptoms of brain dysfunction (e.g., disturbances of motor or of cognitive function) may be seen. In this section, we will review these two kinds of observation as they have appeared in some of the Anglo-American and French literature of the past 100 years, and then review some of the neurological explanations for this syndrome that such observations have prompted.

II. OBSESSIVE-COMPULSIVE DISORDER OCCURRING IN BRAIN DISEASE

A. Epidemic Encephalitis

Epidemic encephalitis is believed to be a viral infection of the nervous system.[11-12] Epidemics are thought to have occurred in Europe every half-century or so since at least the 17th century, the last major epidemic in Britain and Northern Europe occurring at the end of the first world war, and spreading shortly afterwards to North America.[11] Typical of the chronic sequelae of this encephalitis, which may follow the acute stages immediately or after an interval of months or years, are a marked motor disturbance — especially a Parkinsonian picture, with oculo-gyric crises — and protean psychiatric abnormalities, from dementia to less-severe neurotic states.[12-13] Neuropathologically, the chronic phase is characterized by severe, diffuse degeneration of cells in the substantia nigra, although degeneration is also seen in the gray matter of the cerebral cortex and the basal ganglia.[11-12]

Many studies of the mental sequelae of epidemic encephalitis report on the occurrence of obsessive-compulsive phenomena,[8,13-28] albeit often with little clinical description to substantiate their findings. In some studies, however, conclusions are supported by clinical vignettes which allow a more critical appraisal of the relevance of these early observations to present-day descriptions of obsessive-compulsive disorder. In 1928, for example, Delbeke and Van Bogaert[17] studied the oculo-gyric crises occurring in 25 personally observed cases of chronic epidemic encephalitis. In their study they particularly focused on the psychiatric symptoms associated with these oculo-gyric crises, and they described three phases through which such patients seemed to pass: (1) a phase of anxiety, (2) a phase of "psychic inhibition", and (3) a phase characterized by "states of obsessionality". They illustrate their observations of obsessional states with a number of case descriptions. One patient, before the onset of a crisis, reported: "at that moment I am worried about everything, I go twenty times to see if the door has been locked, I realise that it is ridiculous and that it is locked, but to calm my anxieties I go and see, I check the locks, I clear out the stove to make sure it is properly extinguished. When I have done that, when I am no longer worried, I stay there, surprised, sometimes for a long time thinking about the same things and staring at the same objects like a man having a waking dream". The authors comment that the content of such obsessions is often very impoverished, and that such states could be provoked by specific precipitants; for example, in one case by the sight of her lover, in another by the sight of a soldier dressed in bright colors, and in a third by the fact of having to step up onto a pavement.

Jellife[18-19] published a more extensive study of postencephalitic oculo-gyric crises, focusing particularly on the associated psychopathology. In his monograph of 1932 he reports on 4 personally observed cases and on 200 described in the world literature. Jellife notes that scant attention had hitherto been paid in such case reports to the psychopathological features of these crises, and he praises certain authors such as Ewald, Stern, and Skalweit for their particular concern with these mental phenomena. Jellife reports the 6 cases described by Ewald in 1925: one case was that of a man of 24 who, some 2 years following an attack of epidemic encephalitis, developed Parkinsonian signs and frequent oculo-gyric crises sometimes lasting several hours; during these crises consciousness was maintained and the attacks were sometimes accompanied by compulsory ideas and impulses: "Letters had to be counted. Were there 9 or 10? The vowels a,e,i,o,u, did they follow regularly in a row in a sentence? or were they reversed?" At a later date: "he would wait to see if somebody would mispronounce a syllable; he knew it was nonsense but, according to the tone, some portent as to his getting better or worse was forthcoming; then he began a complicated series of ideas regarding esperanto consonants and their proportion to the vowels — $1^1/_2$:1, 6:3, 2:1, etc., which were related to the prognosis of his illness. Sizes of objects — centimeter, decimeter — were also brought into relation with the probable length of his illness; otherwise his personality was unaffected and even during these compulsory idea periods he could laugh at this senselessness."

Jellife pays particular attention to Stern's paper, written in 1927: "Psychical Compulsive Processes and their occurrence in Encephalitic Looking Spasms with Observations upon the Genesis of Encephalitic Looking Spasms", a paper which Jellife regards as approaching "the very centre of the problem" which is his purpose to discuss. Stern describes a number of cases in whom compulsory ideas are associated with the forced eye-movements. One patient "has to fix her ideas and her eyes on a certain point. As in the compulsion neurotic she knows it is nonsense but 'why is the 0 round? How is glass made? How was writing discovered?' She cannot get away from it. Then the eye-cramp sets in . . . " Another patient, a 19-year-old man with postencephalitic parkinsonism, suffered from oculo-gyric crises lasting up to 24 hr, during which he must think "of definite words with distinct meanings. 'Wohen, warum, wozu, was' — these are as phantasied not as written. They repeat themselves on the inside. They press upon his head, as it were; he is conscious of their being outside of him. He would drive them away by a formula: 'Get away Satan, Jesus is mine.' Then he loses his ritual and is lost for hours in the compulsion." Another patient during his forced eye movements has a counting compulsion: "He must count from 1 to 100 very slowly and than backwards to zero. Internally he counts softly and cannot suppress the series unless he can interpose a song when the number compulsion gives way. Only during the eye spasms do such compulsions occur.''

Jellife discusses Skalweit's study of 1928 with his report of three cases in which he dwells ''chiefly upon the central theme of compulsive states in general in postencephalitic cases of which the oculo-gyric crises are but a part of the general picture.'' Skalweit describes the mental phenomena that were associated with oculo-gyric crises in a woman of 28: ''She seems to seek things out which reflect light — water surfaces, metallic objects. For a time she can overcome the compulsion to seek such objects and turn in an opposite direction but the eyes will, in spite of her, turn in this direction. These attacks come on when she goes out and she must look at the ends of grass-blades . . . She compares this feeling to that of the desire to scratch an itching spot which can be denied but it is very uncomfortable. She would like to see a 'white wall' but the shiny objects compel her attention. If she closes the eyes she obtains peace. At the time of the examination she had a typical mild parkinsonian syndrome . . . ''.

Jellife noted that Chlopicki, in a report of six cases in 1931, discussed the problem of compulsive activities in oculo-gyric crises. One of his cases was that of a 30-year-old man who suffered oculo-gyric crises lasting 2 to 8 hr or longer: ''Ten to twenty minutes before the attack he is restless and irritable, his lips are dry, and peculiar ideas arise concerning redress from his director or his wife for injury, or whether the closed window is really closed. He makes certain that the door is closed. In the free interval also he has some of this compulsion. He washes his hands, looks under the bed to see if anyone is hiding there; cannot sit on a chair without blowing off the dust from fear of dirtying himself. If he suppresses the action of closing the door or window or cleaning the chair, he is very restless, and finally must yield to the compulsion.''

Based on his digest of case-reports, Jellife considered that the phenomena seen in oculo-gyric crises were most consistently classifiable under the general rubric of the ''compulsion neuroses'', and in his grouping of the phenomena he was in broad agreement with Delbeke and Van Bogaert in their study described above; Jellife described four elements: (1) the abnormal eye-movements, (2) slowing of thought-movement (bradyphrenia, stickiness or slowing), (3) an emotional state of anxious compulsion, (4) involvement of consciousness (vigilance) from a slight to a marked degree.

Lewis[8] in 1936 described three cases in whom an obessional illness followed epidemic encephalitis. The first case was that of a woman of 28, a severe chronic encephalitic of 10 years standing, who also suffered from oculo-gyric crises, and who was ''obsessed by the ruminative thoughts 'what is what' and 'did you say did I say'. These she has to revolve and rearrange endlessly in her mind, e.g., 'what is what, did you say did I say, what word is that word what, what do the words the word what mean'. Besides this thought, so reminiscent of the literary output of Gertrude Stein, she sometimes sees her obsessing sentences as though spelt, and spelt wrongly . . . The obsessional thoughts occur independently of her oculo-gyric crises.'' The second case was that of a 25-year-old man who suffered encephalitis at the age of 11 and the onset of oculo-gyric crises at the age of 19. Lewis gives an account of his obsessional symptoms which might occur just before or just after a crisis: ''I have to fight against thinking. I keep on continually thinking: 'what's going to become of the country? Where do clothes come from, and electricity and wireless? I can't stop myself. I feel frightened — I feel something terrible is going to happen— buildings will fall; and then I think where cement comes from, where wood comes from, how trees grow? And then I think I'm a murderer — I'm a spy. I know they're silly ideas, but I can't help thinking them. I try to put it out of my mind, but it seems impossible.'' Lewis' last case was that of a 46-year-old man who suffered encephalitis at the age of 29. He was ''now troubled by obsessional pallilalia. 'I can't help repeating things, I try not to. Singing a song for example, I keep on repeating over and over again 'Is it in the trees, is it in the trees, is it in the trees?' It's when I'm agitated, too, I'll keep on saying things 'I'm going to hang that cup up, I'm going to hang that cup up, I'm going to hang that cup up'. I can't stop myself and when I go to wash my face, I keep splashing the water, I can't get my hand to my face. Everything I do seems to be wrong. I used to say 'Damn' all the time, I couldn't help it.'''

Lewis noted that in many postencephalitic cases there were no indications of obsessionality in the premorbid personality, and that the obsessional phenomena might either be part of oculo-gyric crises or independent of them. He also noted that in such patients the subjective compulsions often went hand in hand with "objective compulsions in the field of motor behaviour".

In 1940, Brickner and colleagues[23] attempted to illustrate their ideas on the physiology of obsessional illnesses using descriptions of cases suffering from epilepsy and postencephalitic parkinsonism. Most of the latter were drawn from the already published literature cited by Jellife, but they also described one of their own cases: this was a 33-year-old woman who had contracted encephalitis at the age of 10 and in whom oculo-gyric crises began at the age of 21; during a crisis the patient was "suddenly assailed with the fear that she was unable to pass urine or faeces, though there was no subjective sensation of fullness of the bladder or rectum; she maintained complete insight into the absurdity of the idea, but was powerless to control it . . . she tried vainly to rid herself of her delusive ideas by attempting to divert her thoughts elsewhere. Again and again she was compelled to attempt evacuation." The obsessive-compulsive symptoms disappeared at the end of a crisis, and between attacks she was otherwise free of such symptoms.

In 1942, Rosner[25] reported three cases of chronic epidemic encephalitis, in two of whom oculo-gyric crises were associated with a variety of mental symptoms which included obsessive-compulsive phenomena; the first case was that of a 30-year-old woman who had acute encephalitis at the age of 16 and oculo-gyric crises at the age of 18; during a crisis, in addition to a change of personality with bizarre and aggressive behaviour and speech disturbance, the patient was subject to a number of "obsessive-compulsive ideas": "she is frequently prompted to recall the name of a theatre at which she used to attend as a very young child. Despite all her efforts during which she spends many miserable hours in futile concentration, she has never succeeded. During a recent spell, while staring out of her living room window, she found herself counting the floors of the building opposite. She thought 'there are six floors in that building. Add one makes seven. One, two, three, four, five, six, seven — all good children go to heaven'. The jingle was displaced from consciousness only after considerable effort and voluntary distraction. During spells she is subject to repeated 'false calls' to move her bowels. Ruminative tendencies become particularly marked while at stool: 'I sit for hours thinking about all sorts of things; I think about numbers and letters. I can't get them out of my mind. I add them, subtract them, and multiply them.' On one occasion she was compelled to repeat to herself 'why don't I love my mother as I do my daddy?''' The author noted that between oculo-gyric crises "this patient is pleasant, friendly, physically active and socially competent".

In 1951, Schwab and colleagues,[27] as part of their study of the psychiatric symptoms and syndromes in a group of 200 patients suffering from Parkinson's disease, reported on certain "paroxysmal psychiatric disorders" which may occur in the disease, and they provided brief case-histories of four post-encephalitic patients to illustrate the occurrence of "specific attacks of compulsive thinking, counting and the use of words". The first patient, aged 35, had a 4-year history of postencephalitic parkinsonism and was subject to frequent oculo-gyric crises during which she "is forced to think of the date 'September 2nd, 1950'; during these attacks, which last for half an hour, the patient is "miserable with this obsessive forced thinking, and is only curious and amused by it during the intervals between the attacks".

The second patient, a man of 35 with a 6-year history of Parkinson's disease, was subject to "paroxysmal attacks of anxiety with an increase in his tremor during which he has to count numbers, usually under 20, over and over again". The authors comment that "it is interesting to note that as far as could be determined during an interview when he had one of these attacks the rate of counting was synchronous with his 4/second tremor".

Ajuriaguerra[28] in 1971 summarized some aspects of the obsessive-compulsive states to be seen in the postencephalitic, which are: "characterised by parasitic elements interfering with the train of thought: melodies, ruminations, incongruous thoughts with sexual or anti-religious themes (ideas of sexual assault, incest, ideas of religious obscenity) ideas of aggression towards other people, towards loved ones, ideas of being forced to count words or numbers, to analyse words, sentences or syllables. These phenomena go hand in hand with compulsions, with commission of acts following upon these ideas (kill, kill, hit) with accompanying aggression or with direct action felt as automatic: motor (spit, destroy, hit out; or more simply, whistle, cry out, sing) or verbal (insults, threats, obscenities as in the Gilles de la Tourette syndrome). These states are rarely permanent; they occur most frequently during periods of anxiety or of agitation; the patients live through them as if they were forced upon them; they have a primitive character and are most often seen as manifestations of illness which demand help".

B. Epilepsy

In 1900, Haskovec[29] noted that although Griesinger some years previously had commented on a possible relationship between epilepsy and obsessional illness, there had hitherto been little discussion of this relationship, all subsequent authors refusing to see any connection whatever between epileptic fits and the sometimes paroxysmal nature of obsessional ideas. Haskovec disputed an element considered to be essential in the diagnosis of obsessional illness, namely clarity of consciousness; he felt that clear consciousness was not invariably present, and furthermore that "obsessions and impulsions were very often to be seen in epilepsy", and that "epilepsy sometimes occurred in obsessional patients".

Janet,[2] at about the same time, took a similar view, and also disputed the notion voiced by some authors that any episodes of altered consciousness seen in obsessional states were merely phenomena "secondary" to the obsessions since in many cases they antedated the obsessions by a considerable period. Others pointed to the paroxysmal onset of epilepsy, with its unconsciousness, its total amnesia and its repetitiousness as further features which distinguish epileptic from obsessional states; but Janet again noted that in the former total loss of consciousness or total amnesia are not invariably present, and that in the latter disturbances of consciousness and memory are by no means invariably absent. He reported several cases from his practice to support his thesis, all of whom suffered classical epileptic attacks. One case experienced obsessional ideas about killing her mother, or obsessionally guilty thoughts about having killed her; she believed that she would do harm to everybody and desired to be locked in a hospital which would prevent her from committing such crimes.

Another case was that of a 22-year-old woman who had been epileptic since the age of puberty; following one particular epileptic aura she began to doubt everything — whether what she saw existed or not, whether her mother was her mother, whether she herself existed or not. She doubted everything she was told. This state lasted several days, and was relieved only following the onset of a grand mal fit.

As well as cases in which epilepsy seemed to antedate the appearance of such psychopathology, Janet observed the reverse process in some of his patients. He describes a man of 52 who, since the age of 12, had been subject to scruples and inexplicable questioning ideas such as "Why do men live? How are men made? Why must men die?". These ideas had greatly tormented him for a long time. For some time also, whenever he posed himself such questions, he experienced a particular emotional state during which he would tremble and shake, and feel cold sensations in his legs, breathing difficulties and pounding of the heart; "in a word, since his youth, he had suffered from anxiety attacks at the same time as obsessional ideas and scruples". Janet noted that he appeared to be a typical case of psychasthenia which would pose no diagnostic problem. However, from the age of 50 the character of his illness changed; more and more often he began to lose consciousness during his anxiety attacks; on three occasions these attacks were followed by an irresistible need to sleep for several hours, and on one occasion he was incontinent of urine.

Janet discussed the problem of "mental ruminations and intellectual obsessions" and their similarity to epileptic phenomena; more specifically he noted that epileptic fits are sometimes heralded by a particular idea against which the patient often struggles but experiences as irresistible and which, therefore, have striking features in common with obsessional ruminations. He described, for example, a young male patient whose epilepsy was antedated by rumination; he presented with a number of obsessional ruminations about infinity; he was concerned about the infinity of everything and felt forced to meditate on the infinity of time and space, and on the infinity of happiness and unhappiness; Janet reports that in a letter he received from the patient describing his symptoms it was clear that he was fully conscious of the absurdity of his ruminations; on interview the patient requested Janet not to discuss these ideas of infinity with him as "this caused him a strange discomfort"; believing that he would cause no more than an attack of anxiety of the kind seen in patients with such a "folie du doute et du scrupule", Janet deliberately dwelt "on the stretches of space which link up indefinitely with other stretches of space"; the patient immediately complained of feeling as if "he was losing his head" and of "falling"; then suddenly, he fell backwards and went pale, his eyes rolled up and all his limbs trembled; a few moments later he regained consciousness, but there was a large stain of urine under him; the patient had no recall of what had happened to him and began conversing again, albeit with a certain hebetude, and a retrograde amnesia; Janet felt that this man's epileptic fit had been the end-point of a state of anxiety triggered by his rumination about infinity, and he recorded that he had observed a number of exactly similar cases.

In 1908, Parhon and Goldstein[30] reported the following case under their care: the patient was a 13-year-old girl who had suffered from epilepsy since the age of 10; her epilepsy was treated with bromide, and on high doses, the frequency of her fits diminished, although she became apathetic and lazy with impaired memory and ability to learn at school; her thyroid was noted to be enlarged, and she was thought to be suffering from hypothyroidism; her bromide treatment was stopped and she was started on thyroid extract; she was followed up over the next 3 months and the frequency of her fits diminished to less than half that seen during treatment with bromide; however, after 6 months of treatment a change in her mental state was observed; the patient's mood became changeable and she refused to leave the house; at other times she was ill and sad; there then appeared a picture characteristic of "psychasthenia" in which various phobias and obsessions predominated; she became obsessed with dirt and felt the need to wash her hands with soap 40 times or more a day; she became plagued by the question of whether such and such a person known to her "has a heart in his body or not" and asked her mother about this; she was also obsessed with the idea that she did not love her mother any more and felt the need to tell her constantly that she did; in view of these changes her treatment with thyroid extract was stopped for a 5-week period and replaced by a course of warm baths; all her mental symptoms disappeared and the patient herself laughed over the fact that she could have been tormented by such "absurdities"; a few days after her thyroid treatment was re-instituted, her obsessional ideas returned in exactly the same form, though less intensely. The authors discussed the role of both the thyroid treatment and the background of epilepsy in the genesis of this patient's obsessional state; they hypothesised that the treatment with thyroid extract had diminished the frequency of her epilepsy of which her obsessional state was but an attenuated or "degraded" form.

Brickner and colleagues[23] in 1940, were interested in the similarity between some of the mental features of obsessive-compulsive disorder and the fixed, perseverative and "forced" thinking and motor behavior of some epileptic auras, and they gave a number of case examples to illustrate their ideas. For example, they described a 32-year-old man who suffered from both major and minor fits, and in whom the major fits were associated with fixation of gaze; he would find it impossible to avert his gaze from whatever he was looking at — for example, these were on one occasion the cards in a game of bridge he was playing at the time. Also, he could not divert his thought from whatever it happened to be on; this phenomenon was followed in a few seconds by a convulsion; on other occasions the same mental phenomenon occurred without any overt seizure following. The same patient displayed other compulsive phenomena which occurred when reading or adding a column of figures, and were again independent of any overt seizure; when reading he felt compelled to fix his attention on the first word — or the first letter of the first word — of each line, and he could not progress to the rest of the line without severe discomfort and until a certain interval had elapsed; the same fixation of his attention occurred with each figure in a column he tried to add up, and he could not progress smoothly to the next figure.

In order to illustrate his ideas about the relationship between epilepsy and obsessive-compulsive disorder, Garmany,[31] in 1947, described the case of a 51-year-old man who had suffered from epilepsy since the age of 17, having approximately 5 or 6 fits each year; two years previously, at the age of 49, his fits had ceased, but at about the same time time "he began to feel worried as to whether he had paid his debts; he would go into shops in the village and ask for confirmation that he had paid his bill of a week before, clearly aware that he had done so, but feeling compelled to confirm it; at first one confirmation would suffice, but later several visits might be necessary; at the same time he became fearful of leaving the gates of fields open, and despite the utmost care and repetition in closing a gate, would be tormented by the fear that the catch might have slipped, and he would sometimes return over several fields in order to confirm that the gate was secure; at night he worried lest a passing car might run over someone in the dark, and that the victim would be undiscovered until the morning; during one night he lay awake and on three occasions when he heard a car pass he got out of bed, dressed, and walked two hundred yards down the road looking for a body; his passage through the village was a torment to him, for he had at least three separate compulsions to contend with; he had to go into every shop he passed and confirm that he owed no money there; he had to stop and kick every stone off the pavement lest someone should slip and be injured; and he had to pick up and examine any small piece of paper he saw lest someone might have lost something of value; eventually his anxiety became intense and he felt unable to move out of doors, and he was admitted to the hospital; there he would ask repeatedly if he were entitled to the cigarettes and stamps with which he was issued, at the same time apologising for his 'stupid behaviour' and asking the nursing staff to be as tolerant of him as possible; when his temperature was taken he felt compelled to ask for confirmation that the thermometer had been properly removed and not been allowed to slip under his clothing; after washing his hands, which he did frequently, he returned repeatedly to be sure that he had turned off the taps; he was troubled incessantly by a rumination as to whether in a large hospital someone might be buried prematurely because a doctor was not available to confirm the fact of death". This patient had never before suffered from obsessional symptoms, and though an orderly and tidy man, his premorbid personality was not considered to be abnormal.

Other case reports suggesting similarities between the clinical phenomena of epilepsy and obsessive-compulsive disorder have been provided by, for example, Allen,[32] Heuyer and colleagues,[33] Roberts and colleagues,[34] and Kettl and Marks.[35] Bear and Fedio,[36] pointed to obsessional inter-ictal behavioral characteristics in a group of patients with temporal lobe epilepsy.

Finally, in their comprehensive study of the epilepsies, Marchand and Ajuriaguerra,[37] in 1948, discussed the relationship between epilepsy and obsessional illness. They noted the occurrence in some epileptic states of "impulsions", or phenomena in which "the individual is irresistibly led, despite himself, to commit a certain act. Sometimes he is conscious of it, sometimes not." Such impulsions can be observed during epileptic absences, confusional states or automatisms, and thus constitute merely one element in a variety different clinical pictures. In addition to these forms, the authors noted the occurrence in epilepsy of two other forms of "impulsion" which are conscious and remembered, which impede the normal action of the will and which can be considered as psychic equivalents of epileptic fits: those impulsions which erupt in states of irritability, and those which occur in an isolated way without other associated phenomena; the former may often appear only during treatment of the epilepsy and only as the frequency of the fits diminishes. The nature of these "impulsive acts" varies widely and can take the form, for example, of an impulse to utter obscenities, to commit acts of violence or sexual offences, to steal or to set fires; the impulsion can be entirely intellectual, e.g., the impulsion to perform mental calculations (arithmomania).

C. Tics

Soon after Gilles de la Tourette's description in 1885 of the syndrome which now bears his name,[38] Guinon reported five similar cases and described the occurrence in three of them of obsessive-compulsive symptoms.[39] One case was that of a 38-year-old man who developed tics of the face, upper limbs, and whole body at the age of 8 or 9 years; currently, the patient displayed a number of these abnormal movements, together with compulsive vocalisations, echolalia, and echopraxia considered typical of the syndrome. At the age of 29 he had contracted syphillis and he showed some tabetic features, although his general condition was good. This patient was also prone to a number of what Guinon refers to as "fixed ideas": he felt compelled to count 1,2,3,4,5,6,7 several times before doing almost anything, such as getting out of bed, sitting down, starting to walk, or opening a door; he would make as if to put an object down or turn a door knob five or seven times before actually doing so; he also felt compelled to avoid the number 6 in his counting rituals and, to a lesser extent, any word with the sound "i" in it, and to use certain staircases and not others. These features had worsened recently, and he was now no longer only obliged to count to 5 or 7 before doing anything, but also to turn round and round on himself 5 or 7 times, or do 5 or 7 paces in a circle.

Another case was a 24-year-old man in whom a tic of the head and neck started following some epileptic fits at the age of 2, and had persisted ever since. Shortly before he was seen, and following a severe emotional upset, coprolalia and tics of the whole body, which could be very pronounced and intense, supervened. Despite these involuntary movements, this patient was able to pursue uninterrupted his profession as a dancer, at which he was highly talented. This patient did not display echolalia, but demonstrated coprolalia to a marked degree, and also echopraxia — the imitation of movement occuring either immediately, or after a gestation period of up to several days. For about a year before his admission, this patient had been subject to periods of irrational and nonspecific fear, but during his present illness this fear had become almost continuous and had become associated with suicidal impulses, claustrophobia, a reluctance to be left alone, a compulsion to arrange and re-arrange objects in front of him and a compulsion to ask himself all sorts of unlikely questions, e.g., why he breathed, why his heart beat, why a certain window he was looking at had 6 panes in it and not 5; he would be fully aware of the absurdity of these ideas which would cause him great anxiety.

Guinon expanded on the observation first made by Charcot that "fixed ideas" were a frequent accompaniment of tic disorders in their severer forms, and he understood the convulsive movements, convulsive utterances, and convulsive ideas as manifestations of a common cerebral disturbance varying only in degree of severity.

There followed other similar case-reports and more general descriptions of the association between tic disorders and obsessive-compulsive phenomena;[40-45] e.g., in 1905, Meige[42] devoted a section of his classical monograph on the clinical features, etiology and treatment of tics to the "mental state" to be seen in such patients. He wrote that they are prone to the "appearance and development of fixed ideas, impulsions and obsessions", and that "all types and degrees of these mental disturbances are to be found" in these patients. He observed "doubt, scruples, and all sorts of manias: the exaggerated love of order, arithmomania, onomatomania, 'folie du pourquoi', etc . . . Phobias abound; all can give rise to tics or coexist with tics."

Creak and Guttman,[44] in 1935, were also interested in the relationships between chorea, tic-disorders, and obsessive-compulsive phenomena. Based on their clinical experience, they hypothesised that "the groups of muscles affected by the tic may determine the nature of the accompanying psychic phenomena". They suggested, for example, that it is the respiratory tics which tend to be associated with compulsive utterances as in the Gilles de la Tourette syndrome, whereas the abnormal eye-movements seen in oculo-gyric crisis, which are grouped by some authors alongside the tics, tend to be more associated with obsessional thoughts.

Wexberg,[45] in 1938, commented that the relationship between tics and compulsary neuroses is well known, and that the two disorders cannot be definitely separated. He also mentioned the possibility of some familial relationship between the two, and cited the example of a patient he had seen with severe compulsory neurosis whose father had suffered from tic.

More recent research has tended to confirm and extend these earlier observations, suggesting that as many as 90% of patients with Gilles de la Tourette's syndrome may exhibit obsessive-compulsive symptoms, and pointing to a familial association between obsessive-compulsive symptoms or traits, and tics.[46-51] Guinon's view that obsessions and tics may share common causative cerebral mechanisms has once more become a theme of current debate and speculation.

D. Parkinson's Disease

In 1951, Schwab and his colleagues[27] reported on some of the psychiatric symptoms and syndromes encountered in a group of some 200 patients suffering from Parkinson's disease. From their clinical impressions they classified those patients "who were from a psychiatric point of view essentially normal"into three personality groupings. One of the groups conforms closely to the obsessive-compulsive type of personality and included "suspicious, worrying, demanding, tense people who are often over-concerned with their health. They tend to be obsessive and perfectionistic and they demand from life, their family, and their physician far more than is obtainable."

In a later paper[52] these authors reported on the inability to maintain at the same time two voluntary motor activities, which they comment "is a conspicuous feature of nearly all patients suffering from Parkinson's disease". In their discussion of the clinical and practical consequences of this deficit, they drew attention to similarities with the kind of difficulties which often beset the obsessive-compulsive patient in his attempt to execute the complex sequences of voluntary movement necessary for day-to-day living: "If the Parkinsonian patient cannot perform acts concurrently, he must execute them in consecutive order . . . In patients who have this difficulty to a marked degree, their usual method of dealing with it is to try harder and harder to perform as normal, each time resulting in conspicuous failure leading to increased frustration. Other patients who have a high degree of adaptability learn to accomplish tasks by approaching them in a consecutive manner. There is elimination of concurrent motor activity which leads to excessively ritualistic and over-deliberate reaction patterns suggestive of obsessive-compulsive personalities."

In 1956, Diller and Riklan[53] studied the psychiatric features of 114 patients who had been referred for neurosurgical relief of their Parkinson's disease. Using clinical interviews, ward observations, life history data, and a variety of psychometric tests, 24(21%) were rated as "severely disturbed" — a group which was said to include a number of "severe obsessive-compulsive character disorders".

More recently, Lees[54] has commented on the "somewhat inflexible, obsessional personality traits displayed by a number of Parkinsonian patients". Mouren and his colleagues,[55] in 1983, attempted to study the personalities of 30 patients with idiopathic Parkinson's disease. Using uncontrolled information gained from interviews both with the patient and with the spouse or close relative, the authors attempted to define the characteristics of these patients' personalities both before and after the appearance of their Parkinsonism. They concluded that "meticulousness and perfectionism were present in 77% of cases"; "mental rigidity, uprightness and excessive moralism" in 47%, and an "obsessional tendency towards orderliness" in 37% of cases.

E. Other Brain Diseases

In 1964, Grimshaw[56] attempted to compare the frequencies of past neurological illnesses in a group of 103 consecutive patients with obsessive-compulsive disorder with a control group of 105 patients matched for age and sex who were suffering from other non-obsessional neuroses. In the obsessive-compulsive group 20 (19.4%) had a past history of one or more important neurological illnesses, such as meningitis, encephalitis, Sydenham's chorea, and convulsive disorders. In the control group there were 8 cases (7.6%) with histories of important neurological illnesses, the difference between the two groups being significant at a 5% level.

Reports of obsessive-compulsive disorder in other brain diseases can be divided into three main categories: (1) Basal ganglia disorders, (2) Lesions of the posterior pituitary, producing diabetes insipidus, and (3) Physical injuries to the brain.

1. Basal Ganglia Disorders

In 1981 Laplane and his colleagues[57] reported the case of a man of 53 years who was fit and well up to the age of 41 when he developed an encephalopathy following a wasp sting. Approximately 2 years following his acute illness this patient developed obsessive-compulsive behavior which persisted until he was seen and investigated by the authors. "At night-time I count. I can't stop myself from counting, it's stronger than I am. To stop I have to concentrate, I have to empty my head to get rid of the numbers." At other times it involved movement: "When I see an electric switch I have to operate it. Sometimes I am drawn towards them and have to switch them off and on in multiples of three." Sometimes he has to touch things several times, or put his hand under his thighs and count 1,2,3,4,5,6,7,8,9, up to 12 and then start again from 0. "I have fixed myself at 12 times 12, my hands need it." When asked why 12, he replied "I think it's to stop myself. If I haven't stopped at 12 times 12, I have to start again. It's a force in me which pushes me on, which is stronger than I am." Sometimes he does things in squares of 3 or multiples of 3, for no apparent reason. His wife found him one day on all fours in the street when he complained to her that it was because he couldn't manage to push a stone with his foot because of his unsteady walk, and he was therefore having to do it by hand. "The stones are pushing me towards them. As I stumble I am doing it with my hand. That's all there is to it." If he is occupied, or distracted by a new situation his compulsions go. Nevertheless, even during a medical examination he was noted to be rhythmically rubbing his thumb and index-finger together and when asked about it he said "I am counting". He is able to stop if he is momentarily distracted, or if he is taken sufficiently by surprise by something. In a similar way, the presence of some third party has the power to distract him, which varies according to the social importance that person has for him. If it is his mother, that annoys him and he carries on, whereas his wife, or a stranger, can relieve him of the compulsion. The authors comment that a striking feature of the clinical picture is his mental emptiness, but equally striking is the relative perservation of his intellectual function with a full scale IQ of 114. Though he spends his days doing practically nothing, he complains that this is neither boring nor distressing. When pressed to say what is in his mind he replied: "It's difficult to say — not a lot — I count"; and when questioned about possible depression in response to his disabilities he says he doesn't feel exactly "sad . . . it's like a lack". His premorbid personality was unremarkable.

Neurological examination showed abnormal movements, but most of the choreiform movements which he had earlier shown had gone. Nevertheless, his voluntary movements were somewhat brusque, and his trunk and neck postures and his gait were reminiscent of patients with chorea. Some facial movements were reminiscent of neuroleptically induced dyskinesia. There was a more or less permanent stereotyped movement of the jaw reminiscent of a tic, and there was a rapid rubbing together of the fingers; the authors say that this could be taken for a tremor — except that it accompanied his mental calculations and ceased at the same time. His gait had features reminiscent of Parkinson's disease, and he had an unspecified problem with writing.

Waking and sleep EEGs were unremarkable. However, a CT brain scan showed bilateral areas of low density localized to the antero-medial regions of the lentiform nuclei, involving the putamen and probably adjacent areas of the pallidum, and an area of low density in the head of the right caudate nucleus. The cerebral cortex appeared intact, although previous studies of the brains of patients dying from similar encephalopathies had shown some evidence of some cortical involvement; however, the authors pointed out that in this patient the relative preservation of cortical function on psychometry, the absence of neurological signs other than those associated with basal ganglia dysfunction, and the normal EEGs would suggest that cortical impairment was contributing only slightly to the clinical picture.

The following year the same authors reported a further, similar case of obsessive-compulsive behavior, consisting mainly of counting compulsions, in a man of 25 who had sustained bilateral lesions of the globus pallidus following accidental poisoning with carbon monoxide.[58] At about the same time, Pulst et al.[59] reported the case of a previously healthy 58-year-old man who, following severe accidental carbon monoxide poisoning, developed an illness reminiscent of Gilles de la Tourette's syndrome with some of the compulsive features associated with this syndrome.

Similarly, at about the same time, Ali-Cherif and his colleagues[60] reported two cases of accidental carbon monoxide poisoning in whom profound mental inertia, obsessive-compulsive phenomena, and lesions of the globus pallidus on the CT brain scan were features. One of these cases was that of a woman who had suffered accidental carbon monoxide poisoning at the age of 18, leading to admission to hospital in a coma which lasted 15 days. From the time of her recovery of consciousness there was no abnormality on neurological examination; in particular there were no extrapyramidal signs. Having previously been a very active person, she was now totally inactive. The patient had to be encouraged to get up and attend to her toilet, and had to be prompted in everything she did. One day, for example, her father left her alone to peel some apples and came back 2 hr later to find she was sitting in the same place. Nothing seemed to interest her. Her parents described another change in her behavior: having been previously a rather untidy person she now showed a keenness for order, not only in her personal possessions, but also for things in the house. She used to check that her clothes were put away correctly in her wardrobe and would frequently re-fold them when they were already perfectly done. She reacted with aggression if the order she had put things in were disturbed. She also had developed the habit of collecting certain things. Thus, at the time of her latest admission to hospital, she collected the cardboard tubes from rolls of toilet-paper, piling them up in a corner of her room and refusing to remove them. She expressed no distress over her condition, and when asked what she thought about during her long periods of inactivity she replied "nothing". An EEG was normal. A CT brain scan showed bilateral areas of low density in the region of the globus pallidus, unmodified by contrast injection. The authors commented that in both their cases the dominant feature of the mental state was the mental inertia; they also remarked that the obsessive-compulsive features were notably unassociated with symptoms of doubt or anxiety, and seemed to be "secondary to the mental emptiness and inactivity".

2. Lesions of the Posterior Pituitary

In 1928, Bailey[61] described the case of a 15-year-old boy who was admitted to the hospital for investigation of excessive drinking, polyuria, anorexia, and intermittent vomiting. Physical examination was normal. His course was variable over the ensuing 3 years, and he had periods of relative normality. Partly in view of his poor response to physical treatments such as pituitary extract, he came to be regarded as a case of "compulsion neurosis"; he described his drinking as "coming from thirst", and "in order to clean out his urine", and as being much like an "addiction" that was "very inconvenient"; "he could not go to sleep without water and had to go to the toilet on the top floor to get it; he hated to disturb his mother and father — 'Gee whiz, I don't want to do this — but I have to do it' he would say." He liked to drink from the tall, thin glasses and had bought some special ones for himself. While in hospital it was noted that "he was unreliable and stole water and ice-cream cones whenever he could do so", and that "he became sick when he could not have his way." The polydipsia and polyuria were, for some time, judged to be an "internal cleansing action associated with a sense of sin", and his nausea, vomiting and anorexia to be expressions of a "fear or disgust of self". However, on his last admission to hospital, 3 years after the onset of his symptoms, he had developed clear-cut focal neurological signs, and he had signs of raised intracranial pressure; the patient died suddenly a few days after admission. A postmortem examination of the brain revealed a very large tumor of the pineal which filled the third ventricle and extended into both the lateral ventricles and the fourth ventricle. Sections of the pituitary showed that although the anterior lobe was normal, the posterior lobe was almost completely replaced by tumor.

Barton,[62] in 1954, reported the case of a 29-year-old woman who was admitted to the hospital with a 4-month history of disabling obsessional rituals — in the form of washing rituals and the need to repeat tasks in multiples of 6 — and obsessional ruminations that she smelled badly, and she was unable to complete the performance of tasks satisfactorily. There was a background of mild obsessional symptoms for 1 year, with sporadic urinary incontinence, and a history of excessive thirst and drinking for 15 years. The obsessional symptoms and polydipsia had been worsening for about 4 months, and she also had some subjective hypothermia, transient hypersomnia, and ideas of reference over this time. There was no history of encephalitis. Neurological examination revealed some dysdiadochokinesia in the left arm, which swung less on walking. Investigation showed her to have cranial diabetes insipidus, although no structural abnormality was ever demonstrated. During treatment with pitressin and supportive psychotherapy all her symptoms subsided somewhat, but improvement was not maintained, and depression and anxiety became added to the clinical picture.

In a later paper, Barton reported some further cases in whom obsessive-compulsive phenomena were associated with diabetes insipidus.[63] One patient was a man of 36 who was admitted to hospital with a history of disabling obsessional ruminations and rituals which started at the age of 15 and had prevented him from working. Initially, he had shown excessive concern for his homework. He expressed fears of going mad and a belief he knew to be senseless that he was responsible for the Blackheath murder. Washing his hands and saying his prayers began to take a long time, and after about a month he developed obsessional ruminations in the form of unanswerable questions such as "Is life real?", or "Why is a chair?", and he began to feel he must draw his recently widowed mother into certain sexual practices. He developed the compulsion to do certain things which he knew to be absurd, and which should be resisted, such as going over his handwriting a second time. At the age of 25 he developed a polydipsia which was thought to be obsessional, and some months later his sleep cycle became inverted, and he would sit up all night keeping meticulous notes on the weather. On admission, his shaving, dressing, and washing were disrupted by a compulsion to repeat many acts 9 times. He was drinking up to 20 pints of water daily. The only physical sign of note was a mask-like face. Vasopressin produced a normal diuresis.

Another case was that of a woman of 43 who suffered from the onset of excessive thirst and polyuria 6 weeks following the onset of her first pregnancy which was complicated by hyperemesis. One week later she began to show drinking rituals when she would drink 3, 5, 7, or 9 cupped handfuls of water, sometimes merely to complete the number. The symptoms ceased after a spontaneous abortion at 5 months, but recurred after a few months when she was one month pregnant again. Six years later she began to suffer a compulsion to touch objects that seemed to "protrude", such as hooks on the wall, the top of a door or television set, or the sides of chairs. "I would pretend to bend over to get something as an excuse to touch it." She had to touch her hand against a wall 3 times, or 3 times 3. She had to rinse her face 3, 5, 7, 9 and 3 times in the bath, and might doubt how often she had rinsed it and start again. Twelve years after the onset of her obsessional symptoms, she developed compulsive hand-washing; she also had to take 3 small pieces from a roll of toilet paper before tearing off 3, 7, or 10 sheets. Other rituals consisted of taking three steps on a big paving stone before stepping onto the next one, keeping a thimble in her pocket or wearing it to touch, twisting her head in bed so that she had her left ear touching the pillow before she slept, and combing the parting in her hair 3, 5, and 3 times. This patient also sometimes complained of feeling cold and of shivering. She was sleepy and walked in the streets with her eyes shut, and she was obese (23 stone) having greatly increased in weight following the onset of her obsessive-compulsive symptoms. Neurological investigation was unrewarding; vasopressin produced a normal anti-diuresis, but her response fluctuated over subsequent years.

In discussing his cases, Barton pointed out that the obsessive-compulsive symptomatology has often to be specifically elicited if it is not to go unrecorded in a more routine history-taking.

3. Physical Injuries to the Brain

In his study of the after-effects of head-injury, Hillbom[64] paid close attention to the psychiatric sequelae of such injuries, and he reported a number of patients who had become "coercive-compulsive-anancastics"; however, case-descriptions are too sketchy to allow precise understanding of this term.

In 1977, Capstick and Seldrup[65] investigated the frequency of birth abnormalities in a group of 33 patients with obsessional states, comparing it with that in a matched psychiatric control group in whom none had ever exhibited any obsessional symptoms. Eleven obsessional patients gave a history of abnormal birth such as breech, forceps, protracted labor or Caesarian delivery, in contrast to only 2 in the control group; this was a highly significant difference. The authors expressed the view in their discussion that while these patients "could have suffered from minimal organic brain damage", the psychological influence of separation between mother and child, which was presumed to have occurred during the early hours following the birth due to such factors as the use of anesthetics or maternal fatigue, "could have been more relevant in the later developement of the obsessional states."

In 1983, McKeon and colleagues reported four cases of obsessive-compulsive neurosis following head injury.[66] Three of these formed part of a consecutive series of 25 obsessive-compulsive patients collected for another study. One case was that of a 37-year-old woman who sustained a head injury in a road traffic accident. She was unconscious for about 5 min and had a posttraumatic amnesia of about 12 hr. Neurological examination was normal. When she regained her memory in the hospital following the accident, she began to feel "dirty" and to feel that everything she touched she "contaminated", together with feeling agitated and fearful. Following discharge "this 'dirty' feeling persisted and she became involved in repeated and lengthy cleaning rituals which occupied a large part of her daily activity and prevented her from performing her usual household chores. She realized that these ideas were ridiculous and initially tried to resist them, but became increasingly depressed." Her premorbid personality was unremarkable and her family history negative, except that her son suffered from grand mal epilepsy. Psychometry, skull X-ray, and CT brain scan were normal. An EEG showed bilateral symmetrical 9 to 10 Hz alpha rhythm posteriorly mixed with slow waves. Four years after the injury she was free of pathological obsessions and depression and her EEG was normal. Another case was that of a 24-year-old woman who at the age of 16 was hit on the head by her mother with a hair brush. Though she felt dazed and light-headed afterwards, there was no loss of consciousness or amnesia. "The following day she was anxious and uneasy when she awoke and on her way to school found she was searching for something she had lost. This idea of having dropped her handkerchief or money on the ground resulted in lengthy periods of searching on the streets and in late arrival at school. The compulsive searching later extended to her bedroom and other rooms in the house, although she realized her obsessions were irrational, and resisted them intensely for a few weeks. Eight years later, she was admitted to the hospital in a depressed state, with obsessions and compulsions to search." Her work-life had been seriously disrupted by her rituals which had resulted in poor time-keeping. Her premorbid personality was unremarkable. An EEG, a CT brain scan, and psychometry were normal. Her depression responded to clomipramine, but her rituals were unresponsive to behavior therapy.

III. NEUROLOGICAL ABNORMALITIES IN PATIENTS WITH OBSESSIVE-COMPULSIVE DISORDER

A. Motor Function

In 1938, Schilder referred to the repeatedly observed occurrence of obsessions and compulsions in epidemic encephalitis, and argued for a close relationship between the neurologically based disturbances of motility seen in these patients and their particular psychiatric symptoms.[67] He wrote: "in spite of the fact that such an obvious connection exists between the impulse disturbance on an organic basis and compulsions, the question rarely has been raised whether it isn't possible to find an organic basis for obsession neurosis in general. Naturally in the majority of cases there is no history of epidemic encephalitis. If one examines a large material of obsession neurosis cases, one finds not so rarely organic symptoms identical with those found in chronic encephalitis."

Schilder estimated from his clinical experience that organic signs could be found in about one third of cases, and that in another third the symptomatology aroused the suspicion of an organic background. Schilder was insistent that the patients to which he was referring did not constitute an atypical postencephalitic psychiatric population: "the main group in which I was interested are cases in which with the ordinary neurological examination, the neurological findings might have been neglected. I would furthermore like to say that in these cases the process is not known. The evidence speaks very much against an epidemic encephalitis."

The physical signs which Schilder had observed in his cases were rigidity of the face and mask-like facies, flexor rigidity of the arms, tremor, impaired convergence of the eyes, "great urge to talk and propulsive features in speech", and "hyperactivity and motor urges of higher degree". Schilder then provides some case histories to illustrate his thesis. One case, for example, was that of a man of 26 who had a cat and dog phobia since the age of 5, and since the age of 14 he was continually plagued with the idea that he might not be clean enough, and would spend hours in the bathroom washing his hands again and again; he feared contamination by germs and would wash very carefully; he could not stand any interrruption and, if interrupted, he would begin all over again. Examination showed a man with quick and jerky movements; his speech was incessant, delivered in a loud voice and in a propulsive manner. His face was a little rigid, and there was a slight tremor of the head and fingers. There was no arm convergence, and eye-movements were normal.

In 1950 and 1952, Vujic[68-69] described a number of abnormal motor signs which he had noted in a wide variety of psychoneurotic conditions; he particularly referred to obsessional illness, and felt his signs complemented those described by Schilder. Vujic's additional signs included abnormal postures of the arms and hands, either at rest or when outstretched, narrowing of the palpebral fissure, tonic contraction of the frontal muscle, and loss of arm-swing when walking.

Much more recently, Behar and his colleagues in 1984 studied a group of 17 adolescents suffering from obsessive-compulsive disorder, in 4 of whom a neurodevelopmental examination was also performed.[70] These subjects, who were comparable in age, psychometric performances, EEG, and CT brain scan appearances with the rest of the group, all showed neurodevelopmental anomalies indicating "immaturity", consisting of abnormalities of tone, reflex, and posture.

B. Cognitive Function

In his study of patients with "primary obsessive-compulsive syndrome", Flor-Henry and his colleagues, in 1979, administered a battery of neuropsychological tests to 11 subjects.[71] The group was compared with the stored data on a control file from 11 normal controls who were matched for age, sex, and years of education. In addition to the WAIS, use was made of a modified version of the Reitan test battery which generated 28 neuropsychological variables, and the "performance profiles" obtained were subjected to an analysis which attempted to correlate performance with functional impairment of different regions of the brain. The authors concluded from their results that in 10 of the 11 patients there was evidence of bilateral frontal impairment, with the left hemisphere showing greater dysfunction than the right.

Rapoport and her colleagues, in 1981, administered a variety of psychological tests to a group of nine adolescents with obsessive-compulsive disorder.[72] Patients were compared with a control population of volunteers matched for age, sex, and handedness. Although no details of subtest scores are given, the authors commented that "intelligence and Bender-Gestalt tests gave no evidence of organicity". Tests of language expression and comprehension showed no difference between patients and controls. A dichotic listening test revealed a lack of the usual laterality effect in speech perception in the patient group, though accuracy of reporting was normal. The group also contained 4 patients who were predominantly left-handed in writing, eating, and sport; this proportion (44%) contrasts with the 10% reported for the general population. The authors pointed out that such lack of usual laterality had been reported in patients with other psychiatric conditions.

Insel and his colleagues, in 1983, attempted to replicate the findings of Flor-Henry described above.[73] They studied a group of 18 patients with obsessive-compulsive disorder. All patients were evaluated using the WAIS, and 16 were also tested with the Halstead-Reitan battery. In 9 patients (50%) verbal scores on the WAIS exceeded performance scores by at least 15 points, and although no single subtest score was consistently low, 9 patients

(50%) had scores less than or equal to 8 in one of three performance subtests: picture arrangement, object assembly, and digit symbol. Verbal subtests showed consistently high scores. Although the group mean of the results did not suggest any impairment for the group of 16 patients as a whole in comparison with normative data, 4 patients (25%) nevertheless had "overall patterns suggestive of organic involvement", although the patterns of deficit in these patients were apparently markedly different. Furthermore, 9 patients (56%) were impaired on the tactual Performance Test in comparison with normative data. This last finding was the only result in those tests common to both this study and that of Flor-Henry in which there was agreement between the two studies. In their discussion of the results, the authors pointed to the abnormalities in the performance subtests of the WAIS and in the Tactual Performance Test as possibly suggestive of "impairment in tasks requiring the perception of spatial orientation". They regarded as consistent with this fact that patients with obsessive-compulsive disorder "frequently volunteer that they are clumsy, nonmechanical, or 'never able to get a round peg in a square hole'". However, they commented that "many obsessive-compulsives are markedly slow and indecisive" — features which may have been highlighted by the particular performance subtests of the WAIS and the Tactual Performance Test in which some of the patients were impaired. In consequence, the authors felt uncertain whether such psychometric impairments "reflect psychiatric manifestations of obsessive-compulsive disorder or whether they hint at a right hemisphere deficit"; the possibility that these two perspectives might not be mutually exclusive was not considered in the discussion.

Behar and colleagues,[70] in 1984, enlarged the group of 9 adolescents with primary obsessive-compulsive disorder reported by Rapoport and described above, with a further 8 similar patients, and used additional psychological tests not given to the original cohort. Amongst these tests, the patient group performed significantly worse than the controls in the Stylus Maze Learning and the Money Road Map Test, there being no difference in the remaining tests. The authors comment that since the patients performed similarly to controls in tests of reaction time, two-flash threshold, and decision times, their performance in the above two tests could not be attributed to poor attention or "obsessive style"; rather performance in the Stylus Maze Learning Test implied an impairment of the ability to discern and follow a set of unstated rules and patterns, while performance in the Money Road Map Test implied an impairment in the ability to mentally rotate the self in space — both of which abillities being believed to reflect frontal lobe function, although the authors com mented that "the true localising strength" of their findings is unknown.

V. Electroencephalography

In 1944, Pacella and colleagues[74] studied the EEGs of 31 patients manifesting severe obsessive-compulsive symptoms "as the dominant and leading features of their clinical pictures". Of these patients, 26 were classified as "obsessive-compulsive neurosis"; the remaining 5 patients displayed obsessive-compulsive phenomena in a setting of schizophrenia. Brief clinical details were given for all patients, and the authors also provided detailed illustrative case descriptions for 4 of the patients. EEGs were performed "routinely" and before diagnosis was made or treatment instituted. 20 of the patients (64.5%) were said to have shown mild to severe degrees of EEG abnormality. Fourteen of the abnormal records (45% of the whole group) showed "convulsive-type patterns" consisting of occasional or frequent runs of 2 to 4 cycles per second potentials of high amplitude, in 12 of which the "paroxysmal cerebral dysrhythmia" was noted only during a 2-minute period of hyperventilation. Nine recordings (28.9%)were definitely normal. There appeared to be no correlation between severity of obsessive-compulsive symptoms and the incidence of EEG abnormality. In a study performed by the authors of over 100 clinically normal individuals, "not over 10% exhibited serial, high amplitude 2 to 4 cps waves, even after a period of over-ventilation." In another study of a group of over 400 psychiatric patients of all types, which included the above obsessive-compulsive patients and, in addition, many disturbed and deteriorated patients, the incidence of EEG abnormality was approximately 30% — less than half that noted in the group of obsessive-compulsives alone.

Taking the above study as their starting point, Rockwell and Simons,[75] in 1947, attempted to study the question of whether "personality organisation", as judged by certain qualitative clinical impressions, correlated with the presence or absence of an abnormal EEG in patients suffering from obsessive-compulsive disorder. They studied a group of 24 patients who showed "well-marked obsessive-compulsive symptoms" of varying types. The authors stated that 13 of the total group of 24 patients had abnormal EEGs characterized by the presence of excessive quantities of 3 to 7 cps slow waves, and these occurred predominantly in the group with abnormal personalities; in addition, 2 of the EEGs in this subgroup showed "paroxysmal features" although there was no evidence of epilepsy in the clinical histories of these patients. Out of the 11 patients with normal personalities 10 had normal EEGs.

In 1960, Gibson and Kennedy[76] published a single case-study of a patient with severe obsessive-compulsive disorder in whom they performed serial EEGs over a period of several months. The patient was a 34-year-old woman admitted to the hospital with an acute exacerbation of a long-standing obsessional disorder. For many years she had had washing and toilet rituals and a fear of going out. Fifteen months before admission she began to complain of a compulsion to growl and bark like a dog; later she began to fear that she might harm her mother by splitting her head open with an axe. There was no clinical evidence of epilepsy, and neurological examination was normal. An EEG performed shortly after admission was abnormal and was characterized by fairly frequent paroxysmal discharges of slow waves and spikes over both hemispheres, with varying asymmetry and irregularity; although the amount of abnormal activity varied a good deal, further records over the next 3 months confirmed these findings. While in the hospital the patient developed a urinary tract infection during which it was noted that she became more irritable, aggressive, and disturbed by her obsessional thoughts. An EEG in this phase showed a greatly increased number of paroxysmal discharges.

In the same year, Ingram and McAdam[77] studied 30 patients showing obsessional symptoms of some severity. Of the cases, 22 were "typical" of the disorder according to textbook descriptions. Among these 22 "typical" cases, 3 recordings were "mildly dysrhythmic" and showed slow generalized bursts on hyperventilation; these changes were considered to be minimal and "borderline". Only 1 recording was clearly abnormal, and showed several bursts of synchronous, generalized delta waves. The authors concluded from their findings that "specifically obsessional traits and symptoms have no relation to EEG variables" and that "the relationship between obsessional illness and both epilepsy and EEG abnormality is considered to be a chance one".

Epstein and Baline in 1971,[78] studied the all-night sleep EEGs in three patients with obsessive-compulsive disorder who had shown abnormalities in their waking EEGs, but in whom there was no evidence of epilepsy. Illustrative samples of recordings are provided. The first patient was a 36-year-old woman who for 8 years had suffered a compulsion to search through and check items of rubbish in her rubbish-bin, feeling that something of value may have been lost; this ritual would take 2 to 3 hr to perform, and attempts to resist the compulsion would cause marked unease. This patient suffered from less intense compulsive rituals in which she would examine water draining from dishes or a mop for fear that something of value may be contained in the water. The second patient, a 34-year-old woman, complained of unwanted "bad thoughts" revolving around fear over the safety of her family; these bad thoughts had been present for 12 years and were only relieved by certain rituals such as handwashing, filling a pot with water, rewashing dishes, and folding clothes. The third patient was a 47-year-old woman, who, at the age of 16, developed an obsessional idea that she "had hurt her mother's head and killed her". After persisting for many years it was replaced by a similarly distressing idea that she had set fire to a house. At the time of investigation she felt a compulsion to think "I have not done enough for my mother", and she had a number of rituals, for example, checking the gas-heater; she also feared that she had "left the water on" in the shop where she worked. Sleep EEG abnormalities common to all three patients consisted of theta and spiking becoming more localized to the temporal areas during Stage 1 and REM sleeps, abnormalities which the authors commented were similar to those reported in temporal lobe epilepsy by other workers, although they cautioned that an explanation might be that any generalized EEG abnormality may possibly become more localized during sleep, irrespective of the clinical manifestations associated with such an abnormality.

More recently, Flor-Henry and colleagues[71] in 1979, used spectral analysis to study the EEGs of 10 unmedicated patients out of a group of 11 with obsessive-compulsive disorder. Recordings were compared with EEGs from 23 normal subjects. The authors felt that a disturbance of the dominant hemisphere was suggested by their results.

Rapoport and her colleagues,[72] in their study of 9 adolescents with obsessive-compulsive disorder described above, performed all-night EEGs on their patients and compared them with a group of 15 age- and sex-matched controls. Of 24 EEG variables 8 were significantly different between patients and controls. The authors singled out a "short REM latency" and "total sleep time", and a trend toward increased "sleep latency" and "reduced sleep efficiency" — features which they commented have been previously reported in "middle-aged depressed patients with primary affective disorder". The authors admitted to some puzzlement over this alteration in the sleep structure of obsessional patients, and they speculated that it might reflect some connection, which could be 'psychodynamic or biological', between obsessional and depressive illnesses.

Insel and his colleagues, in their study of 18 obsessive-compulsive patients described above,[73] performed routine EEGs on all subjects. Two recordings (11.1%) were read as abnormal. One showed "non-specific intermittent left temporal sharp wave activity", and the other showed "left hemisphere bursts of rhythmic theta activity". Some nonlocalized increase in theta activity was evident in five other tracings and was interpreted as a "non-specific normal variant". Eleven records were clearly normal.

In 1948, Jenike and Brotman[79] carried out a retrospective survey of the notes of 12 patients suffering from obsessive-compulsive disorder who had undergone an EEG. Of the records 4 (33%) were described as clearly abnormal: in one subject there was "occasional slowing in both posterior temporal regions when drowsy"; in another, "paroxysms of high voltage sharp wave activity more prominent in the fronto-temporal regions"; another had "bursts of right temporal sharp waves", and the last patient had "frequent bursts and runs of sharp slowing bilaterally in the fronto-temporal areas". The authors considered that the findings were so similar to those reported in temporal lobe epilepsy that they gave a trial of anticonvulsants to the four patients with abnormal EEGs, but without obvious benefit.

D. Computerized Tomography of the Brain

CT brain scans have been performed in two recent studies of patients with obsessive-compulsive disorder. Insel and colleagues,[73] in 1983, in their study of 18 such patients described above, administered a CT brain scan to 10 of the patients, including all those who had shown any psychometric or EEG abnormality, and compared them with 10 matched control subjects. Results of the study, which was carefully controlled, showed no significant differences between patients and controls on measures of ventricular-brain ratio, cerebral hemisphere asymmety and cortical atrophy. Behar and colleagues,[70] in their study described above, compared 16 of their group of 17 such patients with 16 matched, normal controls on various CT scan measures. Results were reported as showing a significant difference in ventricular-brain and Evans ratios between patients and controls. Attempts were made to correlate the ventricular-brain ratios with other data gathered in the study, such as severity and duration of illness, EEG and psychometric findings, history of birth injury, etc., but with negative results. In the discussion of their results, the authors cautioned that enlarged ventricles have been reported in other psychiatric disorders, and are not necessarily of an etiological importance that is specific to obsessive-compulsive disorder.

IV. NEUROLOGICAL EXPLANATIONS OF OBSESSIVE-COMPULSIVE DISORDER

In his monograph of 1889 on the psychology of attention, Ribot[80] considered in some detail the nature and cause of "fixed ideas" which were of a "purely intellectual" nature, such as "arithmomania", "onomatomania" and "metaphysical mania". Ribot regarded such phenomena as signs of "degeneracy" and emphasized the role of constitution in their development. "Not everyone can have fixed ideas. There is a primary condition which is required for their development: a neuropathic constitution. This can be hereditary; it can be acquired . . . the fixed idea, even in its simplest form, which is our concern here, and which seems entirely theoretical and confined to the realm of intellectual processes, is, however, not a purely interior event without physical concomitants. Very much to the contrary, the organic symptoms which accompany it indicate a neurasthenia: headaches, neuralgia, feelings of oppression, disturbances of motility, vaso-motor control and sexual function, insomnia, etc. The mental phenomenon of the fixed idea is only the effect, amongst many others, of one and the same cause." Ribot considered that, from a psychological point of view, fixed ideas represented an "hypertrophy of attention", and his notion of the physiological basis of fixed ideas was consistent with his views on the physiology of attention: " . . . attention depends on affective states, affective states can be reduced to tendencies, tendencies are fundamentally movements (or arrests of movements) which are conscious or unconscious. Attention, spontaneous or voluntary, is, therefore, at its very origin linked to the state of the motor system . . . (attention) only exists by virtue of a restriction of the field of consciousness, which is equivalent to saying that, physically, it suggests an activity of a limited part of the brain. Whether one imagines this part as a localised region or, which is more probable, as made up of diverse elements spread throughout the mass of the brain and working in concert to the exclusion of other elements, has no importance. The normal state of consciousness suggest a diffuse state of activity depending on scattered parts of the brain working together. Attention suggests a concentrated state of activity depending on the working of a localised part of the brain."

Concerning the specific problem of fixed ideas, Ribot had earlier in his monograph speculated that: "Physiologically, one can probably think of fixed ideas in the following way: in the normal state the brain works as a whole; it is a diffuse process; it depends on discharges passing from one group of cells to another, which is the objective equivalent of the perpetual changes which occur in consciousness. In the morbid state, only some nervous elements are active, or at least their activity is not transmitted to other elements. Furthermore, it is not necessary for nervous elements to occupy one point or limited region of the brain; they can be scattered, although they must be firmly interlinked and associated to perform their common function. Whatever their position in the brain they are effectively isolated: all available eneregy has accumulated in them and they do not pass it on to other elements, from which arises their dominance and exaggerated activity. There is physiological imbalance, due probably to the nutritional state of the cerebral centres . . . Esquirol called the fixed idea a catalepsy of the intelligence. One could also compare it to a motor phenomenon: contracture. This is a prolonged contraction of the muscles; it depends on an excessive irritability of nervous centres; the will is powerless to check it. The fixed idea has an analogous cause; it consists of an excessive tension, and the will has no control over it."

In a paper of 1894 on "imperative ideas", Tuke[81] provided a variety of case-illustrations to demonstrate the diversity of clinical pictures which may be encountered under this rubric. He described, for example, cases of arithmomania, onomatomania, maladie du doute, delire du toucher and its antithesis, mysophobia. In discussing the physiological basis of such mental abnormalities, Tuke commented that "at certain points these cases seem to be closely allied to epilepsy", and, although he does not develop this observation, he nevertheless attempted to provide a neurological framework within which imperative ideas might be understood. For this he drew on Laycock's ideas of reflex action of the cerebral cortex, and of cerebral evolution and dissolution, particularly in the form in which these ideas had been propounded and developed by Hughlings Jackson. In a paper commenting on Tuke's review, Jackson, while finding himself in accord with Tuke's approach to the specific problem of imperative ideas, also provided a concise summary of his views on the neurology of insanity in general:[82] "For my part I consider that illusions, delusions and all other positive mental symptoms in insanities signify activities of healthy arrangements of the highest cerebral centres (so-called 'organ of mind'). What we call an insane man's illusions are his perceptions, what we call his delusions are his beliefs, and more generally his positive mental symptoms sample the mentation remaining possible to him, a mentation occurring during what is left intact of his highest centres, of what disease has spared. The physical condition for these positive mental symptoms is not caused, using the word cause in its strict scientific sense, by disease, not caused, that is, by a pathological process. Disease is, I submit, answerable only for the co-existing negative mental element of insanity. Here let me remark that, to take one kind of mental symptom, an illusion, a positive mental state implies a co-existing negative mental state; if a man sees a black cat where there is only a black felt hat, not only is there for him a black cat, but this is not for him a felt hat. Similarly, mutatis mutandis, for other positive mental symptoms sampling the positive elements of a patient's insanity. As to the physical. Disease of the highest range of the highest centres producing loss of its function or destroying it, answers to the negative mental element in a case of insanity; whilst the activities of the lower, intact, ranges answer to the positive mental element. But sometimes the positive symptoms are super-positive as in post-epileptic unconsciousness (negative element) with mania (positive element). In these cases the lower ranges are in over-activity, and yet I think that these ranges are healthy, that they are untouched by the pathological process which has produced loss of function of, or has destroyed, the highest ranges. Their over-activity is an hyper-physiological state and is analogous to the heart's beating more frequently after section of the vagi . . . The loss of function or destruction of the highest range of the highest centres is not only a loss (answering to the negative mental element of an insanity), but is also a taking off of control from the lower ranges; the 'taking off' of the higher is a 'letting go' of the lower. The disease may be said to be 'constructively accountable' for the over-activity of the lower ranges, for, had not disease destroyed the highest range, that highest range would have continued to control the lower ranges."

Tuke applied these concepts to the phenomenon of imperative ideas: "I believe that Dr. Jackson has not in any of his writings referred specifically to imperative ideas other than as insane delusions, which according to him 'signify evolution going on in the entire (i.e.) unmutilated nervous system.' There is in short a 'disvolution' of the highest cerebral centres, but as Dr. Jackson points out 'there are all gradations traceable as a consequence of disease beginning in the highest centres, from such slight depths of dissolution as those which permit nearly normal actions, to dissolutions so deep as to afford no evidence of mentation.' Accepting this hypothesis . . . we seem to find an answer to the question which I began with, what are the cerebral conditions accompanying imperative ideas? The most automatic is no longer under the control of the voluntary or least automatic cerebral levels or layers, or in Jacksonese, 'reversals of evolution have occurred, being reductions from the least to the most organised of the highest cerebral centres.''' Tuke, therefore, held that imperative ideas arise as a result of ''weakened higher centres'' (of unknown cause, and without demonstrable structural change in the brain) acting to release lower and more automatic cerebral centers governing such behaviors as counting (in arithmomania), or touching (in delire du toucher), or personal hygiene (in contamination phobia).

Although adopting a primarily psychological approach to obsessional illness, Janet,[2] writing at the turn of the century, nevertheless emphasized the importance of biological factors: ''Many psychaesthenics, preoccupied as they are with their obsessions, their mental compulsions or their phobias, complain only of psychological symptoms and an observer could be disposed at first sight to think simply that they were suffering from a disease of the mind . . . This essential point cannot be insisted upon too much: obsessive patients, by their talkativeness, by their interminable descriptions of their extraordinary ideas, distract the doctor from a physical examination which should never be neglected . . . whatever interpretation one gives to their mental state, it should not be forgotten that they are first and foremost sick people.''

In attempting to describe and classify the psychological phenomena of these illnesses, Janet believed that he was proceeding in the only way that the limitations of contemporary neurological understanding would allow; but if his justification of a psychological approach was based on an awareness of the limitations of neurological knowledge, so also was his view that psychological explanation was, by nature, if not provisional, at least partial: ''In our ignorance of the essential workings of the nervous system and of the causes which determine the ebb and flow of cerebral function, pathogenic theories of mental distrubance can scarcely be more than the most natural classifications of the observed symptoms.'' Later he writes: ''These psychological studies, in analysing the phenomena, and reducing them to their essential elements, are preparing the way for an anatomical interpretation — an interpretation which is not yet possible at the present time.''

Two concepts of importance in Janet's framework of understanding of obsessional states were those of ''hierarchy of mental function'' and of ''derivation'' of mental symptoms — concepts reminiscent of Hughlings Jackson's hierarchical and evolutionary view of cerebral organisation, and of his concept of the ''release'' of more automatic functions in the lower centres in disease-states affecting the highest cerebral centres. Janet wrote: '' . . . we have shown the weakening and disappearance of certain psychological functions in obsessional patients; on the other hand we have seen how other functions were well preserved, and rather seemed exaggeratedly developed; in studying this difference, one is naturally led to conclude that all the different types of mental operation do not have the same facility of expression, and that when there is a weakening of cerebral function, they disappear, not simultaneously, but successively and progressively because of this unequal facility of expression. In a word, mental operations seem to arrange themselves in a hierarchy in which the higher are complex, difficult to reach and inaccessible for our patients, whereas the lower are easy to perform and remain available to them.''

Following directly from this idea was Janet's concept of ''derivation'' of mental symptoms: ''When a force, which at its origin is destined to be expended in the production of a certain function, remains unused because this function has become impossible, then derivations are produced, that is to say that this force is expended in the production of other functions which are unintended and useless.'' Janet applied this idea to the phenomenon, for example, of obsessive rumination: '' . . . mental rumination . . . is simply a derived symptom at a low level of mental operation. The essential feature of mental rumination is the developement of unreal thoughts, with no connection with action, without certainty in relation to the past, the future, the imagination or the present. These thoughts which are foreign to our person, to our present, to our action, seem to have no freedom and to be imposed from outside. They, therefore, have all the features of low-level mental operations needing little energy.''

To explain the loss of higher and the emergence of lower levels of function, Janet invoked the idea of "psychological energy" ("tension psychologique"): "The degree of psychological energy, or the level of mental activity, is manifested by that level in the heirarchy of mental functioning reached by an individual when he is functioning at his highest level. The ability to act upon reality, the attainment of certainty are activities requiring a high degree of energy; dreaming, motor overactivity, emotion requiring much less energy one can consider as low-level functions corresponding to a lower mental level." For Janet, this psychological energy was the expression of a cerebral process which remained to be elucidated.

Janet considered the various forms in which this loss of psychological energy may be seen and considered the clinical and theoretical relationships between obsessional and epileptic states: " . . . there are considerable oscillations of mental level, and lowerings, or collapses of psychological or nervous energy as we have understood the term, which play the principal role in the two kinds of neurosis. But in epilepsy, as far as we can know, this collapse is considerable and momentary. It goes as far as complete loss of consciousness for quite a short time, following which energy is restored, undoubtedly incompletely, but to a degree sufficient for the individual to feel almost back to normal and not to complain of the feelings of incompleteness which are so characteristic of the psychasthenic. I believe that the convulsions of an epileptic fit represent derived symptoms; they are violent and of a very elementary kind, that is to say that the derived symptoms are not in the form of mental ruminations, a conscious process and relatively high in the hierarchy of mental functions, nor in the form of conscious emotional states, nor in the form of semi-purposeful movements such as tic, but in the form of movements of the most elementary kind. As soon as the features of the epileptic fit diminish, as soon as the collapse of energy is less great but more protracted, when the derived symptoms are less elementary, one sees states of epileptic delirium appear which are very similar to psychasthenic phenomena. Some of these epileptic states give rise to feelings of incompleteness, to doubts about reality, the self and the outside world, which are absolutely identical to those seen in obsessional patients. This is only an illustration to show the apparent relationship between the psychoneuroses, but it seems to me that the psychasthenic state is a chronic and attenuated form of epilepsy."

In the closing chapter of his monograph, Janet provided a summary of his views: "The actual obsessions, which are the most striking feature of the condition, are only the end result of a series of much deeper problems . . . In psychasthenia, the collapse of mental energy is much less sudden, less profound and more prolonged than in epileptic attacks; it does not lead to that restriction of consciousness with its fixation on certain points that is seen in hysteria; it seems, in this psychoneurosis, to remain general and to give rise to a reduction of efficiency in all mental processes and of the power of adaptation to reality. The functions which are most disturbed are the functions which put the mind in touch with reality, those which govern attention, will and feelings and emotions adapted to the present. Other functions seem to remain intact and they show in this way that they are on a lower level — these are discursive intelligence and language, exaggerated and disordered emotions, maladaptive and partly automatic movements. This loss of psychological energy gives rise to a mental malaise, a state of anxiety, to feelings of incompleteness, which are stronger the more intelligence is preserved. Under the influence of this disturbing anxiety, and by the fact that higher functions are suppressed, lower-level processes which have been preserved become greatly exaggerated and give rise to tics, to motor overactivity, to emotions of anxiety, and to very varied mental ruminations. Finally, ideas form, according to circumstances, which serve to sum up and interpret all these disturbances, and ideas formed in this way continue to carry features of the previous mental state; they are permanent and obsessional because they sum up and express a permanent state; there is no end to them, and they do not give rise to bizarre beliefs, but they keep the form of anxious emotions and ruminations. The factors which influence the disorder are all the circumstances which cause fluctuations in the level of mental activity and cause a rise or fall in that psychological energy of which a fall is the basis of the whole illness. Treatment consists of using all those physical and moral influences which can cause a rise in the level of mental activity; it is especially necessary, by an education of the powers of attention, even of emotion, to make the brain adopt a habit of more active functioning — something which is not always an impossible demand. Emotional instability, which is a secondary phenomenon, as are tics and ruminations, disappears when the higher functions and the power to act upon reality reconstitute themselves, and the obsessions will fade when they are no longer needed to express a general state."

Meige,[42] a contemporary of Janet, noted in his study of tics their frequent association with obsessional states, and considered that lack of cortical control was important in the development of both phenomena: "The tic is therefore a psychomotor disturbance. It is a motor disturbance because the objective phenomenon has the features of a convulsion, and a disorder of normal muscular contraction; and it is also a psychological disturbance as it is an act which is at its origin goal-directed, but which has become inappropriate, illogical and absurd. This transformation shows that there is an abnormal psychological state. It demonstrates the lack of control of motor action, the weakness of the inhibitory power of the will. The tiqueur neglects to use that surveillance which is necessary if his movements are to be measured and thoughtfully coordinated." Meige though that such control was a cortical function: "The irregularity and insufficiency of cortical control favors the appearance and development of fixed ideas, of impulsions and obsessions. In fact, one finds in tiqueurs all types and severities of these mental disorders. Their coexistence with tics is important to be aware of as inappropriate movements are very often connected with these mental problems."

In his discussion of tic, Wilson in 1927[43] offered a mixture of psychological and physiological ideas to explain the phenomena, although he tended to emphasize the role of mental processes: "No feature is more prominent in tic than its irresistibility. The strain of holding the movement back is as great as the relief in letting go . . . The element of compulsion links the condition intimately to the vast group of obsessions and fixed ideas . . . Behind all tic phenomena lies a psychical predisposition . . . A varying degree of mental infantilism . . . stigmatises the tiqueur; in the language of psychoanalysis, he is narcissistically fixed . . . Grafted on a constitutional basis, tic makes its appearance in some cases (possibly in the majority) as the expression or outward manifestation of an unconcious desire — or better, as I am fully convinced, of a desire that is often but half-hidden from consciousness." Even in cases where at a tic is part of a postencephalitic picture, Wilson insisted on the prime importance of psychological mechanisms: "Explanation of the genesis of these and similar motor phenomena of a post-encephalitic kind is beset with difficulties, and, while it is natural to attribute them to the disease, we must not lose sight of the psychopathic soil in which they may be sprouting, for they are neither universal nor invariable as a sequela." In the discussion which followed Wilson's paper there were some dissenting views, particularly in relation to the importance Wilson had aascribed to psychogenic factors. In connection with the views he expressed on postencephalitic tics, Guillain[83] commented: "It is undeniable that emotion can exaggerate or modify post-encephalitic dyskinesias, just as it can and does exaggerate many organic conditions, such as pyramidal and cerebellar symdromes . . . I consider that the functional element is of little importance in post-encephalitic syndromes; the predominating feature is a disorder of the tonic mechanisms of attitude or of simple motor automatisms." The views of Cruchet[84] also differed from those of Wilson; with respect to the mental state of tiqueurs, Cruchet commented: "It is easy to observe in a great number of tiqueurs an abnormal mental state characterized by mental infantilism, instability of ideas, inconstancy, carelessness, absence of will, impulsiveness, excessive emotivity, capricious affectivity, disorderly imagination and loss of deliberation and judgement. On the contrary, some other tiqueurs are extremely intelligent and possess a strong will. Others again, belong to the mental group of those with obsessions and fixed ideas. Several of my tiqueurs cannot go into their rooms without touching the handle of the door three or four times; they cannot get into their bed without looking underneath it . . . Certain tiqueurs exhibit neither mental heredity nor personal mental defect. I do not believe that a 'habit illness' is always a cause, for in certain cases tics are not executed in spite of the will of the patient but quite independently of it; he is as it were the spectator of them. The power of the will seems normal, yet incapable of restraining mental disorder. One may compare this condition to that which occurs in the fixed state of post-encephalitic Parkinsonism. At the beginning of an act the Parkinsonian subject can execute his movements normally, but little by little the will loses its effect, although local muscular resistance does not exist. The same applies to tiqueurs; the will can restrain muscular activity for a certain period, but its action lasts only for a short time, and the tics return. In both cases the will is normal, but its action is as insufficient to reduce the fixed Parkinsonian state as to prevent the explosion of a tic. The trouble is evidently of cerebral origin, but it is not necessarily a mental one.'' Cloake[85] offered a similar critique of Wilson's approach: "It is of great importance to realise that closely similar movements may arise from disorder of neural mechanisms at very different levels, and that etiological differences are of prime importance in determining the line of treatment to be followed. This is true not only of tics but of various other neurological conditions, and, I believe, also of the respiratory sequelae of epidemic encephalitis, which may in one case have their origin in rigidity of the chest wall and in another in brain lesions at various levels. A view of this kind allows a conception of disordered behavior which is not interrupted by any distinction as between physiogenic and psychogenic. At the highest levels disorders of function manifest themselves as psychical disorders, and correlated physiological and anatomical changes may be unrecognisable.''

Following their report of three cases of postencephalitic obsessional behavior, Claude and his colleagues,[14] in 1927, commented: "Thus, an infectious disease of the brain such as encephalitis lethargica gives rise to more and more complex and new syndromes, signifying a more and more diffuse and less systematised disorder of the nervous system; starting as a mesencephalitic lesion, the encephalitis manifests itself with focal signs; little by little the clinical picture changes to give rise to symptoms which are extremely diverse, fluctuating and variable, ranging from peripheral problems reminiscent of the polyneuritides, and neurovegetative syndromes, up to syndromes classified up till now amongst the psychoneuroses; in this way, one of our patients presented, during the evolution of his illness, with typical emotional crises and, furthermore, we see in our patients a classical picture of obsessional illness. Thus, next to the anatomico-clinical syndromes of a focal nature such as are seen in neurology, and for which one reserves the term organic, one must be taking account of diffuse, fluctuating and widespread processes successively affecting very diverse areas of the brain, and which give rise to dynamic problems which, if they persist, can lead to lasting anatomical changes.''

In the same year, Courtney,[15] in relation to his case of postencephalitic compulsive behavior, stressed the importance of studying the sequelae of epidemic encephalitis for a greater understanding of such psychoneuroses as obsessional states: "I have long been convinced that, in the character, degree and extent of the remaining clinical manifestations of epidemic encephalitis, there are invaluable clues to the pathogenic mechanisms of those phenomena, neurotic and psychotic so-called . . . Even in fatal cases, the pathologic changes in the mesencephalon are not gross. Furthermore, many of the clinical syndromes produced by this disease are not different from those which at present are ineptly considered of purely psychogenic origin. I contend, therefore, that the psychologic interpretation of the syndromes in question should be dropped entirely, and that they should be regarded as due to pathologic changes in the encephalon, identical in degree, at least, with those which exist in certain stages of epidemic encephalitis.''

Similarly, Delbeke and Van Bogaert[17] in the following year perceived the opportunities which a study of the sequelae of encephalitis could offer for the illumination of the cerebral basis of certain neurotic symptoms such as obsessions and anxiety: "Here, as in aphasia, torsion spasms, etc., this question of the interaction of organic and psychic factors presents itself as one of the essential problems of present-day neurology, because one should not ask whether oculo-gyric crises are epiphenomena of a subjective or emotional process in these postencephalitics, but, rather, what can the analysis of those psychic factors which influence authentic organic symptoms such as oculo-gyric crises teach us about the neuroses . . . The role of psychic factors in the onset, the repetition and sometimes the inhibition of these crises with ocular spasms does not imply that they can be considered as emotional in origin. The crisis is accompanied by a series of authentic neurological symptoms, as much cortical as sub-cortical in origin. These disturbances imply cerebral dysfunction, and the examples of chorea and torsion spasm teach us how easy it is for psychic factors to act on a nervous system which has been unbalanced by disease. The fundamental — and unresolved — question is to know why, in certain organic disorders, these psychic factors come into play with such force that they can unleash organic symptoms."

Bailey,[61] in 1928, used his case of diabetes insipidus and compulsion neurosis as the basis of a forthright critique of purely psychological approaches to these problems: "The main interest in this patient . . . is for the theory of the neuroses. One need not quibble as to whether there was a typical case of compulsion neurosis. Any clinical syndrome varies with the causative factors, and no one would maintain that in every case of compulsion neurosis there was a causative tumour of the third ventricle. That there may be in every case a lesion of some kind in the basal region of the brain, is a hypothesis not so easily discarded. The importance of these regions of the brain for the emotional life is every day being clarified by clinical and experimental investigation. However, such an hypothesis is a mere hypothesis at present with little evidence to support it. It is too bad, nevertheless, that careful necropsies are not more often done on neurotic patients. If the present instance means anything, they should often be very illuminating . . . The trouble with most psychologic studies is that a few chapters of the whole story are told. They are like novels without endings . . . The polydipsia and polyuria in this case were not caused by a sense of sin, nor the vomiting by disgust, but the conditional precipitating factor underlying both the disturbance of external behavior and of the water regulation was an insult to the nervous system by a tumour . . . The essential lesson of the case, it seems to me, is to teach us as physicians, while not neglecting the situation, to keep our eyes open in every behavior disturbance for the other conditioning factor — an organic defect of the nervous system or a defect of the bodily economy on which its functioning depends."

One of the most astringent critics of the purely psychological approach to mental illness was Von Economo,[13] on the basis of his experience and studies of encephalitis lethargica. He was also one of the most optimistic adherents of the view that this disease presented unique opportunities for studying the cerebral basis of conditions hitherto considered as purely psychogenic. In his monograph of 1929 he wrote: "The sorry fact of the helplessness of our medical art in chronic cases of encephalitis lethargica has been in some measure compensated amongst the scientifically minded by the immense gain which our knowledge of pathological and normal nervous mechanisms has derived from our acquaintance with this disease, and by the justified hope that, in days not too far ahead, this knowledge will open up new avenues toward the discovery of remedial aid . . . future scientific generations will hardly be able to appreciate our pre-encephalitic neurological and psychiatric conceptions, particularly with regard to so-called functional disturbances. To emphasise the difference of encephalitis from most of the nervous diseases known to the present day, it must be remembered that the majority of the latter are system diseases, as for instance the hemiplegias, tabes, aphasias, etc., with injury to the long tracts and their continuations and association-tracts. All conditions other than those diseases were generally looked upon as functional disorders as, for instance, chorea, tics, abnormalities of psychological disposition, compulsive states, the psychoses, writer's cramp, etc. Now we can, in a similar manner, describe encephalitis lethargica as a functional affection, but on an organic basis. The apparent contradiction which this would have constituted in the past exists no longer. The functional symptoms, organically caused, of this polio-encephalitis arise outside of the long tracts . . . Formerly the term 'functional' described only disturbances originating from previous psychological experiences of the individual . . . The psychoanalytical school considered these psychological experiences as almost exclusively of a sexual nature. Encephalitis now shows us that quite a series of these functional disturbances (urge-like and compulsive actions and thoughts, depressions, abulia, etc.) may not only rest on a psychological or hormonal basis, but also on an neuro-anatomical, organic one . . . The dialectical considerations of those psychologists who claim psychological experience exclusively as the determinant principle of the arrangement and coordination of mental (functional) activity in consequence lose considerable weight.''

In a section on the specific problem of compulsory movement and compulsory thought, and their genesis, Von Economo speculated on the possible role of subcortical structures of the brain: ''The lapse into tics on the one hand and compulsive movements on the other is a gradual one; there is a tendency towards repetitions and with them to iterations and stereotypes. These compulsive movements may also contain psychological elements which associate themselves with the originally organic motor disturbances and increase them to compulsive actions, often of a grotesque kind . . . This combination of a complicated motor disturbance with the process of thoughts and possibly other psychological phenomena is instructive . . . such psychological states of compulsion may also occur without any disturbance of motility as compulsive thoughts, conditions of urge, which, for instance, may lead the patients, as do analogous disturbances, familiar to the student of psychopathology, to the commission of abnormal acts . . . Also in these chronic disturbances of motility we find the patient saying, not, as we have seen in certain acute disturbances of motility, 'I have a twitch in my hand', but rather as a rule, 'I have got to move my hand that way'. The frequent subjectivisation of these processes, experienced as compulsory by the patient . . . is, I believe, one of their characteristic attributes. From this subjectivisation we may deduce that in cases where the condition manifests itself, centres (probably in the diencephalon-mesencephalon) are affected, whose motor function contributes intimately to the constitution of the 'sensation of personality' and even to a greater extent than some parts of the cerebral cortex; because, for instance, in Jacksonian fits affecting the arm as a result of a lesion of the anterior central convolution of the cerebrum, the patient says, 'I have a twitch in my arm', that is, the movement does not become subjectivised, though it has its origin in the cerebral cortex.''

The views of Schilder[67] on the importance of organic factors in obsessional neurosis have already been summarized above. In the discussion which followed Schilder's paper, Wexberg[45] speculated on the possibility of a deficiency in those patients of higher motor function: '' . . . there must be something basically wrong in what might be called the rhythmical function, the higher rhythmical coordination of . . . motor function, which is essential for every motor activity in general. It is a higher function which, for instance, has something to do with building up and finishing a certain coordinated activity, not a movement but an activity . . . It seems to me to be some kind of inability or insufficiency of psychomotor rhythm. This is to be seen in the unjustified repetition, reiteration and perserveration present in practically all cases of compulsion neurosis, particularly the fact that they cannot get away from some point, some idea, and that they stick to it and cannot go on.''

Brickner and colleagues[23] thought that the most fruitful approach to an understanding of obsessive-compulsive states was by considering them as part of a continuum of disorders occurring in nature of which the salient common feature is one in which "repetitiveness or fixedness of action dominate the intellectual or muscular picture"; in their study of 1940 they continued: "Such repetitive or fixed behavior, when clinically observed in the strictly psychological (neuro-intellectual or neuro-emotional) domain is ordinarily thought of as obsessive. In the psychomotor field it is called compulsive, and in the strictly motor sphere it has a variety of names, depending largely on the syndrome in which it appears — perseveration, catatonia, propulsion, iteration, echolalia, pallilalia, stereotypy of movement or thought and others. Indeed, if examples leading from perseveration to obsessive thinking are presented in sequence, it is difficult to make a fundamental distinction between any two of them. The uniform thread of repetitiveness or fixedness which runs through all of these suggests that a common physiological mechanism may underly them all, regardless of the behavior field in which they are clinically manifest."

The authors then present a number of case-illustrations of patients suffering from such apparently diverse disorders as cerebral tumor, epilepsy, and postencephalitic Parkinsonism, with and without oculo-gyric crises, to illustrate the difficulty in making a clear distinction between obsessive-compulsive phenomena understood primarily as "psychogenic" and similar phenomena appearing in conditions understood primarily as brain disorders. In their discussion of the neural basis of obsessive-compulsive phenomena, they emphasized the need for a physiological approach: "Our theory is that the external manifestations of obsessive or compulsive behavior are the reflections of a certain specific kind of neural activity. They occur when a given neurone organisation acts either in a fixed repetitive manner; when those neurones, instead of being activated once by an impulse which then passes on to other neurones, imprisons the impulse so that it either keeps the neurones in a constant state of activity, or it reactivates the same neurones again and again. The theory has appeared useful to us because it permits the alignment with comprehensible physiological occurrences, of behavior which is ordinarily difficult to classify. Psychological matters should be better understood when they can be thought of as obeying the same laws and as following the same patterns as events in the so-called somatic sphere."

In consequence of their views, these authors parted company with some previous writers on the subject, for example, Jellife, who had attempted a primarily psychological understanding of the observable neurological events; e.g., in discussing the phenomenon of the postencephalitic oculo-gyric crisis, of which obsessive-compulsive features were a frequent accompaniment, Brickner and his colleagues stated: "Jellife, not altogether tacitly, distinguished between psychological and physiological functions. Had he not made this separation, it might be that the two viewpoints would approach each other closely. Jellife infers a psychological basis for the crises, whereas we see both the somatic and psychological phenomena of the crisis as the common result of the sudden preponderance of fixing and repeating processes over all others, throughout the brain. To us the content of the thoughts is secondary, the point being that whatever is thought will be obsessively thought. We do not deny the possible 'psychological' conditioning of some attacks and we agree that almost anything that happens to patients may have its content so conditioned. But that is quite another thing than a psychological origin of the attacks themselves. Why may not the 'psyche' be considered incorporated in the somatic processes, instead of the opposite, which is Jellife's implication?"

In 1947, Garmany[31] briefly reviewed some of the literature on the organic basis of obsessional illness before describing his case in whom epilepsy was associated with a severe obsessive-compulsive disorder. Garmany felt that such a combination was unlikely to be very frequent, but he speculated that: " . . . a single pathology may be responsible for both manifestations, and that from some affected extrapyramidal area, a discharge may spread to both hemispheres, producing a generalised seizure. To speculate further would be profitless. It appears certain now that generalised convulsions may ensue from subcortical disease apart altogether from cortical involvement . . . and it is equally certain that obsessional states are associated with ascertained disease of the brain-stem in its upper part. Certain other conditions, such as Gilles de la Tourette's disease, compulsive utterances and tics, and compulsive laughter might be regarded as sharing some of the features of both. It is possible, therefore, that there is a pathological entity which may produce this quite unusual combination of symptoms.''

Barton,[62-63,86] following his report of a series of patients displaying obsessive-compulsive phenomena in association with diabetes insipidus, speculated on the possible anatomical and neurochemical basis of obsessive-compulsive disorder: ''The association of bitemporal hemianopia with diabetes insipidus suggested a single lesion in the region of the optic chiasma, particularly of the neurohypophysis. Similarly, the coexistence of diabetes insipidus and obsessional neurosis may improve understanding of the latter. Diabetes insipidus, however, does not certainly result from a lesion of the supraoptico-hypophyseal system, which cannot be used as a landmark as can the visual-field lesion. The compulsive component is common to micturition, thirst, and obsessional neurosis, but does not explain them. Nevertheless, the syndrome points to an organic cause for obsessional neurosis in impaired function of the hypothalamus and midbrain. Obsessional neurosis and evidence of hypothalamic disturbance should be sought in cases of diabetes insipidus. I am confident that the syndrome is commoner than supposed. Identification of further cases, and adequate examination, including neurohistology and electroencephalography of the third ventricle, may contribute to our knowledge of the anatomy and physiology of appetite, thirst, and compulsion.''

From his EEG studies of three patients with obsessive-compulsive disorder, Epstein[78] in 1971 speculated on the importance of temporal (limbic) structures in the genesis of the syndrome: '' . . . the waking state in the obsessional neurotic need not be qualitatively the same as in the 'normal'. The possibility of an alteration in brain mechanisms, perhaps with a limbic locus, in the waking state (a dissolution in the Jacksonian sense, with release of normally inhibited elements) manifesting itself in the forced appearance of an idea, must be entertained.''

More recently, Laplane and his colleagues, in 1981,[57] introduced their first published case of basal ganglia damage associated with obsessive-compulsive phenomena with a confession of ignorance: ''The biological roots of compulsive phenomena are completely unknown, and the connections which there may be between compulsive behaviour, tics and choreic movements have been, up till now, only vaguely worked out.'' The authors hoped that their case report might throw new light on the problem, and in their discussion they focussed on the biology of the neostriatum, and invoked Janet's concept of psychological energy ('tension psychologique') as the most useful explanatory paradigm: ''For the first time, it seems that a biological basis can be given to the function of self-activation, often invoked by dynamic psychologies . . . The description of this function which would seem to fit best with the characteristics of our patient is that of Janet's 'tension psychologique', of which the highest manifestations are adaptation to reality, in contrast to the inactivity of our patient and the lowest forms, 'the useless muscular movements', of which compulsive activities and tics form a part. The application of this theory to our case would suggest that psychological self-activation would be an essential function of the neostriatum . . . Obsessional behaviours, both physical and mental, can, from a descriptive point of view, be considered as an activity that fills the psychological vacuum. This behaviour can, furthermore, be likened to the stereotyped behaviours of mentally retarded patients who are left to their own devices. Whilst not concealing the uncertainty of the conclusions which can be drawn from a case in which the lesions are only approximately defined, and which seems at present to be unique of its kind, it remains stimulating to note in our patient the succession and coexistence of abnormal movements whose kinship has been discussed for a very long time: choreic movements and tics on the one hand, and tics and obsessional behavior on the other. Such a case seems to give consistency to the idea of a common biological origin to these different aspects of neurological and mental pathology. It is appropriate to stress, in conclusion, that despite the intensity of the compulsive phenomena, our patient did not have an obsessional neurosis, and it would be inappropriate to try and to explain this disorder solely on the basis of the model presented by our patient. However, a deepening of our knowledge of the striatum could well cast some light on such abnormal movements as facial dyskinesias, tics, and on obsessional neurosis as much as on chorea. The double motor and mental aspect of basal ganglia function, furthermore, was very strongly suggested followed Von Economo's epidemic encephalitis.''

In a later paper,[87] they attempted to place their patients in the context of previous reports in the literature: ''It is of interest that in some neurological diseases obsessive-compulsive behavior has been reported and that they were thought possibly to be due to basal ganglia impairment. In numerous cases of Parkinsonism, most of them being post-encephalitic cases, obsessive-compulsive activities were described such as arithmomania, vocalisations or purely obsessional thoughts. These behaviours often occurred in short crises lasting some minutes and could be accompanied by motor disorders such as oculocephalogyric fits and by personality changes such as emotion-aggressivity but mainly anxiety. Some cases, however, are very similar to our patients, since obsessive-compulsions are described as occurring without modifications of the emotional state, as a substitution phenomenon triggered by an inner blocking of thoughts. The association of obsession-compulsions and tics can be present in Gilles de la Tourette syndrome. An organic background at the basal ganglia level has been suspected in this disease on the basis of anatomical, clinical and pharmacological findings.'' Finally, having reviewed some of the clinical and experimental work suggesting the importance of the pallidal area for many aspects of motor, cognitive and other mental function, the authors concluded that: '' . . . a disorder of this area should be considered in research on several psychiatric conditions, such as hebephrenia, obsessional neurosis or severe depression.''

In recent years there has been some renewed interest, particularly in the American literature, in those aspects of obsessive-compulsive disorder that are regarded as indicative in some way of its biological basis; Elkins et al.,[88] Jenike,[89] Lieberman,[90] and Turner et al.,[91] for example, have reviewed research and clinical experience in such disparate areas as stereotactic surgery, pharmacology, genetics, and electrophysiology, in order to support the general hypothesis that brain dysfunction underlies obsessive-compulsive disorder; for the elucidation of the specific brain mechanisms involved, hopes are sometimes pinned on future innovations in techniques of brain investigation: e.g., Insel[73] wrote: ''It seems likely that in the near future new techniques for imaging cerebral glucose metabolism or cerebral blood flow may yield a clearer map of neuropsychological function in obsessive-compulsive disorder.'' This view was echoed by Behar and his colleagues:[70] ''We hope that more sensitive techniques of functional assessment of brain function, such as positron emission tomography, will provide further clues to altered central nervous system states in this disabling disorder.''

V. CONCLUSIONS AND DISCUSSION

Two closely related areas of literature have been surveyed; the first consisted of those reports of obsessive-compulsive phenomena occurring in association with brain disorder; the second consisted of those reports of neurological abnormalities occurring in patients with obsessive-compulsive disorder. Both kinds of report frequently, though by no means always, prompted attempts to understand obsessive-compulsive disorder in neurological terms, and such explanations as have been offered have been reviewed in a separate section. From such a review it is possible to draw three specific conclusions which can be simply stated as follows:

1. Obsessive-compulsive phenomena may arise as manifestations of brain disease; when they are so based, these phenomena are often clinically similar to — if not indistinguishable from — those which constitute the psychiatric syndrome of obsessive-compulsive disorder.
2. The clinical categories of brain disease which may give rise to obsessive-compulsive phenomena are various; there is no one clinical neurological condition which invariably gives rise to this pattern of psychiatric disorder. However, the region of the brain most commonly implicated is subcortical, involving such areas as the basal ganglia and elements of the limbic system.
3. Where these have been specifically looked for, patients with obsessive-compulsive disorder may display a range of signs and symptoms of brain dysfunction, which may be either clinically demonstrable (e.g., abnormalities of movement or cognition), or demonstrable only on laboratory investigation (e.g., EEG or CT brain scan abnormalities). However, there is no single pattern of neurological disability which patients with obsessive-compulsive disorder invariably display.

Given the etiological heterogeneity of neurological syndromes such as epilepsy, tic, and Parkinsonism, and the likely etiological heterogeneity of the psychiatric syndrome of obsessive-compulsive disorder, it is, perhaps, unsurprising that tighter conclusions cannot be drawn about the etiology of obsessive-compulsive disorder from the clinical observations that have been reviewed. Over half a century ago, Mapother[92] perceived and forcefully stated the implications for research of the heterogeneity of clinical neurological and psychiatric syndromes: "A little attention to current principles of neurology would . . . tend to lessen beliefs current among psychiatrists in the fallacies of the single cause, the solitary syndrome, the standard course and the specific cure . . . Neurology illustrates abundantly the following principles:

1. Any one mechanism responsible for a characteristic syndrome may be affected by one of many nocuous factors (as in the epilepsies, the choreas, the Parkinsonisms).
2. Any one nocuous factor may in a single case or in different cases affect several mechanisms, i.e., produce several syndromes (e.g., in epidemic encephalitis, in neurosyphilis, in amyotrophic lateral sclerosis).
3. The influence of one nocuous factor may render a mechanism susceptible to the influence of a second, and such complexity and multiplicity of cause is so common as to be almost the rule. The precipitation of acute symptoms by physical or emotional exhaustion in disseminated sclerosis or by injury in general paralysis are examples.
4. There is such complex reciprocal connection between each neuron system and others (e.g., between pyramidal, striatal, cerebellar and rubrospinal) that a lesion affecting primarily one such system inevitably causes repercussions in a number of others, even in the absence of any direct operation of the nocuous factor on other systems. For the last reason alone, apart from any other, it is hardly conceivable that a neurological syndrome can be pure, and the same is true of psychiatric syndromes. Neglect of the lessons of neurology results, for psychiatry, in a textbook isolation of syndromes that is fantastically remote from anything seen clinically and in a search for unitary causes and unitary cures of such syndromes which is as naive and as futile as the quest for the elixir of life."

Perhaps the most striking general conclusion which can be drawn from such a review of the literature is that neurological understanding of this psychiatric syndrome has advanced very little since the early formulations of, for example, Jackson, Tuke, and Janet at the turn of this century; indeed, where attempts have been made more recently to understand the disorder in terms of clinical and experimental neurology, their concepts re-emerge. Knowledge of the functional specificity of different regions of the brain has undoubtedly increased, but the emphasis placed by recent authors such as Laplane or Epstein on the importance of subcortical areas such as the basal ganglia and limbic system, would not have been unfamiliar to, for example, Von Economo in 1929. Knowledge of the importance of neurotransmitter systems has similarly increased, but specific neuro-chemical hypotheses arising from a coherent theoretical background concerning brain-behavior relationships are still largely absent from current ideas. Furthermore, the related notion of functionally linked groups of neurones spread diffusely within the brain would not have been an unfamiliar concept to such writers as Ribot or Janet. Similarly, in this country at about the same time, Gowers,[93] wrote: "It must also be remembered that the conception of a physiological centre does not necessarily involve that of local limitation. Nerve-cells act together that are far apart, and those that are adjacent are often independent. Diffused through a mass of grey matter may be many separate mechanisms, not necessarily more in one part than in another, and if we call them 'centres' we must not allow mathematical conceptions to govern our physiological ideas."

With the possible exception of the relationship between the "forced thinking" of epileptic states and the ruminations of obsessive-compulsive disorder, there have been few recent attempts to interpret the clinical phenomena of obsessive-compulsive disorder in terms of the phenomena displayed by patients, or experimental animals, with brain-dysfunctions whose mechanisms may be better understood, and specific hypotheses concerning those systems of the brain whose impairment might be given to rise to the phenomena of obsessive-compulsive disorder are lacking from current research in this area. Where a "biological" approach has been adopted, this has tended to consist in the application of, for example, a large battery of neuropsychological tests, or a large battery of "biological" tests on the blood, the electrical activity and gross structure of the brain, the functions of the endocrine system, etc., in a search for "biological markers" which may then generate a hypothesis as to the nature of the brain dysfunction.

If it is true that our understanding of the clinical neurology of obsessive-compulsive disorder is little more advanced than that of the pioneer neurologists and psychiatrists of the end of the 19th century, then a possible implication for further research in this area is that the clinical methods they employed should begin now, in the light of contemporary knowledge of brain function, to be re-applied to the clinical phenomena under scrutiny. Holmes,[94] in 1946, stated that the process of neurological diagnosis must start as an "attempt to interpret symptoms and physical signs in terms of disordered function". Such "disordered function" is expressed as primary, or negative, symptoms and signs arising from loss of function, and as secondary, or positive, symptoms and signs arising from attempts to perform a function differently, or from excessive functional activity. An illustration from everyday neurological practice is given by the phenomenon of hemiplegia, of which the primary, or negative, symptom and sign is loss of movement on one side of the body; the secondary, or positive, symptoms and signs being, for example, the associated disturbance of gait, hypertonicity, hyper-reflexia, and extensor plantar response. This kind of clinical analysis is directly applicable to the phenomena displayed by patients with obsessive-compulsive disorder. Of primary importance is a description of those functions that have been lost (the primary or negative symptoms), e.g., the ability to perform an action with normal speed and fluency, and of those symptoms and signs (secondary or positive) which arise from the attempt to perform the lost function differently, or from excessive functional activity, e.g., the inappropriate repetition of certain actions. Such a clinical approach to the history and signs forms the prelude to an effective use of such techniques as may be currently available for the further investigation of brain function, as it allows very specific questions to be asked about those systems of the brain which are likely to be impaired.

Finally, the history of the attempt to understand the neurological basis of obsessive-compulsive disorder is one which reflects the history of attempts at neurological understanding throughout the rest of psychiatry. In particular, the conflict between the psychological and physical explanatory paradigms, and the potential for their confusion, has sometimes implicitly, but frequently explicitly, been a point of debate between different observers of the same phenomena in this relatively circumscribed area of psychiatry, as it has been in psychiatry as a whole. With respect to the specific problems posed by obsessive-compulsive disorder, Brickner and his colleagues[23] have offered one of the clearest statements of the dilemma: "In any consideration of the physiological processes underlying psychological events, it is essential to maintain a rigid separation between the terminologies of psychology and physiology. Although such a statement may appear trite and unnecessary, it deserves emphasis because of a common tendency to confuse the two vocabularies. As an illustration of the needed distinction, in discussing obsessive and compulsive symptoms, the argument will not even touch upon the customary theory that such symptoms are a compensation for or an expression of anxiety. It is the neural mechanism by which the symptom is produced in which we are interested, whether the precipitating factor is an electrical stimulus or an unconscious anxiety state."

The importance of this debate and its current relevance, can, perhaps, best be illustrated by extracts from two editorials now separated by over half a century; the first is taken from a 1930 issue of the *Journal of Neurology and Psychopathology*:[95] "Notwithstanding advances that are taking place all along the front of medicine, it is a question whether present-day conceptions of what are currently denominated 'neuroses' and 'psychoneuroses' are as serviceable as is often claimed. To classify and differentiate the neuroses has ever been troublesome, mainly though not entirely owing to aetiological uncertainty. Old distinctions between 'functional' and 'organic' types of symptom have proved valueless except for clinical diagnosis, and in this field also their usefulness is over . . . No less deserving of reconsideration is the term 'psychoneurosis', to find proper justification for which would be a matter of no little arduousness. Like 'psychomotor', 'psychosensory', 'psychovisual' and others, it represents a blend of physiological and psychological without any definition of the assumed interaction or interrelation. If it stands for something a little more than neurosis, a little less than psychosis, its descriptive value is outweighed by its nosological impropriety . . . The word 'neurosis' implies disorder of neural function — it is therefore essentially unspecific; if it entails the admission of psychical causation it is unsuitable; and so long as it is imagined to stand for physiological disorder with physiological integrity it is meaningless . . . The more deeply clinical observation penetrates the arcana of the neuroses the more possible does it seem that the class as a whole will suffer dismemberment. Chorea was once a neurosis, and paralysis agitans! Have we finished with telegraphist's cramp when we note its occurrence in a 'neurotic' person, with 'nervous heart', bad dreams, fatiguability, fears and what not? What if these are themselves but the expression of physiological disarray?"

The second extract is taken from a 1982 issue of *Psychological Medicine*:[96] "Neurologists have taken a long holiday from the study of psychiatry in general . . . In partial recognition of the growing similarity of both clinical problems and technologies in neurological and psychiatric research, American neurologists have recently given birth to the new subspecialty of 'behavioral neurology', deliberately avoiding the uncomfortable possibility that this concept might be better called 'psychiatry' or at the very least 'neuropsychiatry'. Pari passu, psychiatrists . . . have rediscovered the importance and excitement of the study of the brain."

PART B: PHARMACOTHERAPY

I. MOOD ELEVATORS AND PSYCHOSTIMULANTS

Rojo-Sierra[97] used solutions of cocaine in doses of 15 to 20 g. The beneficial effects were most pronounced in hand-washing rituals. Baruk et al.[98] used ortedrine for one patient who improved considerably, but became addicted to the drug. Gilberti et al.[99] used amphetamine-like substances but reported only limited benefit. Use of this group of drugs should probably be avoided because of the disastrous side-effects.

II. NEUROLEPTICS

Baruk et al.[100] reported improvement in 50% of patients with levopromazine. Lopez-Ibor[101] used properazine in doses of 500 mg/day with good results. Dally[102] reported some improvement in his patients who were treated with perphenazine. Tapia[103] favored the use of haloperidol, which remains an effective treatment for Gilles de la Tourette syndrome — a syndrome which is thought to have close connections with obsessive-compulsive disorder. However, the use of neuroleptics has not gained universal acceptance, and it has also to be borne in mind that they can themselves induce anxiety and movement disorders.

III. ANXIOLYTICS

In view of the prominence of anxiety symptoms in the disorder, many anxiolytics have been used in its treatment. Breitner[104] reported significant improvement with chlordiaze-poxide in high doses. Bethune[105] and Rao[106] treated patients with diazepam and reported improvement in more than 50%. Orvin,[107] in a double-blind placebo-controlled study, found that nearly 70% of his patients improved on oxazepam. In summary, benzodiazepines have shown an ability to ameliorate symptoms, but none of the studies have been able to demonstrate a preferential reduction of obsessive over anxiety features.

IV. ANTIANDROGENS

Casas et al.[108] have recently tried cyproterone acetate in very resistant obsessive-compulsives and reported dramatic improvement. However, this is the only trial so far to report the efficacy of this drug, and further controlled studies are needed.

V. ANTIDEPRESSANTS

Because of the postulated link between this disorder and depressive illness, it is not surprising that the antidepressants have proved to be by far the most popular pharmacological treatment. Trabucci et al.[109] and Nabarro et al.[110] used imipramine extensively. Later, Vidal and Vidal[111] in a study of 32 obsessive-compulsives reported dramatic improvement in 15 patients, using doses of 150 to 300 mg/day. Their findings were supported by Geisman et al.[112] who used imipramine in doses of 150 to 375 mg/day. Angst et al.[113] reviewed the literature on the use of imipramine, but concluded that the results were inconclusive as the sample sizes were too small. Ananth et al.[114] used doxepin in 13 patients and reported moderate improvement in 9. The dose-range was 75 to 300 mg/day. Perhaps the most popular drug in the treatment of the condition has been clomipramine. As early as 1968, Grabowski[115] reported its efficacy. Later, Lopez-Ibor[116] reported that i.v. clomipramine reduced obsessional symptoms in patients suffering from depressive illness. The first double-blind trial was carried out by Yaryura-Tobias et al.[117] in 1976. They found a highly significant reduction in both obsessions and compulsions over a period of 4 months. More recently, Thoren et al.[118] blindly compared clomipramine and nortriptyline, and found only the former to be better than placebo. Montgomery[119] in another double-blind trial, found clomipramine to be significantly better than placebo.

Around the same time, Marks et al.[120] conducted the largest double-blind trial so far, with 40 patients, and concluded that clomipramine in conjunction with behavior therapy was significantly better than placebo with behavior therapy in a depressed subgroup. Ananth et al.[121] a few months later, published the results of a double-blind trial from which they concluded that clomipramine was better than amitriptyline. Insel et al.[128] compared clomipramine and clorgyline in another double-blind, placebo-controlled trial and found that the former was significantly better than the latter. In summary, it appears fairly convincing that clomipramine does have a place in treatment of this disorder, and that its anti-obsessional and anti-depressant effects seem independent.

Because clomipramine is believed to be predominantly a serotonin re-uptake inhibitor, there has been some speculation that this property may be primarily responsible for its anti-obsessional effects — a hypothesis which is supported by the apparent efficacy of L-tryptophan, a serotonin precursor, in this disorder.[122] Based on this hypothesis, Prasad[123] conducted a double-blind trial comparing zimelidine — another rather more selective serotonin re-uptake inhibitor, — with imipramine, a relatively specific noradrenaline re-uptake inhibitor, and found the former to be significantly superior to the latter. There was no significant difference in the reduction of depression scores. Trazodone, another relatively selective serotonin re-uptake inhibitor, has been tried on an open basis by Prasad[124] and Baxter, both of whom reported significant reduction in obsessional symptoms. However, it was not clear whether this was a consequence of an improvement in depression.

Another category of antidepressants which has been reported of benefit in this disorder is that of the MAOIs. Izikowitz,[125] in 1960, used phenelzine with an encouraging effect. This finding was replicated by Jain et al.[126] in 1970. Jenike[127] used tranylcypromine and reported encouraging and rapid improvement.

VI. CONCLUSION

In conclusion, it seems clear from the evidence that drugs can have a major role to play in the treatment of obsessive-compulsive disorder. In particular, the results with antidepressants, especially clomipramine, seem the most encouraging. Whether clomipramine acts through its selective effect on serotonin is uncertain because its major metabolite, desmethylclomipramine, also has a strong effect on noradrenaline re-uptake. Insel et al.[128] have suggested that an alternative pathway could be an anti-nociceptive effect on opiate receptors, as they found that naloxone exacerbated obsessional doubt. However, the efficacy in this disorder of zimelidine and trazodone — two other antidepressants which are believed to act selectively on the serotonin system — does point to the possible importance of serotonin metabolism in obsessive-compulsive disorder.[129]

REFERENCES

1. **Esquirol, E.**, *Des Maladies Mentales*, Vol. 1, Bruxelles, 1838, chap. 11.
2. **Janet, P.**, *Les Obsessions et la Psychasthénie*, Alcan, Paris, 1903.
3. **Cooper, J.**, The Leyton Obsessional Inventory, *Psychol. Med.*, 1, 48, 1970.
4. **Hodgson, R. J. and Rachman, S.**, Obsessive-compulsive complaints, *Beh. Res. Ther.*, 15, 389, 1977.
5. **Dowson, J. H.**, The phenomenology of severe obsessive-compulsive neurosis, *Br. J. Psychiatry*, 131, 75, 1977.
6. **Stern, R. S. and Cobb, J. P.**, Phenomenology of obsessive-compulsive neurosis, *Br. J. Psychiatry*, 132, 233, 1978.
7. **Murray, R., Cooper, J. E., and Smith, A.**, The Leyton obsessional inventory: an analysis of the responses of 73 obsessional patients, *Psychol. Med.*, 2, 305, 1979.

8. **Lewis, A.,** Problems of obsessional illness, *Proc. Soc. Med.,* 29, 325, 1936.

9. **Lewis, A.,** Obsessional Illness, in *Collected Papers,* Lewis, A., Routledge Kegan Paul, London,, 1967, 157.

10. **Salzman, L. and Thaler, F. H.,** Obsessive-compulsive disorders: a review of the literature, *Am. J. Psychiatry,* 138, 3, 286, 1981.

11. **Greenfield, J. G.,** *Greenfield's Neuropathology,* Blackwood, W. and Corsellis, J. A. N., Eds., Arnold, London, 1976.

12. **Brain, W. R.,** *Brain's Diseases of the Nervous System,* Walton, J. N., Ed., Oxford University Press, Oxford, 1977.

13. **Economo, Von,** *Encephalitis lethargica: its sequelae and treatment,* Newman, O., transl., Oxford University Press, Oxford, 1931.

14. **Claude, H., Baruk, H., and Lamanche, A.,** Obsessions-compulsions consécutives à l'encéphalite épidémique, *L'Encéphale,* 22, 716, 1927.

15. **Courtney, J. W.,** A case of post-encephalitic Parkinson's disease with polydipsia, *Arch. Neurol. Psychiatry,* 19, 188, 1928.

16. **McGowan P. K. and Cook, L. G.,** The mental aspects of chronic epidemic encephalitis, *Lancet,* 1316, 1928.

17. **Delbeke, R. and Bogaert, L. Van,** Le problème générale des crises oculogyres au cours de l'encéphalite épidémique chronique, *L'Encéphale,* 23, 855, 1928.

18. **Jellife, S. E.,** Psychologic components in post-encephalitic oculo-gyric crises, *Arch. Neurol. Psychiatry,* 21, 491, 1929.

19. **Jellife, S. E.,** *Psychopathology of forced movements and the oculo-gyric crises of lethargic encephalitis,* Nervous and Mental Disease Publishing Company Monograph, New York, 1932.

20. **Bromberg, W.,** Mental states in chronic encephalitis, *Psychiatric Quart.,* 4, 537, 1930.

21. **Steck, H.,** Les syndromes mentaux postencéphalitiques, *Arch. Suisses Psychiatr. Neurol.,* 27, 137, 1931.

22. **Wexberg, E.,** Remarks on the psychopathology of oculogyric crises in epidemic encephalitis, *J. Nerv. and Ment. Dis.,* 85, 56, 1937.

23. **Brickner, R. M., Rosner, A. A., and Monro, R.,** Physiological aspects of the obsessive state, *Psychosom. Med.,* 2(4), 369, 1940.

24. **Shaskan, D., Yarnell, H., and Alper, K.,** Physical, psychiatric and psychometric studies of post-encephalitic parkinsonism, *J. Nerv. Ment. Dis.,* 96, 652, 1942.

25. **Rosner, A. A.,** Unit reaction states in oculogyric crises, *Am. J. Psychiatry,* 99, 224, 1942.

26. **Fairweather, D. S.,** Psychiatric aspects of the post-encephalitic symdrome, *J. Ment. Sci.,* 93, 227, 1947.

27. **Schwab, R., Fabing, H. D., and Prichard, J. S.,** Psychiatric symptoms and syndromes in Parkinson's disease, *Am. J. Psychiatry,* 901, 1951.

28. **Ajuriaguerra, J. de,** Etude psychopathologique des Parkinsoniens, in *Monoamines Noyeaux Gris Centraux et Syndrome de Parkinson,* Ajuriaguerra, J. de, Ed., Masson, Paris, 1971.

29. **Haskovec, M. L.,** Contribution à la connaissance des idées obsédantes, in, *13ᵉ Congreès International de Médecine, Comptes Rendus de la Section de Psychiatrie,* 121, 1900.

30. **Parhon, C. and Goldstein, M.,** Etat psychasthénique survenu chez une jeune fille épileptique soumise au traitement thyroidien, disparaissant par la cessation du traitement et réapparaissant par sa reprise, *Revue Neurol.,* 6, 1908.

31. **Garmany, G.,** Obsessional states in epileptics, *J. Ment. Sci.,* 93, 639, 1947.

32. **Allen, I. M.,** Forced thinking as part of the epileptic attack, *N. Zealand Med. J.,* 51, 86, 1952.

33. **Heuyer, G., Lebovici, S., and Bouvier, S.,** Epilepsie et obsessions, *Revue de Neuropsychiatrie Infantile et de Hygiène Mentale de l'Enfance,* 2, 354, 1954.

34. **Roberts, J. K. A., Robertson, M. M., and Trimble, M. R.,** The lateralising significance of hypergraphia in temporal lobe epilepsy, *J. Neurol. Neurosurg. Psychiatry,* 45, 131, 1982.

35. **Kettl, P. and Marks, I. M.,** Neurological factors in obsessive-compulsive disorder — two case-reports and a review of the literature, *Br. J. Psychiatry,* 149, 315, 1986.

36. **Bear, D. M. and Fedio, P.,** Quantitative analysis of interictal behaviour in temporal lobe epilepsy, *Arch. Neurol.,* 34, 455, 1977.

37. **Marchand, L. and Ajuriaguerra, J. de,** *Epilepsies: Leurs Formes Cliniques et Leurs Traitement,* Desclée de Brouwer et Companie, Paris, 1948.

38. **Gilles de la Tourette,** Etude sur une affection nerveuse caracterisée par l'incoordination motrice accompagnée d'echolalie et de coprolalie, *Arch. Neurol. (Paris),* 9, 19, and 158, 1885.

39. **Guinon, G.,** Sur la maladie des tics convulsifs, *Rev. Méd.,* 6, 50, 1886.

40. **Gilles de la Tourette,** La maladie des tics convulsifs, *Semaine Méd.,* 19, 153, 1899.

41. **Grasset, I.,** Leçons sur un cas de maladie des tics et un cas de tremblement singulier de la tête et des membres gauches, *Arch. Neurol. (Paris),* 20, 27 and 187, 1890.

42. **Meige, H.,** *Tics,* Monographies Cliniques, Masson, Paris, 1905.

43. **Wilson, S. A. K.**, The tics and allied conditions, *J. Neurol. Psychopathol.*, 8, 93, 1927.
44. **Creak, M. and Guttman, E.**, Chorea, tics and compulsive utterances, *J. Ment. Sci.*, 834, 1935.
45. **Wexberg, E.**, in Schilder, P., *Am. J. Psychiatry*, (discussion), 144, 190, 1984.
46. **Fernando, S. J. M.**, Gilles de la Tourette syndrome — a report of four cases and a review of published case-reports, *Br. J. Psychiatry*, 113, 607, 1967.
47. **Corbett, J.**, Tics and Gilles de la Tourette syndrome — a follow-up study and critical review, *Br. J. Psychiatry*, 115, 1229, 1969.
48. **Morphew, J. A. and Sim, M.**, Gilles de la Tourette's syndrome: a clinical and psychopathological study, *Br. J. Med. Psychol.*, 42, 293, 1969.
49. **Montgomery, M. A., Clayton, P. J., and Friedhoff, A. J.**, Psychiatric illness in Tourette syndrome patients and first-degree relatives, in *Gilles de la Tourette Syndrome*, Friedhoff, A. J. and Chase, T., Eds., Raven Press, New York, 1982.
50. **Nee, L. E., Polinsky, R. J., and Ebert, M. H.**, Tourette syndrome: clinical and family studies, in *Gilles de la Tourette Syndrome*, Friedhoff, A. J. and Chase, T., Eds., Raven Press, New York, 1982.
51. **Frankel, M., Cummings, J., Robertson, M., Trimble, M., Hill, M., and Benson, F.**, Obsessions and compulsions in Gilles de la Tourette syndrome, *Neurology*, 36, 378, 1986.
52. **Schwab, R., Chafetz, M. E., and Walker, S.**, Control of two simultaneous voluntary motor acts in Parkinson's disease, *Arch. Neurol. Psychiatry*, 72, 591, 1954.
53. **Diller, L. and Riklan, M.**, Psychosocial factors in Parkinson's disease, *J. Am. Ger. Soc.*, 4, 1291, 1956.
54. **Lees, A. J. and Smith, E.**, Cognitive deficits in the early stages of Parkinson's disease, *Brain*, 106, 257, 1983.
55. **Mouren, P., Poinso, Y., Oppenheim, A., Mouren, A., and Nguyen Quang, M.**, La personnalité du parkinsonien — approche clinique et psychométrique, *Annales Médico-Psychol.*, 141(2), 153, 1983.
56. **Grimshaw, L.**, Obsessional disorder and neurological illness, *J. Neurol. Neurosurg. Psychiatry*, 27, 229, 1964.
57. **Laplane, D., Wildocher, D., Pillon, B., Baulac, M., and Binoux, F.**, Comportement compulsif d'allure obsessionelle par nécrose circonscrite bilatérale pallido-striatale, *Rev. Neurol.*, 138(2), 137, 1982.
58. **Laplane, D., Baulac, M., Wildocher, D., and Dubois, B.**, Pure psychic akinesia with bilateral lesions of the basal ganglia, *J. Neurol. Neurosurg. Psychiatry*, 47, 377, 1984.
59. **Pulst, S. M., Walshe, T. M., and Romero, J. A.**, Carbon monoxide poisoning with features of Gilles de la Tourette syndrome, *Arch. Neurol.*, 40, 443, 1983.
60. **Ali-Cherif, A., Royere, M. L., Gosset, A., Poncet, M., Salomon, G., and Khalil, R.**, Troubles du comportement et de l'activité mentale après intoxication oxycarbonée, *Rev. Neurol.*, 140, 401, 1984.
61. **Bailey, P.**, A case of pinealoma with symptoms suggestive of compulsion neurosis, *Arch. Neurol. Psychiatry*, 19, 932, 1928.
62. **Barton, R.**, Diabetes insipidus, obsessional neurosis and ? hypothalamic dysfunction, *Proc. R. Soc. Med.*, 47, 276, 1954.
63. **Barton, R.**, Diabetes insipidus and obsessional neurosis — a syndrome, *Lancet*, 133, 1965.
64. **Hillbom, E.**, After effects of brain injuries, *Acta Psychiatr. Neurol. Scand.*, Suppl. 142, 1966.
65. **Capstick, N. and Seldrup, J.**, A study in the relationship between abnormalities occurring at the time of birth and the subsequent developement of obsessional symptoms, *Acta Psychiatr. Scand.*, 56, 427, 1977.
66. **McKeon, J., McGuffin, P., and Robinson, P.**, Obsessive-compulsive neurosis following head-injury — a report of four cases, *Br. J. Psychiatry*, 144, 190, 1984.
67. **Schilder, P.**, The organic background of obsessions and compulsions, *Am. J. Psychiatry*, 94, 1397, 1938.
68. **Vujic, V.**, Larvate encephalitis and a new extrapyramidal syndrome, *Psychiatr. Neurol.*, 120, 249, 1950.
69. **Vujic, V.**, Larvate encephalitis and psychoneurosis, *J. Nerv. Ment. Dis.*, 116, 1051, 1952.
70. **Behar, D., Rapoport, J. L., Berg, C. J., Denckla, M. B., Mann, L., Cox, C., Fedio, P., Zahn, T., and Wolpman, M. G.**, Computed tomography and neuropsychological test measures in adolescents with obsessive compulsive disorder, *Am. J. Psychiatry*, 141(3), 363, 1984.
71. **Flor-Henry, P., Yeudall, L. T., Koles, Z. J., and Howarth, B. G.**, Neuropsychological and power spectral EEG investigations of the obsessive-compulsive syndrome, *Biol. Psychiatry*, 14(1), 119, 1979.
72. **Rapoport, J. L., Elkins, R., Langer, D., Sceery, W., Buchsbaum, M., Gillin, C., Murphy, D., Zahn, T. P., Laker, R., Ludlow, C., and Mendelson, W.**, Childhood obsessive-compulsive disorder, *Am. J. Psychiatry*, 138(12), 1545, 1981.
73. **Insel, T. R., Donnelly, E. F., Lalakea, M. L., Alterman, I. S., and Murphy, D. L.**, Neurological and neuropsychological studies of patients with obsessive-compulsive disorder, *Biol. Psychiatry*, 18, 741, 1983.
74. **Pacella, B. L., Polatin, P., and Nagler, S. H.**, Clinical and EEG studies in obsessive-compulsive states, *Am. J. Psychiatry*, 100, 830, 1944.
75. **Rockwell, F. V. and Simons, D. J.**, The EEG and personality organisation in obsessive-compulsive reactions, *Arch. Neurol. Psychiatry*, 57, 71, 1947.
76. **Gibson, J. G. and Kennedy, W. A.**, A clinical-EEG study in a case of obsessional neurosis, *Electroencephalogr. Clin. Neurophysiol.*, 12, 198, 1960.

77. **Ingram, I. M. and McAdam, W. A.**, The electroencephalogram, obsessional illness and obsessional personality, *J. Ment. Sci.*, 686, 1960.

78. **Epstein, A. W. and Baline, S. H.**, Sleep and dream studies in obsessional neurosis with particular reference to epileptic states, *Biol. Psychiatry*, 3, 149, 1971.

79. **Jenike, M. A. and Brotman, A. W.**, The EEG in obsessive-compulsive disorder, *J. Clin. Psychiatry*, 45(3), 122, 1984.

80. **Ribot, Th.**, *Psychologie de l'attention*, Felix Alcan, Paris, 1889.

81. **Tuke, D. H.**, Imperative ideas, *Brain*, 2, 179, 1894.

82. **Jackson, H.**, On imperative ideas — being a discussion on Dr. Hack Tuke's paper (Brain, 1894), *Brain*, 318, 1895.

83. **Guillain, G.**, in Wilson, S. A. K., *J. Neurol. Psychopathol.*, (discussion), 8, 93, 1927.

84. **Cruchet, R.**, in Wilson, S. A. K., *J. Neurol. Psychopathol.*, (discussion), 8, 93, 1927.

85. **Cloake, P. C.**, in Wilson, S. A. K., *J. Neurol. Psychopathol.*, (discussion), 8, 93, 1927.

86. **Barton, R.**, Diabetes insipidus and obsessional neurosis, *Am. J. Psychiatry*, 133(2), 235, 1976.

87. **Laplane, D., Baulac, M., Wildocher, D., and Dubois, B.**, Pure psychic akinesia with bilateral lesions of the basal ganglia, *J. Neurol. Neurosurg. Psychiatry*, 47, 377, 1984.

88. **Elkins, R., Rapoport, J. L., and Lipsky, A.**, Obsessive-compulsive disorder of childhood and adolescence — a neurobiological viewpoint, *J. Am. Acad. Child Psychiatry*, 511, 1980.

89. **Jenike, S. E.**, Obsessive-compulsive disorder: a question of a neurologic lesion, *Comprehensive Psychiatry*, 25(3), 298, 1984.

90. **Lieberman, J.**, Evidence for a biological hypothesis of obsessive-compulsive disorder, *Neuropsychobiology*, 11, 14, 1984.

91. **Turner, S. M., Beidel, D. C., and Nathan, R. S.**, Biological factors in obsessive-compulsive disorders, *Psychol. Bull.*, 97(3), 430, 1985.

92. **Mapother, E.**, Tough or tender. A plea for nominalism in psychiatry, *Proc. R. Soc. Med.*, 1687, 1933.

93. **Gowers, W. R.**, *A Manual of Diseases of the Nervous System*, Vol. 2, Churchill, London, 1888, 669.

94. **Holmes, G.**, *Introduction to Clinical Neurology*, Edinburgh, 1946.

95. **Anon.** Editorial, *J. Neurol. Psychopathol.*, October, 163, 1930.

96. **Stevens, J. R.**, Editorial: The Neuropathology of Schizophrenia, *Psychol. Med.*, 12(4), 695, 1982.

97. **Rojo-Sierra, M.**, Tratamiento de la neurosis obsesiva por los derivados del tropano, *Rev. de Psyq. y Psicol. Med. de Europa y Amer. Lat.*, 1, 365, 1954.

98. **Baruk, H. and Joubert, P.**, Action suspensive de l'ortedrine sur certaines obsessions avec toxicomanie consecutive, *Ann. Med. Psychol.*, 109, 69, 1951.

99. **Gilberti, F. and Gregoreti, L.**, Studio farmacopsichiatrico di un caso di psiconevrosi ossessiva, *Sistema Nerv. (Milano)*, 9, 275, 1957.

100. **Baruk, H., Launay, J., Cournut, J., Vallex, A., and Tardy, C.**, Le problème des indications théra-peutiques, des doses, et des incidents de traitment par l'imipramine, *Soc. Moreau de Tours*, 1(8), 14, 1959.

101. **Lopez-Ibor, J. J. and Lopez-Ibor, J. M.**, Tratamiento psicofarmacologico de las neurosis obsessivas, *Act. Luso Esp. Neurol.*, 1(6), 767, 1973.

102. **Dally, P.**, *Chemistry of Psychiatric Disorders*, Logos Press, London, 1967.

103. **Tapia, F.**, Haldol in treatment of children with tics and stutters, *Psychiatr. Quart.*, 43, 647, 1969.

104. **Breitner, C.**, Drug therapy in psychiatric states and other psychiatric problems, *Dis. Nerv. Syst.*, 21, 31, 1960.

105. **Bethune, H.**, A new compound in treatment of severe anxiety states, *N. Zealand Med. J.*, 63, 153, 1964.

106. **Rao, A. V.**, A controlled trial with valium in obsessive-compulsive state, *J. Ind. Med. Ass.*, 42, 564, 1964.

107. **Orvin, G.**, Treatment of phobic obsessive-compulsive patient with oxazepam, *Psychosomatics*, 8, 278, 1967.

108. **Casas, M., Alvarez, E., Duro, P., Garcia-Ribera, C., Udina, C., Velata, A., Abella, D., Espinosa, J. R., Salva, P., and Jane, F.**, Antiandrogenic treatment of obsessive-compulsive neurosis, *Acta Psychiatr. Scand.*, 73, 221, 1986.

109. **Trabucci, C., Zuanazzi, G., and Caceffo, G.**, Noste experienze sullacura con tofranil, *Riv. Sper. Freniatr.*, 83, 328, 1959.

110. **Navarro, F.**, Trattamento delle nevrosi ossissive, *Rassegna Inter. de Clin. Terapia*, 40, 616, 1960.

111. **Vidal, G. and Vidal, B.**, Imipramine et obsessions, *L'Encéphale*, 52, 167, 1965.

112. **Geismann, P. and Kamnerer, T.**, L'imipramine dans la nérvose obsession, *L'Encéphale*, 53(2), 369, 1964.

113. **Angst, J. and Theobald, W.**, Tofranil, *Slampfli et Cie.*, 1970, 31.

114. **Ananth, J., Solyom, L., and Sookman, D.**, Doxepin in the treatment of obsessive-compulsive neurosis, *Psychosomatics*, 16(4), 185, 1975.

115. **Grabowski, J.**, Treatment of severe obsessive depression with G34856, *Follia Med.*, 57, 265, 1968.

116. **Lopez-Ibor, J. M.**, Terapeutica de reactivacion, *Actas Luso-Espanolas Neurol. Psiquiatr.*, 28, 117, 1969.
117. **Yaryura-Tobias, J., Neziroglu, F., and Bergman, L.**, Clomipramine for obsessive-compulsive neurosis — an organic approach, *Curr. Ther. Res.*, 20, 541, 1976.
118. **Thoren, P., Asberg, M., and Cronholm, B.**, Clomipramine treatment of obsessive disorder, *Arch. Gen. Psychiatr.*, 37, 1281, 1980.
119. **Montgomery, S.**, Clomipramine in obsessional neurosis, *Pharm. Med.*, 1, 189, 1980.
120. **Marks, I., Stern, R., and Mawson, D.**, Clomipramine and exposure for obsessive-compulsive rituals, *Br. J. Psychiatr.*, 136, 1, 1980.
121. **Ananth, J., Pecknold, J., and Steen, N.**, Double blind comparative study of clomipramine and amitriptyline in obsessive neurosis, *Proc. Neuropsychopharmacol.*, 5, 257, 1981.
122. **Yaryura-Tobias, J. and Bhagwan, H.**, L-Tryptophan in obsessive-compulsive disorders, *Am. J. Psychiatr.*, 134(11), 1298, 1977.
123. **Prasad, A.**, A double-blind trial of imipramine versus zimelidine in obsessive-compulsive neurosis, *Pharmacopsychiatry*, 17, 61, 1984.
124. **Prasad, A.**, Efficacy of trazodone as an anti-obsessional agent, in *Biological Psychiatry, 1985*, Shagass, C., Josiassen, R., Bridger, W., Weiss, K., Stott, D., and Simpson, G., Eds., Elsevier Press, 1986, 744.
125. **Izikowitz, S.**, Trial of catron therapy in compulsions, *Svensk. Lakad.*, 57, 1993, 1960.
126. **Jain, V., Swinson, R., and Thomas, J.**, Phenelzine in obsessive-compulsive neurosis, *Br. J. Psychiatr.*, 117, 237, 1970.
127. **Jenike, M.**, Rapid response of severe obsessive-compulsive disorder to tranylcypromine, *Am. J. Psychiatr.*, 138, 1249, 1981.
128. **Insel, T.**, Obsessive-compulsive disorder, in *Psychiatr. Clin. N. Am.*, 8(1), 105, 1985.
129. **Sewell, R. and Lee, R.**, Opiate receptors, endorphins and drug therapy, *Postgrad. Med. J.*, 56 (supp. 1), 25, 1980.

Chapter 4

HYSTERIA

Rahul Manchanda and Harold Merskey

TABLE OF CONTENTS

I. INTRODUCTION

Hysteria is known as one of the oldest diagnoses in the history of medicine, yet still somewhat controversial. Lewis[1] pointed out that it is a tough old word which dies very hard and will tend to out-live its obituarists. Every clinician has used the word hysteria at one time or another with the awareness of how little he or others know about its conceptual framework. It is as if we are left with no choice but to abuse the term when we are faced with a symptom complex which does not fit into the so-called framework of an "illness". It was originally supposed to be a physical illness due to the migration of the womb.[2] Sydenham[3] included many conditions under hysteria which we would now suppose to be organic.[4] There was not so much dispute about whether it was an organic disease but about whether it was a disease of the brain or of the uterus. Georget[5] took it to be the former, calling it a "neurosis" by which he meant a physical disease. Brachet[6] and Briquet[7] followed him. Georget was not the first (see References 4 and 8). Yet some of their contemporaries thought otherwise.[9,10] It has been pointed out by Hollender[11] that even to the end of the 19th century, gynecologists tended to favor the uterine explanation of hysteria.

Whereas the psychological accompaniments of hysteria were often recognized, they only began to be seen as primary in the second half of the 19th century. The evolution of these ideas up to the point where they were adopted by Freud has been traced by Veith[8] and by Merskey,[4] as well as others. Ey[12] has pointed out the contrast "Hysteria is essentially disease of the brain" or "Hysteria is essentially not an organic illness". Neither can be completely true. The true evidence for a biological basis for hysteria is scattered and somewhat limited. This is not surprising since it is a common prerequisite for a diagnosis of hysteria that the symptoms not derive from an organic disease. The following is a critical review on the available evidence for an organic basis of hysteria.

Hippocrates named the disease hysteria.[2] The Greek word *hysteria* means the uterus, and he thought the uterus could wander out of the pelvis with hysterical symptoms occurring wherever the womb might lodge. For example, it might attach itself to the liver and could be detached from there by hand.[13] Similarly, it might be lured back to its location by burning fragrant aromas at the vagina or driven back by unpleasant odors at the nostrils. His ideas are said to have derived from even earlier Egyptian sources.[8] Since then the history of changing concepts of hysteria is interwoven with demonology, witchcraft, femininity, sexuality, and emotional conflict. The history of hysteria has been well chronicled by Veith,[8] who reviewed the subject up to the 19th century. More recently, a deliberate parallel title to that historical account further elaborates the ideas which deal with the relationship between feelings, somatic changes, and the state of the brain[4] in hysterical presentations.

For the purpose of the present topic, the historical accounts permit four main conclusions to be drawn.[14] First, it was suggested by Schilder[15] that organic illness might take effect through the repression of a conflict. In other words, the physical changes in the brain could correspond to the psychological changes of repression. This idea, however, has never been proven nor even perhaps tested. Secondly, the same symptoms could have either an organic or a psychological cause. Thirdly, organic disturbances of brain function can provoke neurotic attitudes, and fourthly, organic disease and hysterical symptoms frequently coexist.

II. DEFINITIONS AND DIAGNOSTIC CRITERIA

The definition and diagnosis of hysteria continues to be controversial and debatable, even today. Clinicians have used the term hysteria to describe a symptom, several patterns of illness, attention-seeking behavior and a personality type.[16] Doubts concerning a reliable diagnosis are thus to be expected. Follow-up studies have further supported the notion that the diagnosis of hysteria is fraught with error.[17-19] For the sake of more objectivity, investigators in the field often select aspects of the patient's history and presentation that imply hysteria.[14,16-20] Among the most suggestive features are either a background of nonorganic somatic symptoms or an obvious set of life circumstances that encourage illness for secondary gain. Unfortunately, as we shall note again shortly, these features are quite insufficient to justify a diagnosis of hysteria. The disorder as it stands, finds a place in the International Classification of Disease (ICD-9, 1978).[21] The development of the Diagnostic and Statistical Manual (DSM-III)[22] of the American Psychiatric Association was accompanied by the rejection of the term hysteria. Instead, a group of disorders called somatoform disorders has been introduced. This group includes most of the disorders previously referred to as either hysteria or Briquet's Syndrome, delineated by Perley and Guze. Diagnosis of the latter in this manner yields a 90% diagnostic reliability over a 6 to 8 year follow-up.[23] According to the DSM-III, the essential features of somatoform disorders are physical symptoms suggesting physical disorder for which there are no demonstrable organic findings or known physiological mechanisms and for which there is positive evidence, or strong presumption, that the symptoms are not under voluntary control and are linked to psychological factors or concepts. Whereas the positive symptoms which Head[24] described are generally useful, it should also be recognized that many traditional symptoms employed in neurology, or signs found on examination, are not very reliable and do not properly allow a diagnosis of hysteria to be made. A study of particular note was recently published by Gould et al.[25] Those authors showed that seven features traditionally considered to be indicators of hysteria were frequently found with organic disease and were not diagnostic of hysteria. They were a history suggestive of hypochondriasis, potential secondary gain, la belle indifference, nonanatomical or patchy sensory loss, changing boundaries of hyperalgesia, sensory loss to pinprick or vibratory stimulation, splitting at the midline and give-way weakness. Many of these symptoms represent merely suggestibility on the part of the patient. It is the positive signs found on physical examination which show the patient cannot undertake a specific action of a motor type, which may serve as the best indicator of the classical hysterical symptom. Unfortunately, even the revised version of DSM-III (DSM-IIIR, 1987) also does not provide sufficiently stringent criteria to make an acceptable diagnostic category, at least in respect to conversion symptoms or psychogenic pain disorder. Someone whose symptoms are merely felt to be disproportionate to the physical disorder and who is exposed to emotional stress can be diagnosed as having one or other of these conditions without any other noteworthy restrictions applying on the making of the diagnosis.

A more reliable approach to the diagnosis of hysterical symptoms depends upon demonstrating what was recognized by both Charcot and by Freud to be a key feature of the diagnosis of hysteria. That is to say, that the patients should be possessed of complaints which correspond to an idea in their minds rather than to a physical or pathophysiological state. The proof of this was developed by Head[24] on the basis of what he called positive signs of hysteria. These are signs on neurological examination which demonstrate that the patient is indeed able to use muscles or parts of the body which he or she had thought to be unusable. Slater[17] and Merskey[14] have emphasized the importance of using this approach for the classical conversion symptom and its neurological proof in order to delineate those symptoms which can unequivocally be recognized as having the fundamental features of hysteria, i.e., a complaint of inability to do something which is actually physically feasible.

Much of the confusion regarding the diagnosis of hysteria as a discrete condition is centered around conversion symptoms. Besides the need to diagnose conversion symptoms on a sound basis, it is also the case that they may be found in either sex and among patients suffering from a wide variety of psychiatric, neurological, and medical conditions. DSM-III refers to conversion disorder when conversion symptoms are the predominant disturbance and are not symptomatic of another disorder. Despite the disagreement and confusion surrounding the semantics and diagnostic criteria, varied clinical manifestations from a motor paralysis to complex polysymptomatic somatization disorder continue to be labeled as hysteria. For convenience of description the following presentation patterns of hysteria have been isolated.[14]

1. Hysteria with one or two symptoms, usually motor or dissociative (as in amnesia); sometimes pain
2. Polysymptomatic hysteria, especially hypochondriasis and Briquet's syndrome as described by Guze and his colleagues (now elaborated in DSM-III under the heading of somatization disorder)
3. Hysterical elaboration of organic complaints
4. Symptoms of self-induced illness or self-damage in abnormal personalities ranging from anorexia nervosa to hospital addiction
5. Psychotic or pseudopsychotic disorders (Ganser's Syndrome, hysterical psychosis)
6. Hysterical personality
7. Culturally sanctioned endemic or epidemic hysteria

One of the main difficulties of a comprehensive critical review of hysteria relates to the changing concepts and terminology of the disorder. There is also a lack of substantial research into the biological basis of hysteria. However, for practical purposes, the first three presentations of hysteria (items 1 to 3) mentioned above will serve very well for a consideration of topics for which an organic basis deserves to be considered, whether or not it exists.

III. GENETICS

A. Family Studies

Briquet[7] compiled many statistics on familial aspects of hysteria and these were summarized by Mai and Merskey.[26] Briquet compared the incidence of hysteria in the first degree relatives of 354 hysterical women with the incidence in the relatives of 167 sick, nonhysterical women. Among the hysterical women, 214 out of 1,103 relatives had hysteria (19%), compared with 11 relatives of the nonhysterical women out of 704 (2%). Further, of 100 hysterical mothers who had 220 daughters, hysteria developed in 124 of the daughters. He concluded that persons born of hysterical parents were 12 times as likely to have hysteria develop as those born of nonhysterical parents.

In one of the earliest investigations in the modern period, Kraulis[27] studied 106 hospitalized patients with "hysterical reaction" — defined as a paroxysmal condition with psychogenic disturbances of consciousness such as fits and twilight states. He found that 9.4% of parents, 3.6% of the siblings, and 14.9% of the children had a similar presentation. The risk of hysteria and psychopathy were considerably higher in the families of a subgroup with a personality disorder. Surprisingly, sibs of probands where one parent had a hysterical reaction were not more often similarly affected than those where no parent had been diagnosed. Thus, Kraulis could not deny the role of the environmental factors in hysteria but felt more convinced about the heritability of psychopathic constitution than hysterical reaction. Much lower figures were given by McInnes[28] for the families of 30 hysterical patients. Of the 50 parents, two suffered from anxiety states and two from hysteria; of the 117 sibs, two suffered from anxiety states and none from hysteria. Brown,[29] however, found somewhat higher rates — 19% of the parents and 6% of the sibs of his hysteric group had hysterical symptoms.

Ljungberg's[30] study with 381 probands is by far the largest modern family risk study of hysteria. Unfortunately, no statement is made of the criteria on which a diagnosis of hysteria in the relatives is made. Of the hysterical symptoms in probands, disturbance of gait (47%) and fits (20%) were the commonest. The coincidence of hysteria in the fathers, brothers, and sons of hysterical probands was 2, 3, and 5%, respectively, and in their mothers, sisters, and daughters 7, 6, and 7%, respectively, as against an estimated .05% for the general population. Interestingly, Ljungberg found that 25% of male probands had sustained cerebral injury. The relatives of the male probands who were found to be similarly affected were more often males, but there was an overall increase in the rate of hysteria in the families of male probands as compared to female hysterics. His study, by and large, remained inconclusive, as was a subsequent small study.[31] This clearly supports a multifactorial etiology of hysterical phenomena. Cloninger et al.[32] have put forward the hypothesis that hysteria (referring to Briquet's syndrome) and sociopathy are part of the same disease. They regard male sociopathy, female hysteria, and female sociopathy as increasingly severe manifestations of a multifactorial, polygenic inheritance.

B. Twin Studies

In Slater's twin study,[33] 24 probands had been clinically diagnosed to be suffering from hysteria. None of the 12 monozygotic (MZ) or 12 dizygotic (DZ) co-twins had ever received a diagnosis of hysteria except for one male DZ pair subsequently found to be concordant.[34] Further, concordance was either low or absent in MZ pairs or not significantly higher than in DZ pairs in another study.[35] Recently, Torgersen[36] investigated the contribution of hereditary factors in somatoform disorders as defined in DSM-III. He personally interviewed 14 MZ and 21 DZ index twins and their co-twins. The results showed a concordance of 29% in MZ and 10% in DZ pairs. There was, however, a high frequency of general anxiety disorders in the co-twins of those with somatoform disorders. Similarity in childhood experience seemed to influence the concordance rate, and he concluded that transmission may be environmental rather than genetic.

There is no evidence to suggest any specific genetic factors for hysterical presentation. A low concentration in twins further tends to refute a genetic hypothesis for hysteria, a finding also concluded by Inouye[37] and Shields[34] in their review of genetic studies of hysteria.

IV. PSYCHOPHYSIOLOGY AND NEUROPSYCHOLOGY

Psychophysiological studies on hysterical patients tend to fall into two broad groups: (1) stimulus screening capacity, investigated by arousal, and (2) habituation of the galvanic skin response and the processing of stimuli in the brain, investigated by the evoked-response potentials.

In a controlled but small sample study,[38] hysterical conversion patients (N = 10) reported higher subjective anxiety than the anxious-phobics (N = 71), although the former were rated objectively as less anxious. Psychophysiologically too, the hysteria patients had the highest level of arousal (measured by spontaneous fluctuations in skin resistance, cardiac acceleration, and myogenic activity). However, little emotional distress was communicated by the patient, and this observation is relevant to the question of "la belle indifference", the affective incongruity described characteristically in hysteria. In another study[39] 7 of 11 patients with chronic conversion symptoms had grossly impaired habituation. The patients had greater than normal autonomic activity in the form of higher heart rate, sweat gland activity, and muscle activity. Further investigation concluded that the failure to habituate was not a function of abnormally high levels of arousal but impaired ability to screen out or not attend to irrelevant afferent stimuli.[40]

From the above, it is possible that some symptoms of hysteria may be associated with a defect in the processing of afferent stimuli. Slow habituation is associated with a right hemispheric preponderance and fast habituation with left hemisphere or relative overactivation.[41] A dysfunctional overactivation of the right hemisphere is therefore indicated by the above findings, a lateralization which, in addition, accounts for the affective incongruity of hysteria.[42] Other hypotheses for this lack of concern (la belle indifference) propose intense corticofugal inhibitions causing an attentional dysfunction[19] or a selective depression of awareness of bodily functions.[43] A detailed neuropsychological investigation of hysteria further suggests laterality in hysteria.[44] Ten patients with a stable syndrome of hysteria (Briquet's syndrome) were matched for age and sex and full scale WAIS I.Q. with 10 controls, 10 psychotic depressives, and 10 schizophrenics. All were subjected to an extensive neuropsychological test battery. Compared to the controls, the hysteria group was characterized by an essentially bilateral and symmetrical pattern of anterior cerebral dysfunction, but globally, a greater dysfunction of the nondominant than dominant hemisphere. D-index analysis, (a test which identifies the neuropsychological variables most responsible for the separation between the experimental groups), however, showed that the hysteria group had a greater dysfunction of the dominant hemisphere compared to the normals and depressives. The schizophrenia group, on the other hand, showed greater nondominant hemisphere dysfunction than the hysteria group. Further, a cluster analysis on the 40 subjects produced 3 clusters: normal controls, depressives, and a schizophrenia-hysteria grouping. These findings are interpreted as suggesting that the dominant hemisphere dysfunction is fundamentally related to the syndrome of hysteria and that the dysfunction of the nondominant hemisphere is brought about by associated features: the presence of asymmetrical pain and conversion symptomatology. Conversion symptoms have been found to occur with greater frequency on the left side of the body than on the right, [45] and this left-sided preponderance appears to apply equally to right- and left-handers.[46] Thus, to Flor-Henry et al.[44] the core deficits in hysteria consist of the impairment in verbal communication, the incongruity of affective responsivity, and the defect in understanding endogenous somatic signals. These, in their opinion, are the consequences of altered dominant hemispheric functioning which produces, when it occurs in the female, a secondary disorganization of the contralateral (nondominant) hemisphere which determines the flamboyant facade of female hysteria, but which at the same time masks its fundamental left hemisphere substrate. To date the issue of laterality in hysteria is, however, far from settled.

Investigation of a 15-year-old girl with a glove and sleeve analgesia and thermal anesthesia of the left arm[47] showed that somatosensory-evoked potentials were recorded over the contralateral parietal area when the normal side was stimulated; no definite response was found from the right parietal area when the left affected forearm was stimulated. This led the investigators to conclude that hysterical anesthesia may be the result of an increased inhibition of afferent transmission somewhere along the somatic sensory batteries. Results of further experiments[40,48,49] have implied the following mechanism: (1) a lowering of peripheral receptor sensitivity, and (2) a central mechanism of inhibition along the afferent pathway. Other studies on evoked-response potentials[50-53] have failed to support some of these findings. Normal evoked responses in these studies, as well as in more recent investigations,[54,55] have actually been helpful in establishing the hysterical nature of the neurological deficit in each case. At the present time, it is standard practice in most laboratories to elicit evoked-response potentials by activating peripheral nerves using stimuli well above motor threshold.[56] Using this procedure, no abnormal responses to somatosensory evoked potential testing have been reported in hysterical patients. An abnormal evoked potential response is a reliable indicator of underlying organic disease, if standard testing conditions are maintained and the patient is cooperative and rigorously monitored. However, the converse is not true. The results of evoked potential testing must be interpreted within the context of the clinical setting and with knowledge of other significant laboratory abnormalities. If all other information points to a diagnosis of hysteria, normal evoked potentials may provide important confirmatory evidence.[55]

A recent review[57] on the neuropsychological concept of somatoform disorders focuses on the models suggesting damage to the brain stem structures, modulating influence of the cerebral cortex and of the second somatosensory area. The knowledge is largely theoretical, and adequate empirical evidence is lacking. Miller[57] rightly warns, " . . . neurologizing of this sort must remain in the area of reasonable conjecture with the hope that productive investigation will ensue''. This remains to be seen. He argues in favor of a biopsychosocial approach, incorporating available neuropsychological data, to the understanding of somatoform disorders.

V. ORGANIC BRAIN DISEASE

The association of hysteria with organic disease has been a controversial subject through history. Follow-up studies by Slater and Glithero[17,18] are a significant milestone in the controversy surrounding hysteria. Of the 112 cases seen and diagnosed by them, 85 patients were followed-up 9 years later. Of these 85 patients, 4 had died from suicide and 8 from organic disease; 24 were originally considered to have organic disease and 22 were later found to have organic disease. A further 33 had no organic pathology, but two developed schizophrenia and 8 had affective illness. Finally, there was a nuclear group of only 7 young patients with classical conversion symptoms in response to stress and 14 with chronic personality disorder and multiple symptoms. In sum, 49 (58%) of the patients in whom hysteria had been diagnosed had an underlying organic illness. From this, Slater drew the conclusion so often quoted, that the diagnosis of hysteria was a "delusion and a snare". He remained willing to use the word hysteria as an adjective but not as a noun. In another follow-up study, Whitlock[19] found that 62.5% of his cases, as compared to only 5.5% of a control group of patients with depression and anxiety states, had an accompanying cerebral disorder of a preceding organic brain injury. Merskey and Buhrich[58] reported 67% of 89 patients with motor conversion symptoms had some organic diagnosis, 48% of the total having an organic cerebral disorder or systemic illness likely to affect brain function. Interestingly, 67% of a smaller group (N = 24) compared to 79% of a matched control group had an organic disorder — 50% and 58%, respectively, with organic cerebral disease likely to affect the brain functions. The main relevant organic disease was epilepsy. The authors were not surprised at their findings as their psychiatric service, like Slater's, was primarily for neurologically ill patients. The figures for accompanying cerebral and systemic disease in the studies reported in the literature have ranged from 3%[58] to 66.5%.[59]

A cross-culture comparison shows that motor hysterical symptoms without organic disease still occur frequently in unsophisticated populations and developing countries, suggesting that it is perhaps in the more urbanized developed countries that hysterical symptoms have a definite link with physical illness, especially with organic cerebral disorders.[54] More importantly, it has been recognized that in medical practice the commonest place of presentation of classical hysterical symptoms is the neurological clinic.[14,60,61] Absence of a causative organic disease poses a real dilemma, and Marsden[61] has recently addressed the issue from a neurologist's viewpoint. He proposes the term "neurological hysteria", defined as (1) loss or distortion of neurological function, and (2) not fully explained by organic (visible) or functional (invisible, but perhaps, biochemical) disease, as determined by clinical examination or full investigation.

In a review of the subject,[14] five possible links in the association of physical illness to hysterical symptoms were suggested:

1. An independent emotional stress may lead the patient to elaborate the basic effects of a lesion which has already disturbed function in a part. Perhaps in such a case the most organically affected part is liable to acquire the extra hysterical signs.
2. The unpleasant psychological implications of a physical illness, the discomfort, and the fear attached to it, may lead the patient to elaborate an existing symptom or produce a fresh one.
3. The occurrence of a past or intermittent physical symptom, as in epilepsy, may lead

to a modelling of the hysterical symptoms upon the organic pattern which the patient has experienced. Psychological stress would more easily produce a form of illness of which the patient had some knowledge other than one of which he had no knowledge.

4.   It may happen that physical illness will lead to a regression in behavior and a secondary gain from those around him.

5.   There is the notion that cerebral damage may operate in a more specific way to produce conversion symptoms. If this view is correct, cerebral disease should give rise to more hysterical symptoms than peripheral (nonsystemic) physical illness. A quantitative trend in this direction was shown by Merskey,[62] but the figures, although quite large, were not statistically significant.

From his clinical experience, Marsden[61] further suggests that this association may be due to the patient's desire to convince or help the doctor to determine the real cause of his/her problems. Anxiety leads to elaboration or exaggeration of the real deficit. Also, the patient "may enjoy beating the doctor" i.e., enjoy pitting his/her wits against every effort of the doctor to cure the complaints.

The high prevalence of associated organic illness in hysteria is widely acknowledged and accepted. If it were not for the advances in clinical knowledge that have enabled an organic diagnosis to be made, we might have continued to emphasize the frequency of somatic symptoms primarily of psychological origin.[63] A number of examples are known in which what was formerly considered to be hysterical is now taken to be organic, e.g., facial hemispasm, which was thought to be hysterical by Charcot, and several of the facial dyskinesias which have increasingly been recognized to be related to dopaminergic dysfunction. It is worth emphasizing that when a condition is poorly understood and the patient is referred from one medical department to another, psychological symptoms are likely to develop. These include anger, frustration, irritability, and some criticism of the medical profession. In such a condition, hysteria should not be diagnosed any more in the absence of positive grounds for the diagnosis than in other circumstances where positive grounds are also necessary.

VI. HYSTERIA AND EPILEPSY

A link between hysteria and epilepsy goes back to early writings such as those of Hippocrates[64] and Aretaeus.[65] The French physicians of the last century were particularly prominent in attempting to examine epilepsy and hysteria and the relationship between them, and they introduced the term "hysteroepilepsy".[14] Other terms, e.g., pseudoseizures, simulated epilepsy, psychogenic seizures, nonepileptic seizures, and pseudoepileptic seizures have been used in this context. Unfortunately, many of these terms, especially those which include the word pseudo or simulated, tend to reject the patients who have them as having a genuine illness or problem. Accordingly, such terms are best avoided. Hysterical seizures are quite a frequent manifestation of hysteria, occurring in between 9 and 20% of these patients;[30,67,68] and associated present or past history of epilepsy has been reported in 12 to 65% of these patients.[14,69,70] Previous episodes of epileptic seizures, either observed or personally experienced, provide a model for such hysterical manifestation. Precipitating stress is frequently present. The clinical manifestation of hysterical seizures is variable and has been detailed in several publications.[66,67,70,71] There can, however, be rare and unusual accompaniments of pseudoepileptic seizures that can make diagnosis extremely difficult. Examples include: incontinence, recurrent injury (tongue bite, bruising), twitching, sensory phenomena, and transient global amnesia. For further details, the reader should refer to a recent comprehensive review on the subject.[67]

VII. CONCLUSIONS

The causation of hysterical symptoms is, even today, an intriguing and diverse phenomenon. Considering the variety of manifestations recognized as hysteria, it is hardly surprising that there be different views about its etiology. The issue of whether or not there is a biological basis for hysteria is further complicated by a not-infrequent exclusion criteria of organic basis for a hysterical manifestation. Hysteria, like other neuroses, appears to develop as a response to emotional conflict or stressors in an individual who has a biological and personal predisposition to the development of the disorder. The resulting symptomatology may mimic organic disease, allowing the patient to assume a sick role and relief from the concomitant stress. Organic brain disease, too, may predispose an individual to hysterical manifestation. Conversely, and not uncommonly, hysteria may be the first manifestation of an organic brain disease.

The results of the genetic studies reviewed do not provide any strong support to the genetic hypothesis. On the contrary, environmental factors are probably more significant than a genetic transmission. The neurophysiological model of hysteria emphasizes a hypofunction of the dominant hemispheric systems, leading to a dysfunctional overactivation of the non-dominant hemisphere. There is also the possibility of abnormal interhemispheric relations. This theoretical perspective requires confirmatory evidence. Normal somatosensory evoked potentials provide an important negative finding, suggesting a hysterical phenomenon if the symptom production is not under voluntary control. The presence of organic brain disease appears to facilitate the use of hysterical mechanisms, thus supporting an association between the two conditions rather than etiologically causative role for organic disease in the onset of hysteria.

Advances in clinical knowledge and investigative procedures are demonstrating how the clinical manifestations of certain rare organic disorders may have been mistakenly labeled as hysteria. An understanding of the factors leading up to the clinical manifestation are, therefore, crucial for every case. Until we know more about the disorder, there cannot be a specific treatment technique. A sympathetic, professional, and flexible approach to the disorder, with the aim of securing full psychiatric appraisal, would go a long way in understanding the varied presentation and complex etiology of this fascinating disorder.

REFERENCES

1. **Lewis, A. J.,** The survival of hysteria, *Psychol. Med.,* 5, 9, 1975.
2. **Hippocrates** *Oeuvres Completes d'Hippocrate. Trad. E. Littre,* Vol. 8, J. B. Bailliere, Paris, 1853, pp. 33, 267, 269.
3. **Sydenham, T.,** Discourse concerning hysterical and hypochondriacal distempers. In *Dr. Sydenham's Complete Method of Curing Almost All Diseases, And Description of Their Symptoms. To Which Are Now Added Five Discourses Of The Same Author Concerning The Pleurisy, Gout, Hysterical Passion, Dropsy, And Rheumatism,* 3rd ed., Newman and Parker, London, 1697, 149.
4. **Merskey, H.,** Hysteria: the history of an idea, *Can. J. Psychiat.,* 28, 428, 1983.
5. **Georget, M.,** *De La Physiologie Du Systeme Nerveux,* Vol. 2, J. B. Bailliere, Paris, 1821, 259.
6. **Brachet, J. L.,** *Traite de l'Hysterie,* J. B. Bailliere, Paris, 1847.
7. **Briquet, P.,** *Traite Clinique et Therapeutique de l'Hysterie,* J. B. Bailliere et Fils, Paris, 1859.
8. **Veith, I.,** *Hysteria. The History of a Disease,* University of Chicago Press, Chicago, 1965.
9. **Dubois D'Amiens, E. F.,** *Histoire Philosophique de l'Hysterie,* Paris, Deville Cavellin, 1833.
10. **Landouzy, H.,** *Traite Complet de l'Hysterie,* J. B. Bailliere, Paris, 1846.
11. **Hollender, M. H.,** Conversion hysteria — a post-Freudian reinterpretation of XIXth century psychological data, *Arch. Gen. Psychiatry,* 26, 311, 1972.
12. **Ey, H.,** *History and Analysis of the Concept in Hysteria,* Roy, A., Ed., John Wiley & Sons, New York, Chap. 1, 1982.

13. **Hippocrates,** *Oeuvres Completes d'Hippocrate. Trad. E. Littre,* Vol. 8, J. B. Bailliere, Paris, 1853, 271.
14. **Merskey, H.,** *The Analysis of Hysteria,* Bailliere Tindall, London, 1979.
15. **Schilder, P.,** *The Image and Appearance of the Human Body,* International Universities Press, New York, 1935.
16. **Kendell, R. E.,** A new look at hysteria, *Medicine,* 30, 1780, 1972.
17. **Slater, E.,** Diagnosis of ''hysteria''. *Br. Med. J.,* 1, 1395, 1965.
18. **Slater, E. and Glithero, E.,** A follow-up of patients diagnosed as suffering from ''hysteria'', *J. Psychosom. Res.,* 9, 9, 1965.
19. **Whitlock, F. A.,** The aetiology of hysteria, *Acta Psychiat. Scand.,* 43, 114, 1967.
20. **Roy, A.,** *Hysterical Neurosis, Hysteria,* Roy, A., Ed., John Wiley & Sons, New York, Chap. 8, 1982.
21. World Health Organization, *International Classification of Diseases,* 9th ed., (ICD-9) WHO, Geneva, 1978.
22. American Psychiatric Association, *Diagnostic and Statistical Manual,* Vol. 3 (DSM-III) APA, Washington, D.C., 1980.
23. **Perley, M. and Guze, S. B.,** Hysteria: the stability and usefulness of clinical criteria. A quantitative study based upon a 6-8 year follow-up of 39 patients, *N. Engl. J. Med.,* 266, 421, 1962.
24. **Head, H.,** An address on the diagnosis of hysteria, *Br. Med. J.,* i, 827, 1922.
25. **Gould, R., Miller, B. L., and Goldberg, M. A.,** The Validity of Hysterical Signs and Symptoms, *J. Nerv. Ment. Dis.,* 174, 593, 1986.
26. **Mai, F. and Merskey, H.,** Briquet's concept of hysteria: a historical perspective, *Can. J. Psychiat.,* 26, 57, 1980.
27. **Kraulis, W.,** Zur Vererbung der hysterischen Reaktionsweise, Zeitschrift fur die gesamete Neurologie und Psychiatrie, 136, 174, 1931; cited in Shields, J., *Genetical Studies of Hysterical Disorders in Hysteria,* Roy, A., Ed., John Wiley & Sons, New York, Chap. 5, 1982.
28. **McInnes, R. G.** Observations on heredity in neurosis, *Proc. R. Soc. Med.,* 30, 895, 1937.
29. **Brown, F. W.,** Heredity in the psychoneuroses, *Proc. R. Soc. Med.,* 35, 785, 1942.
30. **Ljungberg, L.,** Hysteria: a clinical, prognostic and genetic study, *Acta Psychiat. Neurol. Scand.,* 32, suppl 112, 1957.
31. **Ey, H. and Henric, E.,** Hérédite et nervoses. *Évolution Psychiatr.,* 24, 287, 1959.
32. **Cloninger, C. R., Reich, T., and Guze, S. B.,** The multifactorial model of disease transmission: III. Familial relationship between sociopathy and hysteria (Briquet's syndrome), *Br. J. Psychiat.,* 127, 23, 1975.
33. **Slater, E.,** The thirty-fifth Maudsley lecture: 'Hysteria 311', *J. Ment. Sci.,* 107, 359, 1961.
34. **Shields, J.,** *Genetical Studies of Hysterical Disorders in Hysteria,* Roy, A., Ed., John Wiley & Sons, New York, Chap. 5, 1982.
35. **Scarthy, L. G. and Kay, D. W. K.,** Protracted cardiac neurosis with congenital heart disease in one of identical twins, *Br. Heart J.,* 31, 404, 1969.
36. **Torgersen, S.,** Genetics of somatoform disorders, *Arch. Gen. Psychiatry,* 443, 502, 1986.
37. **Inouye, E.,** Genetics aspects of neurosis: a review, *Int. J. Health,* 1, 176, 1972.
38. **Lader, M. H. and Sartorius, N.,** Anxiety in patients with hysterical conversion symptoms, *J. Neurosurg. Psychiat.,* 31, 490, 1968.
39. **Meares, R. and Horvath, T.,** 'Acute' and 'chronic' hysteria, *Br. J. Psychiat.,* 121, 653, 1972.
40. **Horvath, T., Friedman, J., and Mears, R.,** Attention in hysteria: a study of Janet's hypothesis by means of habituation and arousal measures, *Am. J. Psychiat.,* 137, 217, 1980.
41. **Gruzelier, J. H.,** Hemispheric imbalance masquerading as paranoid and non-paranoid syndromes, *Schizophr. Bull.,* 7(4), 662, 1981.
42. **Flor-Henry, P.,** Hysteria, *Handbook of Clinical Neurology,* Vol. 2, (46): Neurobehavioural Disorders, Fredericks, J. A. M., Ed., Elsevier Science, Chap. 32, 1985.
43. **Ludwig, A. M.,** Hysteria: a neurobiological theory, *Arch. Gen. Psychiat.,* 20, 771, 1972.
44. **Flor-Henry, P., Fromm-Auch, D., Tapper, M., and Schopflocher, D.,** A neuropsychological study of the stable syndrome of hysteria, *Biol. Psychiat.,* 16(7), 601, 1981.
45. **Galin, D., Diamond, R., and Braff, D.,** Lateralization of conversion symptoms: more frequent on the left, *Am. J. Psychiat.,* 134, 578, 1977.
46. **Stern, D. B.,** Handedness and the lateral distribution of conversion reactions, *J. Nerv. Ment. Dis.,* 164, 122, 1977.
47. **Hernandez-Peon, R., Chavez-Ibarra, G., and Aguilar-Figueroa, E.,** Somatic evoked potentials in one case of hysterical anaesthesia, *Electroencephalgr. Clin. Neurophysiol.,* 15, 889, 1963.
48. **Levy, R. and Behrman, J.,** Cortical evoked responses in hysterical hemianaesthesia, *Electroencephalogr. Clin. Neurophysiol.,* 29, 400, 1970.
49. **Levy, R. and Mushin, J.,** Somatosensory evoked responses in patients with hysterical anaesthesia, *J. Psychosom. Res.,* 17, 81, 1973.
50. **Halliday, A. M.,** Computing techniques in neurological diagnosis, *Br. Med. Bull.,* 24, 253, 1968.

51. **Behrman, J.**, The visual evoked response in hysterical amblyopia, *Br. J. Ophthal.*, 53, 839, 1969.
52. **Behrman, J. and Levy, R.**, Neurophysiological studies on patients with hysterical disturbance of vision, *J. Psychosom. Res.*, 14, 187, 1970.
53. **Moldofsky, H. and England, R. S.**, Facilitation of somatosensory average evoked potentials in hysterical anaesthesia and pain, *Arch. Gen. Psychiat.*, 32, 193, 1975.
54. **Kaplan, B. J., Friedman, W. A., and Gravenstein, D.**, Somatosensory evoked potentials in hysterical paraplegia, *Surg. Neurol.*, 23, 502, 1985.
55. **Howard, J. E. and Dorfman, L. J.**, Evoked potentials in hysteria and malingering, *J. Clin. Neurophysiol.*, 3(1), 39, 1986.
56. **Lesser, R. P., Koehle, R., and Lueders, H.**, Effect of stimulus intensity on short latency somatosensory evoked potentials, *Electroencephalgr. Clin. Neurophysiol.*, 47, 377, 1979.
57. **Miller, L.**, Neuropsychological concepts of somatoform disorders, *Intl. J. Psychiat. Med.*, 14(1), 31, 1984.
58. **Merskey, H. and Buhrich, N. A.**, Hysteria and organic brain disease, *Br. J. Med. Psychol.*, 48, 359, 1975.
59. **Kligerman, M. J. and McKegney, F. P.**, Patterns of psychiatric consultation in two general hospitals, *Psychiat. Med.*, 2, 126, 1971.
60. **Trimble, M. R.**, *Neuropsychiatry*, John Wiley & Sons, Chichester, 1983, 79.
61. **Marsden, C. D.**, Hysteria — a neurologist's view, *Psychol. Med.*, 16, 277, 1986.
62. **Merskey, H.**, The mode of action of organic lesions in promoting hysteria, *Br. J. Med. Psychol.*, 48, 373, 1975.
63. **Merskey, H.**, The importance of hysteria, *Br. J. Psychiat.*, 149, 23, 1986.
64. Hippocrates, *Oeuvres Completes d'Hippocrate. Trad. E. Littre*, Vol. 2, J. B. Bailliere, Paris, 1853, 523.
65. **Adams, F.**, (Transl.) On the Causes and Symptoms of Acute Diseases, Book 2, 1840, Chap. 11, 285, in *The Extant Works of Aretaeus, the Cappadocian*, New Sydenham Society, London, 1856.
66. **Trimble, M. R.**, Pseudoproblems: pseudoseizures, *Br. J. Hosp. Med.*, 29(4), 326, 1983.
67. **Fenton, G. W.**, Epilepsy and Hysteria, *Br. J. Psychiat.*, 149, 28, 1986.
68. **Reed, J. L.**, The diagnosis of 'hysteria', *Psychol. Med.*, 5, 13, 1975.
69. **Fenton, G. W.**, *Hysterical Alterations of Consciousness in Hysteria*, Roy, A., Ed., John Wiley & Sons, Chichester, 1982, Chap. 16.
70. **Lesser, R. P.**, Psychogenic Seizures, in *Recent Advances in Epilepsy*, No. 2, Pedley, T. A. and Meldrum, B. S., Eds., Churchill Livingston, Edinburgh, 1985.
71. **Roy, A.**, Hysterical fits diagnosed as epilepsy, *Psychol. Med.*, 7, 271, 1977.

Chapter 5

THE BIOLOGY OF ANOREXIA NERVOSA

Janet Treasure

TABLE OF CONTENTS

I. HISTORICAL INTRODUCTION

The clinical features of anorexia nervosa were described over a hundred years ago. Sir William Gull, a physician at Guys hospital, recognized the association between psychopathology and emaciation in his 1868 account,[1] and he was able to compare his experience with that of Lasègue, who had published an account of the disorder simultaneously. Gull observed that the illness occurred in young women between the ages of 15 and 23 and was characterized by extreme emaciation in the absence of any signs of visceral disease. The lack of appetite was thought to result from a morbid mental state. Despite severe wasting, these women did not complain of pain or malaise, but were often "singularly restless and wayward". "A change in domestic circumstances" with regular small meals, ignoring the patient's "inclinations about eating" was the recommended method of treatment. At the meeting of the clinical society in London, during which Gull and Lasègue's papers were presented, several similar cases were discussed, including a girl who induced vomiting by "fixing her thoughts upon putrid cat pudding" when coerced into eating. Lasègue noted that these patients did not recognize themselves as thin which he attributed to denial. Before the syndrome was clearly delineated, clinical analogues of anorexia nervosa were either admitted to asylums or to infirmaries; Parry Jones[3] has surveyed the archives of two general hospitals and one lunatic asylum in England and identified several possible cases of anorexia nervosa. There have also been several retrospective case reports of women with probable anorexia nervosa in whom the psychological nature of the disorder was unrecognized.[4-6]

The next major step in our understanding of the eating disorders came when Russell, in 1979, gave a detailed case description of 30 cases with a distinct syndrome, bulimia nervosa.[7] Although these cases shared many of the psychopathological features which Russell had described as typical of patients with anorexia nervosa (such as the morbid preoccupation with body weight and dread of fatness), instead of the decreased food consumption and diminished appetite described by Gull and Lasegue, these patients periodically overate but attempted to avert the "fattening" effect of such binges by stringent weight control measures.

Although bulimia nervosa was not recognized as a distinct syndrome before 1979, epidemiological studies have shown that the disorder is present in about 1 in 100 young females attending their general practitioner,[8] whereas despite the incidence of anorexia nervosa doubling over the last two decades[9] it remains less common and occurs in about 1 in 600 students.[10]

Russell[11] has argued that this increased incidence of eating disorders may be attributed to pathoplastic causal factors, such as the recent changes in female roles and expectations and the current fashion for a slender female form. Such factors may have led to the change in psychopathological content of anorexia nervosa so that the emaciation of the illness is rationalized by expressing a fear of fatness; but also there has been a change in the form of neurotic illness, with the development of a new psychiatric syndrome, bulimia nervosa characterized by overvalued ideas about the importance of body weight.

II. PROBLEMS OF CLASSIFICATION

In this chapter we will use the term anorexia nervosa to describe patients who exhibit Russell's criteria of: (1) self-induced starvation, (2) a morbid fear of fatness, and (3) an abnormality in reproductive hormone functioning, indicated by amenorrhoea in females, or decreased sexual drive and function in males. However, in an attempt to describe a heterogeneous group these patients will be subdivided.

Several systems of subclassification have been proposed. Dally[12] differentiated three groups: an obessional group (group O), a hysterical group (group H), and a group of mixed etiology (group M). The obsessional group retained appetite and had specific clinical features including bulimia, vomiting, and labile mood. Beumont[13] differentiated anorexics into two groups, depending upon whether they lost weight by calorie restriction and were thence termed dieters, or relied upon vomiting and laxative abuse and were termed purger-vomiters.

In this chapter we will subdivide patients with anorexia nervosa into two groups; those who periodically overeat, bulimic anorexia nervosa, and those who have never shown such behavior, restrictor anorexia nervosa. The validity of this division has been substantiated by research from several centers[14,15,16,17] which has confirmed Russell's contention that the symptom, bulimia, is a core feature, and hence should be integral to any system of classification of the eating disorders. Dally's Group O and Beumont's purger-vomiters would probably correspond best with the bulimic subgroup of anorexia nervosa.

It is unfortunate that the criteria used to deliniate the DSM-III syndrome, bulimia, and bulimia nervosa, are so disparate that they have led to confusion and difficulties comparing research findings between centers. DSM-III syndrome, bulimia, is a broadly defined syndrome and so includes patients without the specific overvalued ideas about weight or the morbid fear of fatness. In practice the DSM-III definition has also led to weight being used as a threshold to divide DSM-III, bulimia, from anorexia nervosa. In the UK, bulimic anorexia nervosa would be classified as bulimia nervosa in most centers, whereas in many centers in the US it would be classified as anorexia nervosa.

III. CLINICAL FEATURES

A. Eating Disturbance

In the original reports,[1,2] a lack of appetite was considered to be the primary disturbance in anorexia nervosa, but it is now frequently assumed that this was incorrect and that the appetite is normal but is suppressed as the desire to lose weight is so intense. However, it is now recognized that the eating behavior in anorexia nervosa is heterogeneous[12] as is implicit in the symptoms of the restrictor and bulimic subgroups. Unfortunately, the majority of research into eating behavior in anorexia nervosa was done before it was recognized that the symptom, bulimia, was such a key variable. On empirical grounds, it is probable that the restrictor subgroup have a disturbance in appetite, whereas the bulimic subgroup might have a disturbance in satiety.

The subgroup of patients with anorexia nervosa who consistently restrict their intake frequently deny hunger, although they may accurately perceive the gastric sensations of emptiness or fullness which correlate with hunger in normals.[18] However, a disturbance in satiety rather than appetite was later found in a group of 11 patients with anorexia nervosa, but the balance of restrictor to bulimic patients in this population was not detailed.[19] Emaciated restrictor anorexic patients and normal controls showed similar changes in hunger and fullness over a meal, whereas the bulimic subgroup experienced a rapid rebound of hunger which occurred before the end of the meal and which continued after the meal was finished.[20] After weight gain, the restrictor subgroup appeared to be no longer able to distinguish between hunger and fullness. These results suggest that only when the physiological drive to eat is maximal, such as occurs during starvation, is the restrictor anorexic able to accurately perceive hunger, and this facility decreases after weight gain. This finding could explain the divergence of opinion in the literature as the patient with anorexia nervosa may experience hunger only when weight is decreased.

Taste preference differs between the subgroups of patients with anorexia nervosa; the restrictor subgroup consistently rated all sucrose and fat stimuli as more unpleasant than did bulimics, who showed an enhanced liking for intensely sweet sucrose solutions.[20] In animals, food preference is controlled by a balance of neurotransmitters within the hypothalamus, and the divergent taste preferences between the two subgroups of patients with anorexia nervosa suggest that there may be a specific underlying neurochemical disorder in each group.

B. Depression

Features of depression are common in anorexia nervosa; these include irritability, depressed mood, crying spells, suicidal thoughts, and sleep disturbance.[21,22] Many of these depressive symptoms remit with refeeding alone,[23] thereby suggesting that depression is related to the effects of starvation. These depressive symptoms are more pronounced in patients with DSM-III bulimia, who often fulfill the Research Diagnostic Criteria for major depressive disorder.[24-26]

The shared abnormalities in certain biological variables such as short REM latency, raised basal cortisol, and abnormal dexamethasone suppression tests and thyrotrophin releasing hormone tests, have been used to support the argument which states that anorexia nervosa is an atypical variant of depression.[27,28] However, the validity of such tests as biological markers for depression is contentious, as similar abnormalities have been found in starvation[29] or when the sleep pattern of a patient with depression is adopted by a healthy volunteer.[30]

C. Endocrine Disturbance

The hypothesis that anorexia nervosa arose as a result of pituitary dysfunction followed Simmonds' publication on pituitary cachexia in 1914.[31] This resulted in interest in the endocrinology of the condition and the use of hormonal treatment. It is now recognized that many of the endocrine abnormalities are a result of malnutrition; however, the abnormalities in the hypothalamic, pituitary, adrenal, and gonadal axes, are possible exceptions to this generalization.

Amenorrhea is an invariable feature of severe anorexia nervosa, and the onset either before or immediately following weight loss[21,32,33] and the delay in resumption of menses despite weight restoration suggest that it may be of possible pathogenic significance. Weight loss results in a reduction in the amplitude and frequency of gonadotropin releasing hormone release from the hypothalamus; this leads to a decrease in the absolute amount, and in the ratio of LH/FSH released from the pituitary. As a consequence, the ovaries decrease in size and contain many small cysts, and have been defined by ultrasonographic features as multifollicular ovaries.[34,35] These small antral follicles produce little estradiol, and the organs in which estrogens produce a trophic response, e.g., uterus and breasts, decrease in size. The delay in the resumption of menses after weight restoration may be partially explained by individual variation in the constitutional set point necessary for normal menstrual function, but there may be other confounding factors such as excessive exercise, an inappropriate lean/fat ratio, abnormal dietary constituents, or persistent anorexic attitudes which may also contribute.

Patients with anorexia nervosa secrete an excess of cortisol and have raised basal cortisol levels.[36] Although basal cortisol is also increased in starvation, this is probably secondary to changes in cortisol metabolism, as the production rate is diminished. This increase in plasma cortisol could result either from increased secretion of corticotrophin releasing hormone (CRH) or ACTH from the hypothalamus or pituitary, respectively, or be due to impaired cortisol feedback at either of these two sites. The response to a CRH test differs between patients with anorexia nervosa, and those with Cushing's disease; the ACTH response is attenuated in anorexia nervosa either as a result of the negative feedback of cortisol or receptor down-regulation.[37-39] These results suggest that the defect in the normal control of cortisol secretion results from a hypothalamic abnormality, but normal control is resumed after recovery from the illness.

D. Abnormal Perception of Body Image

The apparent inability of a patient with anorexia nervosa to acknowledge that her body is thin was first described by Lasègue in 1876,[2] and was then given as an example of denial, but Bruch[40] considered it to be one of the fundamental features of the illness and an example of the one of the perceptual abnormalities seen in these patients. In the majority of studies using the distorting image technique, patients with anorexia nervosa have been found to overestimate their size.[41,42] The patients with the most severe distortion of body image often have low self-esteem and more generalized psychiatric symptoms, and a poorer outcome.[43] When the bulimic subgroup were compared with the restrictor subgroup, they demonstrated an exaggerated estimation of their body and expressed a desire to be at a weight lower than this perceived weight, whereas the restrictor subgroup were satisfied with their perceived size.[16]

IV. ANIMAL MODELS

Animal models provide a valuable experimental tool and can be used to investigate the biological facets of an illness, but it has been difficult to find an animal model which incorporates all the clinical features of anorexia nervosa.[44] Rats with lateral hypothalamic lesions initially lose weight, but their weight restabilizes at a lower level, suggesting that the set point has been lowered. There is no evidence which implies that the set point is lowered in anorexia nervosa, indeed one of distubing facets of the condition is that without any intervention, weight loss continues until death or physical damage occurs. One suggestion is that the physiological weight set point may be poorly defended in women who are predisposed to develop anorexia nervosa.[45] An animal model in which weight loss continues with a similar malignant course is produced if rats are allowed access to food for only 1 hr/day, but are given free access to an exercise wheel within their cage; such rats eat less than rats given the same feeding schedules without exercise wheels in their cages.[46] Some animals on the former regimen died. Hyperactivity is characteristic of anorexia nervosa and is one of the few symptoms which is specific to the illness and is not a feature of starvation; thus it is possible that this experimental paradigm is an analogue of anorexia nervosa.

V. THE NEUROCHEMISTRY OF FEEDING

The regulation of eating is a result of a complicated interplay between the peptide, amino acid, prostaglandin, and amine neurotransmitters (summarized in Table 1). Morley[47] suggested that the primary central feeding system is that of food-seeking and ingestion and that fine control is mediated by a cascade of inhibitory neurotransmitters, some of which feed back information from the gastrointestinal tract.

Several methodological difficulties arise in this area of research, as a drug which produces increased food consumption in animals may stimulate appetite but its primary effect may be to increase the metabolic rate or change the nutrient balance, rather than to specifically influence the central control of feeding. Similarly, drugs which decrease feeding may do so as a result of decreased arousal, or be noxious and produce generalized aversion, rather than specifically influence satiety. The animal's state of repletion at the time of testing also influences the effect of any drug.

A. Drugs Which Increase Feeding

The most potent substances which increase feeding are neuropeptide Y and peptide YY; their effect summates with the maximal physiological drive to feed[48,49,50] and can override the peripheral satiety system and hence result in marked gastrointestinal distension. Animals given these substances specifically increase their carbohydrate intake.[51] It is thought that their action is independent of the noradrenergic system but they are antagonized by naloxone, suggesting that there is some interrelationship with the opioid system.[51] The endogenous opiates, dynorphin and alpha-neo-endorphin, (acting on the kappa receptor), stimulate feeding (specifically of fat but to a small extent protein) in animals[52-54] and opiate antagonists

Table 1
THE CONTROL OF FEEDING IN ANIMALS

Substances which stimulate feeding		Substances which inhibit feeding	
Agent	Nutrient	Agent	Nutrient
Noradrenergic α_2 receptor (corticosterone and the vagus have a permissive effect)	Carbohydrate	Monoamines	—
		Dopamine	Carbohydrate
		Serotonin (5HT)	Carbohydrate
		Noradrenergic adrenergic β receptors	Protein
Opiods	Fat (protein)	Amino Acids	
		GABA	
		Adenosine	
Pancreatic polypeptides (neuropeptide Y, polypeptide YY)	Carbohydrate	Prostaglandins	
Gamma amino butyric acid		PGE$_2$	
GABA		PG$_2$	
Growth hormone releasing hormone (GHRH)		Peptides	
		Cholecystokinin	
		Neurotensin	
		Calcitonin	
		Somatostatin	
		Glucagon	
		Bombesin	
		Corticotrophin releasing hormone	
		Thyrotropin releasing hormone	
		Vasoactive intestinal peptide	
		Insulin growth factor	
		Sauvagine	
		Glucagon	
		Caerulin	

have been found to decrease fat intake in man,[53] although tolerance develops if naltrexone is administered continuously. Preliminary studies suggest that growth hormone releasing hormone (GHRH)[55] produced by neurons in the ventromedial nucleus of the hypothalamus, stimulates feeding after intracerebral injection.

Noradrenaline, acting on alpha adrenergic receptors, stimulates feeding in animals, but its effect depends upon an intact vagus and corticosterone.[56] Noradrenaline is thought to decrease the firing of a neuron which inhibits feeding within the paraventricular nucleus.[57] Gamma amino butyric acid (GABA) injected into the hypothalamus also increases feeding, and benzodiazepines which increase feeding in animals may interact with this receptor.[47]

B. Neurotransmitters Which Inhibit Feeding

Several of the multiplicity of substances which inhibit feeding are shown in Table 1; however the specifity and site of action of most of these compounds have yet to be clarified. Serotonin and corticotrophin inhibit feeding by an effect within the paraventricular or ventromedial nucleus of the hypothalamus.[58,59] The prostaglandins have also been found to inhibit feeding and are possible intermediaries in the suppressive effect of calcitonin.

VI. THE BIOLOGY OF ANOREXIA NERVOSA

A. Genetics

A genetic susceptibility to any illness suggests that biological factors may be pertinent.

The concordance rate for anorexia nervosa between identical twins has been reported from several studies to lie between 44 to 50%, depending upon how stringent the criteria are for determining zygosity (this subject has been recently reviewed by Scott.[60]) In the St. Georges/ Maudsley Hospital collaborative study (with 34 pairs of twins), the concordance rate between MZ twins was 55%, and it was 7% between DZ twins.[61] The latter figure is similar to that given as the incidence in the sisters of a proband with anorexia nervosa.[32,62,63] Although the sample size was relatively small, Holland[61] suggested that their findings indicated that there was a genetic predisposition to anorexia nervosa. The biological substrate for this vulnerability is unknown; several possibilities have been proposed including: personality type; a predisposition to obesity or other psychiatric illness such as depression.

The incidence of both eating disorders and affective disorders in the relatives of probands with anorexia nervosa was above that seen in the families of the control group.[64-66] However, the converse, i.e., an excess of eating disorders, has not been found in the relatives of probands with affective disorders[66] suggesting that the two conditions are not inherited in combination. It is probable that the genetic vulnerability to depression and anorexia nervosa do interact and predispose to the complete expression of anorexia nervosa. The incidence of eating disorders in the relatives of probands with the restrictor and bulimic subtypes of anorexia nervosa is also of great interest; although both sets contained an equal number of patients with partial syndromes, cases of anorexia nervosa were only found in the relatives of the restrictor subgroup, whereas the eating disorder in the relatives of those with bulimic, anorexia nervosa took the form of bulimia nervosa.[66]

B. Brain Structure and Function

In many animals food intake and reproductive function are integrated within the paraventricular nucleus of the hypothalamus. Both of these functions are disturbed in anorexia nervosa and although it is possible that the common intermediary is weight loss, a dissociation between weight and menstruation has frequently been found (see endocrine features), which suggests that there may be a common disturbance in hypothalamic function, as proposed by Russell.[67] Additional evidence in support of this theory is the fact that the illness usually occurs at puberty, a time of considerable flux within the hypothalamus when, in women, cyclic reproductive activity becomes established within the supraoptic and paraventricular nuclei. The less complex adjustments in hypothalamic function at puberty in the male, may account for marked female preponderance of the disorder. There have been several case reports of women presenting with the clinical features of anorexia nervosa who have been found to have tumors within the hypothalamus,[70-72] but this level of association is possibly no more than would be expected by chance, and no structural abnormalities within the hypothalamus have been reported in the large number of patients who have had transmission computerized tomography (CT scans) of their brains. The most frequent abnormality noted in the latter investigation is "pseudoatrophy" with widening of the cerebral and cerebellar sulci,[73-76] occasionally associated with enlarged third and lateral ventricles.[77] These changes are most common in the severely malnourished, bulimic subgroup who have raised cortisol levels,[76] and preliminary evidence suggests that they are reversed after weight gain.[76-78] There have been some reports of positron emission tomography (PET scanning) in patients with anorexia nervosa; however, there are technical difficulties if 2 deoxyglucose is used, as ketone bodies rather than glucose, are metabolized by the brain during starvation. Patients with ketone bodies thus have to be excluded from these investigations. The preliminary reports have found no difference in the absolute metabolic rates between patients with anorexia nervosa and normal controls; however, there were differences in the relative metabolic rates (ratio of metabolic rate to individual whole brain metabolism) with lower rates in the cortex and cerebellum, but these may be artifacts arising from the pseudoatrophy.[76]

Table 2
GASTROINTESTINAL HORMONES IN ANOREXIA NERVOSA

	Fasting response		Postprandial response
	Anorexia nervosa	Controls	
Glucose	D	—	—
Insulin	D	—	I
Gastrin	—	—	—
N-GLI	—	—	—
C-GLI	D	—	—
Secretin	—	—	—
GIP	—	—	I
PP	—	—	I, P
VIP	—	—	—

Note: D, decrease; I, increase; P, prolonged.

Table 3
NEUROTRANSMITTERS IN THE EATING DISORDERS

Plasma	Emaciated anorexia nervosa	Weight-recovered anorexia nervosa	Bulimics
NA	D	N	—
DAβ hydroxylase	N	N	—
β endorphin	—	—	I
Trytophan	D	N	—
Urine			
HVA	D	N	—
HMPG	D	D→N	—
CSF			
MHPG	I	—	—
GABA	N	—	—
HVA	D	N	—
5HIAA	D	I	N
Tyrosine	D	N	—
NA	N	N	—
Opiod	I	N	—
β endorphin	N	N	—
Choline	D	—	—
Platelet			
MAOI	D	—	—
α_2 adrenoceptor	D	N	D
5HT uptake	N	N	—
Imipramine binding site	D	?	—

Note: N, normal; D, decrease; I, increase.

C. Neurochemistry

Several of the amine and opiate groups of neurotransmitters which are recognized to be important for the regulation of feeding in animals have been measured in patients with anorexia nervosa (see Table 2), and the preliminary findings from studies in which gastro-intestinal peptides have been measured are shown in Table 3. Unless specifically indicated, the following studies have used patients with anorexia nervosa who have not been divided into the bulimic or restrictor subgroups.

1. Measurement of Substances which Stimulate Feeding

If the hypothalamic control of eating is normal in anorexia nervosa, then an increase in the levels of substances which have been found to stimulate eating (see the neurochemistry of feeding above) would be expected. Opioid acitvity in the cerebrospinal fluid (CSF) has indeed been found to be increased.[79] Also the raised levels of pituitary growth hormone found in patients with anorexia nervosa[80-82] suggests that hypothalamic growth hormone releasing hormone (GHRH) secretion is elevated. Similarly, plasma levels of pancreatic polypeptide were found to be increased after a test meal in one study[83] but this has not been confirmed in other preliminary studies.[84-86] It would be interesting to measure levels of pancreatic polypeptide in patients with bulimic anorexia nervosa, as when this substance is given to animals grossly increased consumption leading to gastrointestinal distension ensues; this pattern of eating resembles the clinical features of bulimia nervosa.

The careful dissection of the noradrenergic system which has established that alpha and beta receptors produce opposing effects upon feeding in animals has been difficult to replicate in man. However, there is evidence for an increase in noradrenaline turnover in patients with anorexia nervosa at low weight; the metabolite of noradrenaline, MHPG, but not noradrenaline itself, has been reported to be increased within the CSF[87,88] and hypothalamic alpha 2 adrenoreceptors appear to be increased — based upon the response of growth hormone (GH) to clonidine.[89] Any noradrenergic effect upon appetite would be potentiated by the raised levels of cortisol found in these patients at low weight.[36]

2. Putative Neurotransmitters of Satiety

If the mechanisms controlling satiety in anorexia nervosa were functioning in a manner appropriate to the physiological state, then a decrease in dopamine, serotonin, and beta noradrenergic activity would be expected, and this has indeed been found.[87,88] However, there are differences in serotonergic metabolism between the restrictor and bulimic subgroups; after weight restoration, the restrictor subgroup have increased concentrations of the serotonin metabolite 5 hydroxyindolacetic acid (5 HIAA) within the CSF when compared with controls and the bulimic subgroup.[90] Moreover, hypothalamic CRH is also increased (see endocrine abnormalities in anorexia nervosa, above) which is the converse of the predicted change. This has led to the hypothesis that psychic stress releases CRH which in turn inhibits feeding, leading to the development of anorexia nervosa. Morley et al.[59] has suggested that the penultimate inhibitory neuron, in the central control of feeding, secretes corticotrophin releasing hormone (CRH) neuron and that the inhibitory effects of noradrenaline and serotonin are mediated via the release of CRH. Thus it is possible that the increased serotonin turnover in restrictor anorexia nervosa is increasing CRH release, thereby decreasing food intake.

D. Drug Treatment

The theoretical rationale for the use of antidepressants to treat anorexia nervosa has been partially based upon the hypothesis that anorexia is a variant of depression.[91-93] If the neurochemistry of feeding is taken into consideration, then those antidepressants which increase the concentration of noradrenaline might be expected to increase food consumption. There are several theoretical caveats to this premise. First, although alpha receptors increase feeding in animals, the beta receptors exert the opposite effect. Second, many antidepressants have subsidiary effects upon serotonin which decreases feeding.

The tricyclic antidepressant, amitriptyline[94] appeared promising in open trials; however, it was not superior to placebo when tested in two recent double-blind trials.[91,93] Clomipramine (a drug which is thought to exert a more pronounced effect on serotonin reuptake than amitriptyline) also did not increase weight gain in a double-blind trial.[95] In addition, lithium has also been found ineffective in a double-blind trial.[96]

Cyproheptadine is a serotonin antagonist and theoretically should be expected to increase feeding. In the controlled double-blind trial referred to above,[93] cyproheptadine was compared to amitriptyline and placebo. Although there was no significant effect in the heterogeneous group of patients with anorexia nervosa, cyproheptadine increased weight gain in the restrictor subgroup but impaired treatment efficacy in the bulimic subgroup.[93] Fenfluramine (which is believed to act by increasing serotonergic transmission[97]) has been used in a controlled trial to treat patients with bulimia nervosa, and although the study was marred by a high drop-out rate, this drug was effective in reducing the amount of overeating and vomiting (Russell, personal communication).

As dopamine has been found to decrease feeding in animals, a dopamine antagonist might be expected to increase food intake. Chlorpromazine was thought to be effective in promoting weight gain in patients with anorexia nervosa in open trials,[98] but controlled trials of pimozide[99] and sulpiride[100] have not confirmed this finding. Although, on theoretical grounds extrapolated from animal studies, an opiate agonist would be expected to increase feeding, it has been the opiate antagonist, naloxone, which has been used in an open trial in patients with anorexia nervosa with inconclusive results.[101]

In conclusion, no pharmacological agent has been found to improve upon the weight gain attained by standard inpatient management. However, it is possible that the rate of weight gain achieved in such specialized units is maximal; therefore a more appropriate model may be to test a drug during outpatient managment. Of the drugs which have been tested in controlled trials, it is those which have serotoninergic effects which hold most promise; but careful selection of patients into the restrictor or bulimic categories is needed.

IX. CONCLUSIONS

Research into the biology of anorexia nervosa is at a most exciting stage, as the nosological differentiation of the illness into restrictor and bulimic subgroups has dispersed some of the confusing findings and has allowed some consistent patterns to develop. The family histories of patients with anorexia nervosa suggests that the bulimic and restrictor subgroups have a different pattern of inheritance.[66] Interestingly, the physiology and biochemistry of the restrictor subgroup diverges maximally from control groups after weight restoration. Thus Halmi[20] has reported that such patients appear to be unable to differentiate between hunger and fullness. These patients also appear to have an increase in central serotonin turnover[90] (a neurotransmitter which in animals acts in the hypothalamus to decrease feeding), and weight gain is promoted by cyproheptadine, a serotonin antagonist.[93] Such abnormalities are not apparent when the patient with restrictor anorexia nervosa is starving.

The bulimic subgroup appear to have an unusual perception of hunger and fullness over a range of weight.[20] The taste preference of bulimic anorexia nervosa for nutrients with a high sucrose concentration suggest that there may be an increase in the neurotransmitters which increase the appetite for carbohydrate, such as neuropeptide Y and noradrenaline. Also, the clinical features of bulimic anorexia nervosa resemble the behavior of animals after the administration of neuropeptide Y. The symptoms of patients with bulimia nervosa diminish with fenfluramine, which suggests that serotonin may inhibit feeding in these patients as it does in animals, perhaps by antagonizing the effects of excessive levels of neuropeptide Y.

Thus, several strands of evidence suggest that there is some biological validity in dividing anorexia nervosa into the bulimic and restrictor subgroups, and it can be hoped that further studies upon more homogeneous subgroups of patients will increase our understanding of the biological nature of the disorder.

REFERENCES

1. **Gull, W. W.**, Anorexia nervosa (apepsia hysterica anorexia hysterica), *Br. Med. J.*, 7, 527, 1873.
2. **Lasègue, E. C.**, On hysterical anorexia, Medical Times Gazette 1873, 2, 265, 367; Translated in, *Anorexia nervosa, A paradigm, Evolution of Psychosomatic Concepts*, Kaufman, M. R. and Heiman, M., Eds., Hogarth Press, London, 1965, 141.
3. **Parry Jones, W.**, Archival exploration of anorexia nervosa. *J. Psychiat. Res.*, 19, 95, 1985.
4. **Bell, R. M.**, *Holy Anorexia*, University of Chicago Press, Chicago, 1985.
5. **Rampling, D.**, Ascetic ideals and anorexia nervosa, *J. Psychiat. Res.*, 19, 89,1985.
6. **Silverman, J. A.**, Anorexia nervosa in Seventeenth Century England as viewed by Physician Philosopher and Pedagogue, *Int. J. Eat Dis.*, 5, 847, 1986.
7. **Russell, G. F. M.**, Bulimia nervosa: an ominous variant of anorexia nervosa, *Psychol. Med.*, 9, 1, 1979.
8. **King, M. B.**, Eating disorders in general practice, *Br. Med. J.*, 293, 1412, 1986.
9. **Szmukler, G. I., McCance, C., McCrone, L., and Hunter, D.**, Anorexia nervosa: a psychiatric case register study from Aberdeen, *Psychol. Med.*, 16, 49, 1986.
10. **Button, E. J. and Whitehouse, A.**, Subclinical anorexia nervosa, *Psychol. Med.*, 11, 509, 1981.
11. **Russell, G. F. M.**, Bulimia revisited, *Int. J. Eat Dis.*, 4, 681, 1985.
12. **Dally, P.**, *Anorexia Nervosa*, William Heineman Medical Books, London, 1969.
13. **Beumont, P. J. V., George, G. C. W., and Smart, D. E.**, Dieters and vomiters and purgers in anorexia nervosa, *Psychol. Med.*, 6, 617, 1976.
14. **Fairburn, C. G. and Cooper, P. J.**, Binge eating, self induced vomiting and laxative abuse: a community study, *Psychol. Med.*, 14, 401, 1984.
15. **Garner, D. M., Garfinkel, P. E., and O'Shaughnessy, M.**, Clinical and psychometric comparison between bulimia in AN and bulimia in normal weight women, in Understanding AN and Bulimia: Report of the 4th Ross Conference on Medical Research, Columbus, Ross Laboratories, 1983.
16. **Garner, D. M., Olmsted, M. P., and Garfinkel, P. E.**, Similarities among bulimic groups selected by weight and weight history, *J. Psych. Res.*, 19, 129, 1985.
17. **Fairburn, C. G. and Garner, D. M.**, The diagnosis of bulimia nervosa, *Int. J. Eat Dis.*, 5, 403, 1986.
18. **Silverstone, J. T. and Russell, G. F. M.**, Gastric hunger contraction in anorexia nervosa, *Br. J. Psych.*, 113, 257, 1967.
19. **Garfinkel, P. E.**, Perception of hunger and satiety in anorexia nervosa, *Psychol. Med.*, 4, 309, 1974.
20. **Halmi, K.**, Satiety and taste in eating disorders, in *Disorders of Eating Behaviour: A Psychoneuroendocrine Approach*, Advances in the Biological Science, Vol. 60, Ferrari, E. and Brambilla, F., Eds., Pergamon Press, New York, 1986, 199.
21. **Morgan, H. G. and Russell, G. F. M.**, Value of family background and clinical features as predictors of long term outcome in anorexia nervosa: four year follow-up study of 41 patients, *Psychol. Med.*, 5, 355, 1975.
22. **Halmi, K., Goldberg, S. C., Eckert, E., Casper, R., and Goldberg, S.**, Catecholamine metabolism in anorexia nervosa, *Arch. Gen. Psychiat.*, 35, 458, 1978.
23. **Eckert, C. D., Goldberg, S. C., Halmi, K. A., Casper, R. C., and Davis, J. M.**, Depression in Anorexia Nervosa, *Psychol. Med.*, 12, 115, 1982.
24. **Hendren, R. L.**, Depression in anorexia nervosa, *J. Am. Acad. Child Psych.*, 22, 59, 1983.
25. **Strober, M.**, The significance of bulimia in juvenile anorexia nervosa: an exploration of possible aetiological factors, *Int. J. Eat Dis.*, 1, 28, 1981.
26. **Hudson, J., Pope, H. G., Jonas, J. M., and Todd, D.**, Family history study of anorexia nervosa and bulimia, *Br. J. Psychiat.*, 142, 133, 1983.
27. **Katz, J. L., Kuperberg, A., Pollack, C. P., Walsh, B. T., Zumoff, B., and Weiner, H.**, Is there a relationship between eating disorder and affective disorder? New evidence from sleep recordings, *Am. J. Psychiatry*, 141, 753, 1984.
28. **Herzog, D. B.**, Are anorexic and bulimic patients depressed? *Am. J. Psychiatry*, 141, 1594, 1984.
29. **Fichter, M. M. and Pirke, K. M.**, Hypothalamic function in starving healthy subjects, in *The Psychobiology of Anorexia Nervosa*, Pirke, K. M. and Ploog, D., Eds., Springer Verlag, Berlin, 1984, 124.
30. **Mullen, P. E., Linsell, C. R., and Parker, D.**, Influence of sleep disruption and calorie restriction on biological markers for depression, *Lancet*, ii, 1051, 1986.
31. **Simmonds, M.**, Ueber embolische Prozesse in her Hypophysis, *Arch. Pathol. Anat.*, 217, 226, 1914.
32. **Theander, S.**, Anorexia nervosa: a psychiatric investigation of 94 female patients, *Acta Psychiat. Scand.*, suppl. 214, 1970.
33. **Kay, D. W. and Leigh, D. A.**, The natural history, treatment and prognosis of anorexia nervosa based on a study of 38 patients, *J. Ment. Sci.*, 100, 411, 1954.
34. **Adams, J., Franks, S., Polson, D. W., Mason, H. D., Abdulwahed, N., Tucker, M., Morris, D. V., Price, J., and Jacobs, H. S.**, Multifollicular ovaries: clinical and endocrine features and response to pulsable gonadotrophic releasing hormone, *Lancet*, 1375, 1985.

35. **Treasure, J., Gordon, P. A. L., King, E. A., Wheeler, M., and Russell, G. F. M.,** Cystic ovaries: a phase of anorexia nervosa, *Lancet,* 1379, 1985.
36. **Walsh, B. T.,** The endocrinology of anorexia nervosa, *Psychiatr. Clin. N. Am.,* 3, 299, 1980.
37. **Gold, P. W., Loriaux, D. L., Royn, A., Kling, M. A., Calarese, J. R., Kelliner, C. H., Nieman, C. K., Post, R. M., Pickar, D., Gallucci, W., Avgerinos, P., Paul, S., Oldfield, E. H., Cutler, G. B., and Chrousos, G. P.,** Responses to corticotropin releasing hormone in the hypercortisolism of depression and Cushing's disease, *N. Engl. J. Med.,* 314, 1329, 1986.
38. **Gold, P. W., Gwirtsman, H., Avgerinos, P. G., Nieman, L. K., Gallucci, W. T., Kaye, W., Jimerson, D., Ebert, M., Rittmaster, R., Loriaux, L., and Chrousos, G. P.,** Abnormal hypothalamic-pituitary-adrenal function in anorexia nervosa, *N. Engl. J. Med.,* 314, 1335, 1986.
39. **Cavagnini, F., Invitti, C., Passamonti, M., and Polli, E. E.,** Impaired ACTH and cortisol response to CRH in patients with anorexia nervosa, in *Disorders of Eating Behaviour. A Psychoneuroendocrine Approach.* Advances in the BioScience, Vol. 60. Ferrari, E. and Brambilla, F., Eds., Pergamon Press, New York, 1986, 229.
40. **Bruch, H.,** Perceptual and conceptual disturbances in anorexia nervosa, *Psychosom. Med.,* 24, 187, 1962.
41. **Garfinkel, P. E. and Garner, D. M.,** Perceptions of the body in anorexia nervosa, in *The Psychobiology of Anorexia Nervosa,* Pirke, K. M. and Ploog, D., Eds., Springer Verlag, Berlin, 1984, 136.
42. **Slade, P.,** A review of body-image studies in anorexia nervosa and bulimia nervosa, *J. Psychiat. Res.,* 19, 255, 1985.
43. **Garfinkel, P. E., Moldofsky, H., and Garner, D. M.,** Prognosis in anorexia nervosa as influenced by clinical features, treatment and self perception, *Can. Med. Assoc. J.,* 117, 1041, 1977.
44. **Mrosovsky, N.,** Animal models: anorexia Yes, nervosa No, in *The Psychobiology of Anorexia Nervosa.* Pirke, K. M. and Ploog, D., Eds., Springer Verlag, Berlin, 1984, 22.
45. **Epling, W. F., Pierce, W. D., and Stefan, C.,** A theory of activity based anorexia, *Int. J. Eat Dis.,* in press.
46. **Kanarek, R. B. and Collier, G. H.,** Self-starvation: a problem of overriding the satiety signal, *Physiol. Behav.,* 30, 307, 1983.
47. **Morley, J. E.,** The neuroendocrine control of appetite, the role of the endogenous opiates, cholecystokinin, TRH gamma aminobutyric acid and diazepam receptors, *Life Sci.,* 27, 355, 1980.
48. **Clarke, J. J., Kalva, P. S., Crowley, W. L., and Klava, S. P.,** Neuropeptide Y and human pancreatic polypeptide stimulate feeding behaviour in rats, *Endocrinology,* 115, 427, 1984.
49. **Stanley, B. G., and Liebowitz, S. F.,** Neuropeptide Y: stimulation of feeding and drinking by injection into the paraventricular nucleus, *Life Sci.,* 33, 2635, 1984.
50. **Morley, J. E. and Levine, A. S.,** Neuropeptide Y potently induces ingestive behaviour after central administration, *Dig Dis. Sci.,* 29, 538, 1984.
51. **Morley, J. E. and Mitchell, J. E.,** Neurotransmitter/neuromodulator influences on eating, in *Disorders of Eating Behaviour-A Psychoneuroendocrine Approach.* Advances in the Neurosciences, Vol. 60, Ferrari, E. and Brambilla, F., Eds., Pergamon Press, New York, 1986, 11.
52. **Morley, J. E. and Levine, A. S.,** Dynorphin (1-13) induces spontaneous feeding in rats, *Life Sci.,* 29, 1901, 1981.
53. **Levine, A. S., Morley, J. E., Gosnell, B. A., Billington, C. J., and Bartness, T. J.,** Opiods and consummatory behaviour, *Brain Res. Bull.,* 14, 663, 1985.
54. **Morley, J. E., Elson, M. K., Levine, A. S., and Shafer, S. R. B.,** The effects of stress on central nervous system concentrations of the peptide dynorphin, *Peptides,* 3, 901, 1983.
55. **Vaccarino, F. J., Bloom, F. E., Rivier, J., Vale, W., and Koob, G. F.,** Stimulation of food intake in rats by centrally administered hypothalamic growth hormone-releasing factor, *Nature,* 314, 167, 1985.
56. **Liebowitz, S. F.,** Neurochemical systems of the hypothalamus. Control of feeding and drinking behaviour and water-electrolyte excretion, in *Handbook of the Hypothalamus,* Raven Press, New York, 1980, 299.
57. **Liebowitz, S. F., Hammer, N. V., and Chang, K.,** Feeding behaviour induced by central norepinephrine injection is attenuated by discrete lesions in the hypothalamic paraventricular nucleus, *Pharmacol. Biochem. Behav.,* 19, 945, 1983.
58. **Morley, J. E. and Levine, A. S.,** Corticotropin releasing factor, grooming and ingestive behaviour, *Life Sci.,* 31, 1459, 1982.
59. **Morley, J. E., Levine, A. S., Gosnell, B. A., and Billington, C. J.,** Neuropeptides and appetite: contribution of neuropharmacological modelling, *Fed. Proc.,* 43, 2903, 1984.
60. **Scott, D. W.,** Anorexia nervosa: a review of possible genetic factors, *Int. J. Eat Dis.,* 5, 1, 1986.
61. **Holland, A. J., Hall, A., Murray, R., Russell, G. F. M., and Crisp, A. H.,** Anorexia nervosa: a study of 34 twin pairs and one set of triplets. *Br. J. Psychiatry,* 145, 414, 1984.
62. **Hudson, J. I., Pope, H. G., Jonas, J. M., and Yurgelin-Todd, D.,** Family history study of anorexia nervosa and bulimia, *Br. J. Psychiatry,* 142, 133, 1983.
63. **Crisp, A. H., Hsu, L. K. G., Harding, B., and Hartshorn, J.,** Clinical features of anorexia nervosa, *Br. J. Med. Psychol.,* 24, 179, 1980.

64. **Gershon, E. S., Schreiber, J. L., Hamont, J. R., Dibble, E. D., Kaye, W., Nurnberger, J. I., Anderson, A., and Ebert, M.,** Clinical findings in patients with anorexia nervosa and affective illness in their relatives, *Am. J. Psychiatry,* 141, 1419, 1984.

65. **Biederman, J., Rivinus, T., Kemper, K., Hamilton, D., MacFadyen, J., and Harmatz, J.,** Depressive disorders in relatives of anorexia nervosa patients with and without a current episode of nonbipolar major depression, *Am. J. Psychiatry,* 142, 1495, 1985.

66. **Strober, M., Morrell, W., Bulrough, S., Salkin, B., and Jacobs, C.,** A controlled family study of anorexia nervosa, *J. Psych. Res.,* 19, 239, 1985.

67. **Russell, G. F. M.,** Anorexia nervosa: its identity as an illness and treatment, in *Modern Trends in Psychological Medicine,* Price, Ed., Butterworth, London, 1970, 131.

68. **Weller, R. A. and Weller, E. B.,** Anorexia nervosa in a patient with an infiltrating tumor of the hypothalamus, *Am. J. Psychiatry,* 139, 824, 1982.

69. **Heron, G. B. and Johnston, D. A.,** Hypothalamic tumor presenting as anorexia nervosa, *Am. J. Psychiatry,* 133, 580, 1976.

70. **Lewin, K., Mattingly, D., and Mills, R. R.,** Anorexia nervosa associated with a hypothalamic tumor, *Br. Med. J.,* 2, 629, 1972.

71. **Enzman, D. R. and Lane, B.,** Cranial computed tomography findings in anorexia nervosa, *J. Comput. Assist. Tomogr.,* 1, 410, 1977.

72. **Heinz, E. R., Martinez, J., and Haenggeli, A.,** Reversibility of cerebral atrophy in anorexia nervosa and Cushing's syndrome, *J. Comput. Assist. Tomogr.,* 1, 415, 1977.

73. **Nussbaum, M., Shenker, I. K., Marc, J., and Klein, M.,** Cerebral atrophy in anorexia nervosa, *J. Pediatr.,* 96, 867, 1980.

74. **Sein, P., Searson, S., Nicol, A. R., and Hall, K.,** Anorexia nervosa and pseudo-atrophy of the brain, *Br. J. Psychiatry,* 139, 257, 1981.

75. **Datloff, S., Coleman, P. D., Forbes, G. B., and Kreipe, R. E.,** Ventricular dilation on CAT scans of patients with anorexia nervosa, *Am. J. Psychiatry,* 143, 96, 1986.

76. **Kreig, J. C., Emnch, H. M., Backmund, H., Pirke, K. M., Herholz, K., Paulik, G., and Heiss, W. D.,** Brain morphology (CT) and cerebral metabolism (PET) in anorexia nervosa, in *Disorders of Eating Behaviour, A Psychoneuroendocrine Approach,* Advances in the Biological Science, Vol. 60, Ferrari, E. and Brambilla, F., Eds., Pergamon Press, New York, 247, 1986.

77. **Artman, H.,** Reversible and non-reversible enlargement of cerebrospinal fluid spaces in anorexia nervosa, *Neuroradiology,* 27, 304, 1985.

78. **Kohlmeyer, K., Lehmkuhl, G., and Poustka, F.,** Computed tomography in patients with anorexia nervosa, *Am. J. Neurol.,* 4, 437, 1983.

79. **Kaye, W. H., Pickar, D., Naker, D., and Ebert, M. H.,** Cerebrospinal fluid opiod activity in anorexia nervosa, *Am. J. Psychiatry,* 139, 643, 1982.

80. **Brown, G. M., Garfinkel, P. E., Jeuniens, C. N., Modolfsky, H., and Stancer, H. C.,** Endocrine profiles in anorexia nervosa, in *Anorexia Nervosa,* Vigersky, R. A., Ed., Raven Press, New York, 1977, 123.

81. **Casper, R. C., Davis, J. M., and Pandey, G. N.,** The effect of the nutritional status and weight changes on hypothalamic function tests in anorexia nervosa, in *Anorexia Nervosa,* Vigersky, R. A., Ed., Raven Press, New York, 1977, 137.

82. **Sherman, B. M. and Halmi, K. A.,** Effect of nutritional rehabilitation on hypothalamic-pituitary function in anorexia nervosa, in *Anorexia Nervosa,* Vigersky, R. A., Ed., Raven Press, New York, 1977, 211.

83. **Alderdice, J. A., Dinsmore, W. W., Buchanan, K. D., and Adams, C.,** Gastrointestinal hormones in anorexia nervosa, *J. Psych. Res.,* 19, 207, 1985.

84. **Tovoli, S., Boletti, G. F., Patrono, D., Severi, D., and Gasbarini, G.,** Release of pancreatic hormones in anorexia nervosa, *Dig. Dis. Sci.,* 29, A28, 1984.

85. **Besterman, H. S., Sarson, D. L., Hsu, G., Crisp, A. H., and Bloom, S. R.,** Gut hormones in thyrotoxicosis and anorexia nervosa, *Proc. Soc. Endocrin.,* 8, 143, 1979.

86. **Nelson, J. D. and Solyom, L.,** Response of the hormones of the gastroentero pancreatic axis to a test meal in anorexia nervosa, bulimia and normal controls, *Disorders of Eating Behaviour: A Psychoneuroendocrine Approach,* Advances in the Biological Science, Vol. 60, Ferrari, E. and Brambilla, F., Eds., Pergamon Press, New York, 1986, 239.

87. **Gerner, R. H., Cohen, D. J., Farbanks, L., Anderson, G. M., Young, J. G., Scheinin, M., Linnoil, M., Shaywiotz, B. A., and Hare, T. A.,** CSF neurochemistry of women with anorexia nervosa and normal women, *Am. J. Psychiatry,* 141, 1441, 1984.

88. **Kaye, W., Ebert, M. H., Raleigh, M., and Lake, R.,** Abnormalities in CNS monoamine metabolism in anorexia nervosa, *Arch. Gen. Psychiatry,* 41, 350, 1984.

89. **Kaye, W. M., Gwirtsman, H. E., Lake, R., Suver, L. J., Jimerson, D. C., Ebert, M. H., and Murphy, D. L.,** Disturbances of norepinephrine metabolism and adrenergic receptor activity in anorexia nervosa: relationship to normal state, *Psychopharmacol. Bull.,* 21, 419, 1985.

90. **Kaye, W. H., Ebert, M. H., Gwirtsman, H. E., and Weiss, S. R.,** Differences in brain serotonergic metabolism between non-bulimic and bulimic patients with anorexia nervosa, *Am. J. Psychiatry,* 141, 1598, 1984.

91. **Bierderman, J., Herzog, D. B., Rivinus, T. M., Harper, G. P., Ferber, R. A., Rosenbaum, J. F., Harmak, J. S., Tondorf, R., Orsular, P. J., and Schildkrank, J.,** Amitriptyline in the treatment of anorexia nervosa: a double blind placebo controlled study, *J. Clin. Psychopharmacol.,* 5, 10, 1985.

92. **Kennedy, S. H., Pira, N., and Garfinkel, P. E.,** Monoamine oxidase inhibitor therapy for anorexia nervosa and bulimia: a preliminary trial of isocarboxaside, *J. Clin. Psychopharmacol.,* 5, 279, 1985.

93. **Halmi, K. A., Eckert, E., LaDu, T., and Cohen, J.,** Anorexia nervosa: treatment efficacy of Cyproheptadine and amitriptyline, *Arch. Gen. Psychiatr.,* 43, 177, 1986.

94. **Mills, I. V.,** Amitriptyline therapy in anorexia nervosa, *Lancet,* 2, 687, 1976.

95. **Lacey, J. H. and Crisp, A. H.,** Hunger, food intake and weight: the impact of clomipramine on a refeeding anorexia nervosa population, *Postgrad. Med. J.,* 56, 79, 1980.

96. **Gross, H. A., Ebert, M. H., and Faden, V. B.,** A double blind controlled trial of lithium carbonate in primary anorexia nervosa, *J. Clin. Psychopharmacol.,* 1, 376, 1981.

97. **Garattini, S.,** Effects of fenfluramine on eating disorders, in *Disorders of Eating Behaviour: A Psychoneuroendocrine Approach,* Advances in the Biological Science, Vol. 60, Ferrari, E. and Brambilla, F., Eds., Pergamon Press, New York, 1986, 327.

98. **Dally, P. J. and Sargant, W. A.,** A new treatment of anorexia nervosa, *Br. Med. J.,* 1, 1770, 1960.

99. **Vandereycken, W. and Pierloot, R.,** Pimozide combined with behaviour therapy in the short term treatment of anorexia nervosa. A double-blind placebo controlled cross over study, *Acta Psych. Scand.,* 16, 445, 1982.

100. **Vandereycken, W.,** Neuroleptics in the short term treatment of anorexia nervosa. A double-blind placebo controlled study with sulpinde, *Br. J. Psychiatry,* 144, 288, 1984.

101. **Moore, R., Millis, I. H., and Forster, A.,** Nalaxone in the treatment of anorexia nervosa: effect on weight gain and lipolysis, *J. R. Soc. Med.,* 74, 129, 1981.

Chapter 6

BIOLOGICAL ASPECTS OF POST-TRAUMATIC STRESS DISORDER

Jonathan Davidson

TABLE OF CONTENTS

I. HISTORICAL CONSIDERATIONS

In 1864 Mitchell, Moorehouse, and Keen[1] described the myriad effects of combat in veterans of the American Civil War. These ailments were variously described as soldier's heart, Da Costa's syndrome (effort syndrome), and neurasthenia. Interest in the effect of war upon human behavior and mental health was rekindled during World War I, in which a new diagnostic entity, "shell shock", was postulated by Mott.[2] In 1919, Wearn and Sturgis[3] used a biological challenge strategy to explore psychophysiological aspects of soldier's heart. This is an interesting study in that, so far as is known, it represents the first such attempt to explore an anxiety disorder. The authors paid good attention to the scientific method; in a controlled study, they challenged a group of patients with soldier's heart by means of i.m. administration of 0.5 cc 1:1000 epinephrine. The authors' original premise was that soldier's heart represented a variant of hyperthyroidism, and they reasoned that the associated hypermetabolic state could be unmasked by an epinephrine challenge. As it turned out, thyroid tests were normal but the soldier's heart subjects responded abnormally to epinephrine, with a rise in systolic pressure above 10 mm, increased pulse rate in excess of 10 beats per minute, and precordial pain, nervousness, palpitations, pallor, tremor, flushing, and sweating. The control group of patients without soldier's heart failed to manifest these abnormal reactions. In addition, the soldier's heart patient group also received i.m. administration of a placebo, to ensure that these abnormal reactions were not due to the procedure itself. Their response to placebo was found to be different from the response to epinephrine. From this landmark study, the authors concluded that soldiers heart was associated with a psychophysiological instability of the autonomic nervous system. Careful reading of the clinical descriptions in this report suggest that not all patients would be diagnosed as posttraumatic stress disorder (PTSD) by current nomenclature, but they would almost certainly have met criteria for generalized anxiety disorder. Without doubt, however, a number of the patients were experiencing a post-traumatic stress reaction.

Armed conflict is not the only stressor which can evoke a maladaptive post-traumatic reactions: civilian accidents were recognized during the 19th century, e.g., railway spine.[4] As also occurred with the classification of phobic disorder, a number of post-traumatic reactions were classified in terms of the original stress. In reality, these variants were probably manifestations of one common condition, which we now know as posttraumatic stress disorder. Kardiner and Spiegel[5] originally identified the commonality of pathological reactions to military and civilian trauma, and Kardiner[6] described their characteristic manifestations as including a preoccupation with the traumatic event, hypervigilance, heightened susceptibility to perceived threat, startle reactions, irritability, a tendency toward explosiveness, nightmares, and a general constriction of interpersonal and social activity. Many of these features are now incorporated in DSM-III[7] (Table 1).

Perhaps to a greater extent than ever before, contemporary life brings us face-to-face with human tragedy and the effects of disaster. Modern technology, increased population mobility, the concentration of large population groups, and the occurrence of two major world wars have set the stage for the occurrence of extreme traumatic experience to millions of people. An idea of the magnitude of the problem is given by the statistic of unnatural deaths, which account for 152,000 (8%) of annual deaths in the U.S. Assuming an average of 4 surviving family members, then 600,000 citizens would undergo bereavement of an unnatural death each year.[8]

While there is good evidence to support the validity of PTSD, the diagnosis is also clouded by some related issues, such as the desirability of achieving compensation, other forms of secondary gain, the occurrence of factitious PTSD,[9] the lack of objectively verifiable diagnostic features, and its frequent coexistence with other diagnoses. Although self-report forms the cornerstone of psychiatric diagnosis, this is especially true for post-traumatic stress

Table 1
DIAGNOSTIC CRITERIA FOR PTSD (DSM-III)

A.   Existence of a recognizable stressor that would evoke significant symptoms of distress in almost everyone.
B.   Re-experiencing trauma by means of at least one of the following:
1.   Recurrent and intrusive recollections of the event
2.   Recurrent dreams of the event
3.   Sudden acting or feeling as if the event were recurring, because of an association with an environmental or ideational stimulus
C.   Numbed responsiveness to, or reduced involvement with the external world, beginning sometime after the trauma, as shown by at least one of:
1.   Markedly diminished interest in one or more significant activities
2.   Detachment or estrangement from others
3.   Constricted affect
D.   At least two of the following symptoms which were not present before the trauma:
1.   Hyperalertness or exaggerated startle response
2.   Disturbed sleep
3.   Guilt over survival when others have not, or about behavior required for survival
4.   Impaired memory or concentration
5.   Avoidance of activities that arouse recollection of the traumatic event
6.   Intensification of symptoms by exposure to events that symbolize the traumatic event.

disorder, and the opportunity to verify such reports is not always present. Mindful of these considerations, the importance of furthering our biological understanding of PTSD becomes clearer.

This chapter considers the genetic and family history in PTSD, the relevance of learning theory, neuroendocrine and neurotransmitter studies, sleep studies, and psychopharmacological treatments.

II. FAMILIAL ASPECTS OF POST-TRAUMATIC STRESS DISORDER (PTSD)

It is pertinent to know whether PTSD has a genetic basis. Do some individuals have a predetermined vulnerability to developing maladaptive reactions after extreme stress? During World War II and thereafter, a number of descriptive studies reported on the family history in cases of stress reaction or neurotic breakdown. Craigie[10] noted a family history of psychiatric disorder in one third of veterans who suffered acute stress responses in World War II. In 1943, studies by Slater[11] and Symonds[12] concluded that if hereditary factors were strong, then neurotic breakdown was likely to occur under mild stress, whereas in veterans with a negative family history, severe stress was needed to bring out symptoms.

Unfortunately, it is difficult to extrapolate the findings of these early studies to our present concept of PTSD, as it is quite possible that a number of these patients were suffering from other forms of neurosis besides PTSD. It is also conceivable that the stresses which were experienced, and which were acknowledged by the authors as often being mild, would not qualify for the DSM-III requirement that the stress should be overwhelming and likely to elicit symptoms of distress in practically everyone. Furthermore, the family histories were poorly characterized. When DSM-III criteria for PTSD were formulated, no information on family characteristics was available. Because of this, we undertook a family history study of World War II, Korean, and Vietnam veterans with chronic PTSD.[13] By applying Family History — Research Diagnostic Criteria (FH RDC),[14] we found that 66% of the 36 patients gave positive family histories of psychiatric disorder. Sixty percent of the whole sample (92% of the subgroup with a positive family history) had a history of alcohol or drug abuse, 22% had a history of anxiety disorder, and 20% each had histories of depression and other disorders. Although depression occurred commonly in the family, it appeared unlikely to be of the melancholic form, since only 6% of the group gave a family history of completed suicide, and none had a history of a relative receiving electroconvulsive therapy (ECT).

The morbidity risk for each illness was assessed in parents and siblings. Among parents, the morbidity risks for alcoholism, anxiety, depression, psychosis, and other illness were as follows: 18.0, 8.3, 5.5, 2.7, and 1.3%, respectively. Among siblings, the morbidity risks for these illnesses were as follows: 12.0, 1.4, 2.8, 3.5, and 2.8%. We concluded that in patients with PTSD arising out of military combat, there was a high incidence of familial psychopathology in first-degree relatives, especially alcoholism. It appeared that PTSD did not bear close relationship either to melancholic depression or to panic disorder, but these findings will need to be assessed more rigorously in future studies. A limitation of our findings is imposed by the use of family history rather than family interview. The sensitivity of FH RDC has been evaluated with respect to a family interview method,[15] and while the sensitivity of the FH RDC is acceptable, its performance falls below the family interview method.

It is clear that more work needs to be done in two areas: adoption of the family study technique to replicate our earlier report on combat veterans with PTSD, and an investigation of family history among first-degree relatives of individuals who have developed PTSD in a civilian setting.

III. PTSD AS A CONDITIONED RESPONSE

From his observations of first World War veterans, Freud[16] recognized that neuroses following combat assumed a different character to those normally seen in his clinical practice. To distinguish it from the latter, Freud used the term "actual neurosis". This view was also shared by Kardiner,[6] who regarded the condition as a "physioneurosis" and suggested that abnormal conditioning might be an important etiological factor. To the extent that the patient was experiencing a neurotic conflict, the source of the conflict was perceived as being the patient's own physiological instability and pathological arousal.

Subsequent studies by Wilson and Dobbs,[17] Blanchard et al.,[18] and Malloy et al.[19] have all endorsed the view that chronic PTSD can be related to a conditioned abnormality of the fear response. Kolb and Multalipassi[20] have suggested that there remains a persistent potential for abnormal arousal in response to any perceived bodily threat. They suggest this response may be mediated by the noradrenergic system, and can lead to a related unpleasant emotional state and altered self-perception.

The first study of autonomic dysfunction in relation to anxiety in war veterans has already been described:[3] it concluded that there was a persistent hypersensitivity and hyperreactivity of the adrenergic system. In a later study,[17] Dobbs and Wilson examined eight decompensated World War II combat veterans 15 years post-combat, along with a control group of 13 age-matched compensated combat veterans and 10 younger university students. Subjects were asked to undergo exposure to a sequence of combat and artillery sounds, as well as photic stimulation with strobe lights. Measures included heart rate, respiratory rate, and EEG alpha rhythm. A brief alerting response was shown in the student control group who manifested no subsequent distress. Among the compensated veterans, there was evidence of substantial physiological arousal, but the majority of patients could tolerate unpleasant stimuli through to completion of the test. Of the decompensated group, 63% were sufficiently aroused that they asked to terminate the stimulus, and manifested marked increases of heart rate and EEG alpha activity.

In the study by Blanchard et al.,[18] Vietnam veterans with PTSD were compared to control groups of Vietnam combat veterans without PTSD, Vietnam era veterans who had not experienced combat, and another group of psychiatric patients. A number of stressors were presented, including combat sounds and the nonspecific stressor of mental arithmetic. PTSD patients showed significant elevations of heart rate, systolic blood pressure, and EMG activity (indicative of muscle tension). Discriminant function analysis showed that the increased heart rate alone could correctly assign 95% of the patients to a diagnostic group. The specificity of psychophysiological arousal was demonstrated by the absence of a difference in arousal measures between PTSD and controls in response to mental arithmetic. The authors concluded that a stimulus which elicits no response in a control group can evoke substantial increase in heart rate among those who had been initially conditioned to the stimulus. Furthermore, this heightened physiological reactivity was evident many years later.

Malloy et al.[19] exposed 3 patient groups to videotapes of 9 helicopter scenes and 9 scenes of a family trip to a shopping mall. The sound intensity progressively increased with scene presentations. Heart rate and sweat gland activity were measured in 3 patient groups: 10 with PTSD, 10 noncombat controls, and 10 psychiatric impatients without PTSD. The PTSD group manifested elevations of heart rate and sweat gland activity, and asked that the combat stimuli be terminated early, after a mean of 6 scenes. No such distress was evoked in the control groups, nor among the 3 groups when viewing the shopping mall scenes. By means of a discriminant function analysis of psychophysiological, behavioral, and self-report responses, it was possible to correctly classify 100% of the PTSD group, and 100% of the combined controls.

Further support for the conditioning model comes from animal studies, such as the report by Solomon et al.[21], who trained dogs to avoid traumatic shock (the unconditioned stimulus) by jumping a hurdle into a safe compartment. A warning signal (conditioned stimulus) immediately preceded shock onset and the criteria for acquiring a conditioned avoidance response (CR) was the achievement of 10 successive avoidance responses to the conditioned stimulus. The animals subsequently were exposed only to the conditioned stimulus, yet failed to extinguish their CR in over 200 trials. From this study it might be concluded that marked resistance occurs towards the extinction of an avoidance response when the original stimulus is extremely noxious or life threatening. In fact, both the instrumental behavior[21] and the autonomic aspects of the response[22] are quite resistant to extinction. By analogy one can therefore understand how in PTSD there is persistence of phobic avoidance, emotional numbing, and constriction of affect (i.e., instrumental anxiety-reducing behaviors) and also the persistence of startle, vigilance, insomnia, shaking, and anxiety (autonomic arousal).

IV. INESCAPABLE SHOCK IN PTSD

Many of the principles and findings with respect to conditioning have been incorporated in the inescapable shock model of PTSD. The impetus for many of these ideas originated in the work of Weiss et al.[22,23] and have been taken into account in subsequent formulations of PTSD.[24] It has been possible to draw some intriguing connections between the inescapable shock model and alteration of catecholamine function.

Animals exposed to severe inescapable physical stress will ultimately develop chronic distress, sleep deficits, decreased motivation for learning new behavior, and deficiencies in learning to escape from other aversive situations.[24] Such behavior is not unlike the learned helplessness which has been described as a model for depression by Seligman et al.[25] It has been suggested that the learned helplessness is due not so much to the shock as to the lack of control in its termination.[26]

Inescapable shock affects neurotransmitters in a number of ways. Initial exposure to a noxious stimulus produces increased norepinephrine turnover, along with depletion of brain norepinephrine and increased production of MHPG.[22,23,27,28,29,30] Depletion of dopamine and serotonin, and elevation of acetylcholine have also been reported following initial exposure to an inescapable shock (IS).[31,32] Furthermore, shock which had no effect on naive animals, produced a lowering of norepinephrine and deficits in escape behavior in animals with previous exposure to IS.[30] Similar changes have been noted following chemical depletion of norepinephrine and dopamine.[33,34] The importance of IS is illustrated by the fact that when the shock is escapable, no reduction of norepinephrine levels occurs.[33]

The behavioral analogy between animals exposed to inescapable shock and patients who have PTSD is clear, although obviously the two situations should not be thought of as identical. One might therefore reason that similar catechol changes also occur in PTSD, i.e., chronic norepinephrine depletion and resultant chronic receptor supersensitivity. The negative symptoms of PTSD (withdrawal, constriction of affect, and reduction of motivation and interest) broadly correspond to the depletion state, while the positive symptoms (nightmares, arousal, sweating, startle reaction, intrusive recollections, irritability, and hypervigilance) could be mediated by overactive or hypersensitive adrenergic receptors. With evidence to support disturbed catecholamine activity in PTSD, it is natural to ask whether locus coeruleus dysregulation also occurs in the disorder. The locus coeruleus is involved in regulating sleep,[34] and its activity may be higher in sleep which is accompanied by autonomic arousal. Kramer et al.[35] have shown that PTSD nightmares are often precipitated by autonomic arousal in stage two sleep. The nightmares are characteristically true-to-life (i.e., eidetic) and are less likely to have the dream-like (oneiric) characteristics of usual nightmares. This suggests a role for the locus coeruleus, as it also subserves the facilitation of memory retrieval.[36] In further support of a connection between locus coeruleus abnormality and PTSD is the report that clonidine and propranolol may ameliorate some of the autonomic aspects of the disorder.[37]

V. NEUROTRANSMITTERS AND NEUROENDOCRINOLOGY IN RELATION TO PTSD

Only recently has the neuroendocrinology of PTSD been investigated. Many of the standard techniques which have been applied to the study of other psychiatric illnesses, e.g., dexamethasome suppression test, TRH infusion, cortisol measurements, and CRF infusion, have yet to be applied systematically to PTSD.

A. The Dexamethasone Test

In a study of the dexamethasone test (DST), Kudler et al.[38] examined 28 patients with the diagnosis of chronic PTSD. Patients received 1 mg dexamethasome at 11 P.M., and a blood cortisol sample was obtained at 4 P.M. the next day. All except 2 patients had been free of psychotropic drugs for a minimum of 7 days. The overall incidence of nonsuppression as defined by a 4 P.M. cortisol level of 5 mcg/dℓ was 6 of 28 patients (22%). Ten patients exhibited a concomitant major depression. When the group was subdivided into those with (n = 10) and those without (n = 18) major depression, the percentage of nonsuppressors dropped to 6% (1 of 18) among patients with PTSD alone, whereas the incidence rose to 50% in patients with a major depression (5 out of 10). In the only other report known to us, Evans et al,[39] described two inpatients with PTSD who were suppressors of cortisol. From this limited evidence, we conclude that PTSD without major depression is associated with normal DST results; the incidence of an abnormal test (6%) is comparable to that found in the normal population. One might still suspect the existence of an HPA axis abnormality, but different strategies will be needed for its detection.

B. Cortisol and Catecholamine Activity

Mason et al.[40] found an increased ratio of urinary norepinephrine:cortisol (NE:C) among inpatients with PTSD. The mean NE/C ratio of 2.54 was more than twice as high as that found in other diagnostic groups, i.e., endogenous depression, mania, and paranoid and undifferentiated schizophrenia. By use of the ratio, it was possible to achieve a diagnostic sensitivity of 78% and diagnostic specificity of 94% in the patient sample. Since norepinephrine and cortisol usually respond together in an acute stress state, a finding of dissociation between the two in chronic PTSD is of considerable interest. The authors relate their results to constructs of anger and denial, drawing an association between high norepinephrine and the tendency of many PTSD patients to externalize their anger. Low cortisol, on the other hand, may relate to their propensity for denial and projection. The basis for these formulations had been laid out previously by Funkenstein et al.[41] and Mason et al.[42] Whether the use of a particular defense mechanism determines the hormonal pattern, or whether the relationship is the other way around still needs clarification, but these associations are of considerable interest. In that cortisol levels were low, PTSD resembled paranoid schizophrenia, both of which are disorders characterized by denial and projection. Low cortisol, which persisted throughout treatment in spite of clinical improvement, may possibly be a trait phenomenon in this group.

PTSD shares many phenomenological features in common with grief reactions, and is also frequently triggered by a bereavement. It is therefore interesting to observe that both norepinephrine and cortisol are elevated in patients who are undergoing prolonged and intensive grief reactions following bereavement, but who do not suffer from PTSD.[43] Perhaps the higher cortisol in a bereavement reaction is associated with less denial on the part of the grieving individual.

Kosten et al.[44] noted elevation of urinary epinephrine and norepinephrine in PTSD patients relative to other diagnostic groups, consistent with the postulate that chronic sympathetic overarousal exists in PTSD. On the other hand, it needs to be reconciled with the antithetical model of inescapable shock and its associated chronic catecholamine depletion state. It also needs to be aligned with the finding of lowered platelet monoamine oxidase (MAO) in chronic PTSD. Without a control group of non-psychiatric patients, it is hard to fully interpret the significance of the norepinephrine:cortisol findings. A priori one might expect that catecholamine depletion is characteristic of several states including depression and schizophrenia, but just happens to be greater in PTSD. One would need a normal control group to assess this possibility.

C. Corticotropin-Releasing Factor

Consideration can be given to relationships between catecholamine function and HPA axis disturbance in PTSD. There is now evidence that CRF not only plays an important role in the peripheral response to stress,[45] but that it serves as a central modulator of the stress response. CRF exists at moderate concentrations in the locus coeruleus,[46] and central administration of CRF to rats increases locus coeruleus firing rates. Concomitant EEG changes occur which are compatible with increased arousal. Moreover, higher doses of CRF administered over a 3 to 7 hr period can produce seizure activity indistinguishable from the activity which follows kindling of the amygdala.[47] Chappell et al.[48] have found that acute and chronic stress in rats can lead to elevated CRF concentrations at the locus coeruleus.

One logical hypothesis would be to postulate that tonic overactivity of locus coeruleus leads to chronic depletion of norepinephrine accompanied by chronic elevations of CRF, to which the pituitary corticotroph eventually becomes densensitized. This will result in decreased production of ACTH, which in turn results in lower cortisol output. Such a chain of events could provide one explanation for the findings described by Mason and Giller. It is clear that more study is required to establish the connection between CRF and chronic maladaptive stress responses.

D. Monoamine Oxidase

The finding that platelet MAO is reduced in chronic PTSD is also of interest,[49] since a number of studies had previously found an association between low platelet MAO activity and a spectrum of psychopathological states,[50,51,52] to which PTSD may now perhaps be added. Secondly we found that patients with a history of alcohol abuse during the course of their PTSD had the lowest MAO values, which raises the question whether PTSD might be subdivided into at least two types according to the presence or absence of alcohol abuse. Lowered MAO activity is consistent with the catecholamine depletion hypothesis described earlier, and has been found to correlate with personality traits of sensation-seeking[53] and poor impulse control,[54] which are also found frequently in PTSD.

E. Tribulin

Glover et al.[55] identified an endogenous inhibitor of both MAO and benzodiazepine receptor binding, which they named tribulin because of its demonstrated association with states of high arousal, anxiety, and agitation. Because such conditions as alcohol withdrawal and panic attacks[56,57] are associated with increased production of tribulin, we decided to investigate the association between tribulin levels and PTSD.

Our unpublished findings[79] indicate that patients with PTSD cannot be distinguished from a group of normal controls on the basis of tribulin activity. We did find that agitated patients with PTSD exhibited significantly higher endogenous MAO inhibition than nonagitated patients. Furthermore, a relationship was observed between endogenous MAO inhibition and platelet MAO activity, whereas no correlation was found between benzodiazepine inhibition and platelet MAO activity.

F. Sleep Studies

The literature is replete with allusions to sleep disturbance following battle. For example, Shakespeare has written in *Henry IV* of the nightmares, physical arousal, and battle reenactments during sleep. It is therefore natural that the sleep of PTSD patients should be studied. Greenberg et al.[58] studied REM sleep patterns in Vietnam combat veterans with PTSD. Reduction of REM latency to less than 40 min occurred over 14 nights of sleep recordings in all 9 patients. Since complaints of nightmares were prominent in most patients, it was surprising that the authors could only identify two clear nightmares during the many nights of sleep recordings. There was also poor dream recall in the patients who were awakened from REM sleep. The authors concluded from their study that their patients had "much greater than normal pressure to begin to dream."

Lavie et al.[59] reported a sleep study in 11 veterans of the Yom-Kippur War. All patients had PTSD and complained of persistent difficulties more than two years following traumatization. They had been hospitalized for treatment of PTSD, claiming that sleep patterns were normal before military experiences. All-night sleep studies revealed sleep onset insomnia, dream interruption insomnia, nocturnal myoclonus, and pseudo-insomnia. Three of the 5 cases of pseudo-insomnia, whose sleep was adequate on the basis of EEG records, exhibited increased body movement and complained of marked fatigue on waking, which often persisted through the day. Decrease in REM percent and a surprising increase in REM latency, which the investigators could not account for, were also noted in the study.

Van der Kolk et al.[60] studied patients whose nightmares were secondary to combat and a group of patients who had experienced life-long nightmares. Compared to the control group, PTSD patients developed nightmares earlier in sleep at approximately 1 to 3 A.M., their sleep was accompanied by gross body movement, nightmares were not restricted to REM sleep, and they were more responsive to psychotropic medication, such as benzodiazepines: 67% of PTSD nightmares were drug responsive as compared to 22% of life-long nightmares. Another difference was that the PTSD nightmare content tended to exactly replicate the trauma, whereas the life-long nightmare was less eidetic.

VI. PHARMACOTHERAPY OF PTSD

The disorder can occur as an acute or chronic condition, and its onset may either immediately follow the stressor, or be delayed by a variable period of time. While the delay is usually not longer than a year, some case reports have described individuals whose PTSD emerged as long as 30 years after the trauma,[61,62] perhaps being triggered by a life change. In such cases it is often found that previously successful coping mechanisms (e.g., employment, maintenance of physical fitness) can no longer be used. If alternatives of successful coping behaviors are not readily available, then this can set the stage for development of PTSD in susceptible individuals. While we think that such delayed reactions are very rare, we have occasionally found that previous PTSD symptoms can also become reactivated in older patients, either in the form of pure PTSD diagnosis, or as a concomitant aspect of some other illness, such as depression.

PTSD is not always an intermittent illness: it can pursue a chronic course, whereas at other times it may be followed by full recovery. It may also intensify with increasing age.[63] Yet another outcome is that an initial diagnosis of PTSD can lead into other diagnoses such as panic disorder, while the PTSD itself recedes. Concurrent diagnoses of alcoholism, substance abuse, depression, and generalized anxiety are often seen, and considerable work needs to be done to separate these various disorders from one another. In fact, chronic PTSD rarely exists in isolation. It stands to reason, therefore, that the treatment of PTSD may involve addressing many other issues besides the core condition.

A variety of factors appear to determine the eventual outcome in PTSD. Among these are the availability of social support networks,[64] prompt access to treatment and counseling,[65] and the ability to integrate and give meaning to the painful intrusive images and feelings that occur.[66] Other factors that may determine severity of illness and response to treatment are the magnitude of the stressor measured in terms of its duration and its dislocating effect upon a patient's life.[67] It makes a difference whether or not trauma is experienced in isolation or as part of a shared community experience, and the community's response to trauma may also determine an individual's outcome.[68] The importance of treatment is great, since untreated or poorly treated PTSD will often result in social isolation, the dissolution of attachment bonds, impairment in forming new attachments, problems with trust, substandard work performance, diminished earning capacity, poor impulse control, and proneness to violence and suicide, not to mention the other psychiatric diagnoses alluded to above.

In the overall scheme of things, pharmacotherapy probably occupies a critical position, yet has been scarcely studied. Such reports as exist are all uncontrolled open trials, and these will be summarized. Essential features of these trials are displayed in Table 2, and the key issues which in this author's opinion need addressing in the pharmacotherapy of PTSD are then presented.

A. A Review of Pharmacotherapeutic Studies in PTSD

An early reference to the use of somatic treatment-methods for PTSD describes an investigation by Freud into the use of electrical therapy by Wagner-Jauregg.[69] Whether or not this was primarily used for therapeutic or punitive purposes remains open to question. Abreactive techniques have also been used for managing acute PTSD, but until recently there has been almost no systematic study of somatic and drug therapy of chronic PTSD.

Although a number of anecdotal case reports have appeared in the literature, we do not know of any double-blind or comparative pharmacotherapy trials in PTSD. While a number of such studies are now being undertaken, the uncontrolled reports are based on relatively small samples. The available publications known to this author are listed in Table 2. Clinical trials have provided support for the use of phenelzine,[70-72] doxepin,[73-75] amitriptyline,[75] imipramine,[76,77] desipramine,[75] clonidine,[37] propranolol,[37] and carbamazepine.[78] In their recent study, Lipper et al.[78] found benefit for carbamazepine in the treatment of intrusive symptoms, hostility, and other positive manifestations of PTSD in a group of Vietnam veteran inpatients. This finding is particularly interesting in light of the kindling effect from high CRF doses in animals studies,[48] and suggests a line of investigation in understanding

Table 2
SUMMARY OF PHARMACOTHERAPY STUDIES IN PTSD

Drugs	Dose (mg)	Duration (weeks)	No. pts.	IP/OP	Outcome	Ratings	Comments	Ref.
MAOI								
Phenelzine	45—90	≥12	5	IP/OP	All improved	None used	Drug permitted abreaction and enhanced psychotherapy	71
Phenelzine	1 mg/kg	8	10	OP	6 patients improved	Zung scale Global rating	Rapid improvement	72
Phenelzine	45—60	6—10	11	IP/OP	7 patients improved	Hamilton scales PTSD rating Impact of Events	Side-effects, limited benefit of drug	70
Tricyclics and tetracyclics								
Doxepin	25—100	?	18	OP	Good effects	None used	Some patients took benzodiazepines; World War II vets; no previous pharmacotherapy; sleep improvement marked	73
Imipramine	50—300 (mean 260)	2—3	10	OP	Good effect on intrusion; no effect on avoidance	Impact of Events	Non-veteran, private practice sample	76
Amitriptyline	150—250	6—8	10					
Desipramine	150—200		4	IP/OP	14/17 improved	Global rating	Drug selection process not given	75
Imipramine	200—250		2					
Doxepin	100		1					
Amitriptyline	139	24	14	?	12/14 improved			77
Doxepin	100	7			4/7 improved			
					11/12 improved			
Other drugs								
Propranolol	120—160	24	12	?	8/9 improved		Improvement on explosiveness, nightmares, sleep, intrusive thoughts, startle, and psychosocial function in both cases	37
Clonidine	0.2—0.4	24	9	?				
Carbamazepine	200—1000 (mean 780)	5	10	IP	7/10 improved	Numerous scales	Side effects: headache and tremor in 4 patients; blood levels measured; dreams, flashbacks, recollections, and hostility responded	78

the biological mechanisms associated with PTSD. Support for lithium, benzodiazepines, and neuroleptics is almost entirely anecdotal.[67] Problems with many of the studies include the absence of rating scales, or the use of somewhat rudimentary rating procedures. Other problems include failure to control for co-existing psychiatric illness, reliability of the diagnostic process, and the fact that all these studies were open and uncontrolled. Some studies report "dramatic" improvement, which in this author's clinical experience is quite rare in veterans with chronic PTSD. This raises the important question of how much improvement can one reasonably expect from pharmacotherapy of the disorder. It also makes one ask whether even modest improvement on a drug is better than what might be accomplished with placebo, psychotherapy, or spontaneous fluctuations in the illness. Another drawback of the cited studies is that all except one[76] have been conducted in veterans, so that we cannot necessarily generalize their results to other PTSD populations. There are several key issues which need to be explored by means of larger double-blind and/or controlled comparison:

1. What treatments are superior to placebo?
2. On what symptoms and aspects of illness is a drug superior to placebo?
3. How does pharmacotherapy compare with, and interact with, psychotherapy in this disease?
4. Do variants of PTSD exist, some of which may be more responsive to one particular active drug as compared to another?
5. What is the time course for the onset of action of a particular treatment?
6. What possible disadvantages may be associated with a given treatment modality?
7. What is the importance of other concomitant diagnoses for treatment outcome?

VII. CONCLUSIONS

Acute PTSD is a normal response to an abnormal stress.[64] Study of this reaction may not therefore be seen as the study of a particular disease state, but rather the study of a universal human reaction. Prolonged and/or delayed reactions, which result in definite morbidity, occur in only a minority. The majority of acute reactions subside, leaving behind a relatively intact survivor. From the biological perspective, three categories of question may be identified: What predisposes a person to morbid stress reaction? What mediates such a reaction? How may intervention/treatment be best accomplished? These three questions can be yoked to a fourth: How do biological factors dovetail with environmental factors?

Both human and animal models serve to expand our biological conceptualization of PTSD. In this chapter, an effort has been made to summarize the salient work accomplished to date in this area, using the three approaches outlined above. Growing understanding of PTSD as a diagnostic entity must be matched by an appreciation of its differences from related disorders: work has scarcely begun here. While they are sure to undergo revision, the establishment of diagnostic criteria, as in DSM-III, has represented an essential first step. A diagnosis which has not been without controversy has at last become legitimized. Figley[64] has pleaded eloquently for greater societal commitment to recognizing and helping the victims of abnormal trauma. The imperative for biological study of PTSD remains strong.

REFERENCES

1. **Mitchell, S. W., Moorehouse, C. R., and Keen, W. S.,** *Gunshot Wounds and Other Injuries of Nerves*, Lippincott, Philadelphia, 1864.
2. **Mott, F. W.,** *War Neuroses and Shell Shock*, Oxford Medical Publications, London, 1919.
3. **Wearn, J. T. and Sturgis, C. C.,** Studies on epinephrine: Effects of the injection of epinephrine in soldiers with "irritable heart", *Arch. Intern. Med.*, 24, 247, 1919.
4. **Trimble, M. R.,** *Post Traumatic Neurosis. From Railway Spine to the Whiplash*, Wiley, Chichester, 1981.
5. **Kardiner, A. and Spiegel, H.,** *War Stress and Neurotic Illness*, Paul B. Hoeber, London, 1941.
6. **Kardiner, A.,** *The Traumatic Neurosis of War*, Basic Books, New York, 1941.
7. *Diagnostic and Statistical Manual of Mental Disorders*, 3rd ed., American Psychiatric Association, Washington, D. C., 1980.
8. **Rynearson, E. K.,** Psychological effects of unnatural dying on bereavement, *Psychiatric Annals*, 16, 272, 1986.
9. **Sparr, L. and Pankranz, L. D.,** Factitious post-traumatic stress disorder, *Am. J. Psychiatry*, 140, 1016, 1983.
10. **Craigie, H. B.,** Two years of military psychiatry in the Middle East, *Br. Med. J.*, 2, 109, 1944.
11. **Slater, E.,** The neurotic constitution, *J. Neurol. Psychiatry*, 6, 1, 1943.
12. **Symonds, C. P.,** The human response to flying stress, *Br. Med. J.*, 2, 703, 1943.
13. **Davidson, J. R. T., Storck, M., Swartz, M., Krishnan, R., and Hammett, E.,** A diagnostic and family history study of PTSD, *Am. J. Psychiatry*, 142, 90, 1985.
14. **Endicott, J., Andreasen, N., and Spitzer, R. L.,** *Family History — Research Diagnostic Criteria*, Biometrics Research, New York State Psychiatric Institute, New York, 1975.
15. **Andreasen, N. C., Rice, J., Endicott, J., Reich, T., and Coryell, W.,** The family history approach to diagnosis. How useful is it? *Arch. Gen. Psychiatry*, 43, 421, 1986.
16. **Freud, S.,** Introduction to psychoanalysis and the war neurosis (1919), in *Complete Psychological Works*, Vol. 17, Strachey, J. (Transl. and Ed.), Hogarth Press, London, 1959.
17. **Wilson, W. P. and Dobbs, D.,** Observations on persistence of war neurosis, *Dis. Nerv. Syst.*, 21, 40, 1961.
18. **Blanchard, E. B., Kolb, L., Pallmeyer, T. P., and Geraldi, R. J.,** A psychophysiological study of post-traumatic stress disorder in Vietnam veterans, *Psychiatric Quart.*, 4, 220, 1982.
19. **Malloy, P. F., Fairbank, J. A., and Keane, T. M.,** Validation of a multimethod assessment of post-traumatic stress disorder in Vietnam veterans, *J. Consult. Clin. Psychol.*, 51, 488, 1983.
20. **Kolb, L. C. and Multalipassi, L. R.,** The conditioned emotional response: a subclass of the chronic and delayed post-traumatic stress disorder, *Psychiatric Ann.*, 12, 979, 1982.
21. **Solomon, R., Kamin, L. J., and Wynne, L. C.,** Traumatic avoidance learning: the outcomes of several extinction programs in dogs, *J. Abnorm. Social Psychol.*, 48, 291, 1953.
22. **Weiss, J. M., Stone, E. A., and Harrell, N.,** Coping behavior and brain norepinephrine levels in rats, *J. Comp. Physiol. Psychiatry*, 22, 153, 1970.
23. **Weiss, J. M., Glazer, H. I., Pohorecky, L. A., et. al.,** Effects of chronic exposure to stressors on subsequent avoidance-escape behavior and on brain norepinephrine, *Psychosomat. Med.*, 37, 522, 1975.
24. **Van der Kolk, B., Greenberg, M., Boyd, H., and Krystal, H.,** Inescapable shock, neurotransmitters, and addiction to trauma: toward psychobiology of post-traumatic stress, *Biol. Psychiatry*, 20, 314, 1985.
25. **Seligman, M. E. P., Maier, S. F., and Geer, J.,** The alleviation of learned helplessness in the dog, *J. Abnorm. Psychol.*, 73, 256, 1968.
26. **Maier, S. F. and Seligman, M. E. P.,** Learned helplessness: theory and evidence, *J. Exp. Psychol.*, (Gen) 105, 3, 1976.
27. **Schildkraut, J. J. and Kety, S. S.,** Biogenic amines and emotion, *Science*, 156, 21, 1967.
28. **Anisman, H. L. and Sklar, L. S.,** Catecholamine depletion in mice upon re-exposure to stress: mediation of the escape deficits provided by inescapable shock, *J. Comp. Physiol. Psychol.*, 93, 610, 1979.
29. **Anisman, H. L., Grimmer, L., and Irwin, J.,** Escape performance after inescapable shock in selectively bred lines of mice: response maintenance and catecholamine activity, *J. Comp. Physiol. Psychot.*, 93, 229, 1979.
30. **Anisman, H. Ritch, M., and Sklar, L. S.,** Noradrenergic and dopaminergic interactions in escape behavior: analysis of uncontrolled stress effects, *Psychopharmacol. Bull.*, 74, 263, 1981.
31. **Usdin, E., Kvetransky, R. and Kopin, I. J.,** Stress and Catecholamines, Pergamon, New York, 1976.
32. **Anisman, H.,** Neurochemical changes elicited by stress: Behavioral correlates, in *Psycopharmacology of Aversively Motivated Behavior*, Anisman, H. and Bignami, G., Eds., Plenum Press, New York, 1979.
33. **Weiss, J. M., Glazer, H. I., and Pohorecky, L. A.,** Coping behavior and neurochemical changes: An alternative explanation for the "learned helplessness" experiments, in *Animal Models in Human Psychobiology*, Serban, G. and Kling, A., Eds., Plenum Press, New York, 1976.

34. **Ramm, P.,** The locus coeruleus, catecholamines, and REM sleep: A critical review, *Behav. Neurol. Biol.,* 25, 415, 1979.

35. **Kramer, M., Schoen, L. S., and Kinney, L.,** The dream experience in dream disturbed Vietnam veterans, in *Post Traumatic Stress Disorder: Psychological and Biological Sequelae,* Van der Kolk, B. A., Ed., American Psychiatric Press, Washington, D. C., 1984.

36. **McNaughton, N. and Mason, S. T.,** The neuropsychology and neuropharmacology of the dorsal ascending noradregergic bundle: A review, *Prog. Neurobiol.,* 14, 157, 1979.

37. **Kolb, L. C., Burris, B. C., and Griffiths, S.,** Propranolol and clonidine in treatment of the chronic post-traumatic stress disorders of war, in *Post-Traumatic Stress Disorder: Psychological and Biological Sequelae.* Van der Kolk, B. A., Ed., American Psychiatric Press, Washington, D. C., 1984, 97.

38. **Kudler, H., Davidson, J. R. T., Meador, K., Lipper, S., and Ely, T.,** A study of the dexamethasone suppression test in post-traumatic stress disorder, *Am. J. Psychiatry,* 144, 1068, 1987.

39. **Evans, D. L., Burnett, G., and Nemeroff, C.,** The dexamethasone suppression test in the clinical setting, *Am. J. Psychiatry,* 140, 586, 1983.

40. **Mason, J. W., Giller, E. L., Kosten, T., Ostroff, R. B., and Podd, L.,** Urinary free cortisol levels in post traumatic stress disorder patients, *J. Nerv. Ment. Dis.,* 174, 145, 1986.

41. **Funkenstein, D. H., King, S. H., and Drolette, M.,** The direction of anger during a laboratory stress-inducing situation, *Psychosom. Med.,* 16, 404, 1954.

42. **Mason, J. W., Giller, E. L., Kosten, T., Ostroff, R. D., and Harkness, L.,** Elevation of urinary norepinephrine:cortisol ratio in post-traumatic stress disorder, *Psychoneuroendocrinology,* in press.

43. **Jacobs, S. C. and Mason, J. W.,** Neuroendocrinologic aspects of bereavement. Paper presented at American Psychiatric Association Meeting, Washington, D.C., May, 1986.

44. **Kosten, T. R., Mason, J. W., Giller, E. L., Ostroff, R., and Harkness, L.,** Sustained urinary nor-epinephrine and epinephrine elevations in post-traumatic stress disorder, *Psychoneuroendocrinology,* in press.

45. **Plotsky, P. M. and Vail, W.,** Hemorrhage-induced secretion of corticotropin releasing factor-like im-munoreactivity into the hypophysio-portal circulation and its inhibition by glucocorticoids, *Endocrinology,* 114, 164, 1984.

46. **Swanson, L. W., Sawchenko, P. E., Rivier, J., and Vale, W.,** Organization of ovine corticotropin-releasing factor immunoreactive cells and fibers in the rat-brain: an immunohistological study, *Neuroen-docrinology,* 36, 165, 1983.

47. **Ehlers, C. L., Henriksen, S. J., Wang, M., Rivier, J., Vale, W., and Bloom, F. E.,** Corticotrophin releasing factor produces increases in brain excitability and convulsive seizures in rats, *Brain Res.,* 278, 332, 1983.

48. **Chappell, P. B., Smith, M. A., Kilts, C. D., Bissette, G., Ritchie, J., Anderson, C. M., and Nemeroff, C. B.,** Changes in CRF-LZ in microdissected rat brain regions with acute and chronic stress, *J. Neuroscience,* 6, 2908, 1986.

49. **Davidson, J. R. T., Lipper, S., Kilts, C. D., Mahorney, S., and Hammett, E.,** Platelet MAO activity in post traumatic stress disorder, *Am. J. Psychiatry,* 142, 1341, 1985.

50. **Haier, R. J., Buchsbaum, M., Murphy, D. L., Gottesman, I. L., and Coursey, R. D.,** Psychiatric vulnerability, monoamine oxidase, and the average evoked potential, *Arch. Gen. Psychiatry,* 37, 340, 1980.

51. **Buchsbaum, M. S., Coursey, R. D., and Murphy, D. L.,** The biochemical high risk paradigm: behavioral and familial correlates of low platelet MAO activity, *Science,* 194, 339, 1976.

52. **Sullivan, J. L., Stanfield, C. N., Maltbie, A. A., Hammett, E., and Cavenar, J. O.,** Stability of low blood platelet monoamine oxidase activity in human alcoholics, *Biol. Psychiatry,* 13, 391, 1978.

53. **Fowler, C. J., Von Knorring, L., and Oreland, L.,** Platelet monoamine oxidase activity in sensation seekers, *Psychiatry Res.,* 3, 273, 1980.

54. **Zuckerman, M., Buchsbaum, M., and Murphy, D. L.,** Sensation seeking and its biological correlates, *Psychol. Bull.,* 88, 187, 1980.

55. **Glover, V., Reveley, M. A., and Sandler, M.,** A monoamine oxidase inhibitor in human urine, *Biochem. Pharmacol.,* 29, 467, 1980.

56. **Bhattacharya, S. K., Die, W. D., Glover, V., Sandler, M., Clow, A., Topham, A., Bernadt, M., and Murray, R. M.,** Raised endogenous monoamine oxidase inhibitor output in post-withdrawal alcoholics: effects of L-dopa and ethanol, *Biol. Psychiatry,* 167, 829, 1982.

57. **Sandler, M., Glover, V., Clow, A., and Elsworth, J. D.,** Tribulin: An endogenous monoamine oxidase inhibitor/benzodiazepine receptor ligand, *Prog. Clin. Biol. Res.,* 192, 359, 1985.

58. **Greenberg, R., Pearlman, C., and Gampel, D.,** War neuroses and the adaptive function of REM sleep, *Br. J. Med. Psychol.,* 45, 27, 1972.

59. **Lavie, P., Hefez, A., and Halperin, G.,** Long-term effects of traumatic war-related events on sleep, *Am. J. Psychiatry,* 136, 175, 1979.

60. **Van der Kolk, B. A., Blitz, R., Burr, W., Sherry, S., and Hartmann, E.,** Nightmares and trauma: A comparison of nightmares after combat with lifelong nightmares in Veterans, *Am. J. Psychiatry,* 141, 187, 1984.

61. **Christenson, R. M., Walker, J. I., Ross, D. R., and Maltbie, A. A.,** Reaction of traumatic conflicts, *Am. J. Psychiatry,* 138, 984, 1985.

62. **Pary, R., Turns, D. M., and Tobias, C. R.,** A case of delayed recognition of post-traumatic stress disorder, *Am. J. Psychiatry,* 1431, 941, 1985.

63. **Archibald, H. C. and Tuddenham, R. D.,** Persistent stress reaction after combat, *Arch. Gen. Psychiatry,* 12, 475, 1986.

64. **Figley, C. R.,** From Victim to Survivor: Social Responsibility in the Wake of Catastrophe, in *Trauma and Its Wake,* Figley, C. R., Ed., Brunner Maazel, New York, 1986, 398.

65. **Fowlie, D. G. and Aveline, M. O.,** The emotional consequences of ejection, rescue, and rehabilitation in Royal Air Force aircrew, *Br. J. Psychiatry,* 146, 609, 1985.

66. **Horowitz, M. J.,** *Stress Response Syndromes,* New York, Aronson, New York, 1976.

67. **West, L. J. and Coburn, K.,** Post-traumatic anxiety, in *Diagnosis and Treatment of Anxiety Disorders,* Pasnau, R., Ed., American Psychiatric Press, Washington, D.C., 1985, 81.

68. **Quarantelli, E. L.,** An Assessment of Conflicting View on Mental Health: The Consequences of Traumatic Events, in *Trauma and Its Wake,* Figley, C. R., Ed., Brunner-Maazel, New York, 1986, 173.

69. **Jones, E.,** in *The Life and Work of Sigmund Freud,* Vol. 2, Basic Books, New York, 1970, 197.

70. **Davidson, J. R. T., Walker, J. I., and Kilts, C. D.,** A pilot study of phenelzine in post-traumatic stress disorder, *Br. J. Psychiatry,* 150, 252, 1987.

71. **Hogben, G. L. and Cornfield, R. B.,** Treatment of traumatic war neurosis with phenelzine, *Arch. Gen. Psychiatry,* 38, 440, 1981.

72. **Milanes, F., Mack, C. N., Nemson, J., and Slater, V. L.,** Phenelzine treatment of post-Vietnam stress syndrome, *V.A. Practitioner,* 6, 40, 1984.

73. **White, N. S.,** Post-traumatic stress disorder, *Hosp. Commun. Psychiatry,* 34, 1061, 1983.

74. **Van der Kolk, B.,** Psychopharmacological issues in post-traumatic stress disorder, *Hosp. Commun. Psychiatry,* 34, 683, 1983.

75. **Falcon, S., Ryan, C., Chamberlin, K., and Curtis, G.,** Tricyclics: possible treatment for post-traumatic stress disorder, *J. Clin. Psychiatry,* 461, 385, 1985.

76. **Burstein, A.,** Treatment of post-traumatic stress disorder with imipramine, *Psychosomatics,* 25, 681, 1986.

77. **Bleich, A., Siegel, B., Garb, B., and Lerer, B.,** Post-traumatic stress disorder following combat exposure: Clinical features and psychopharmacological treatment, *Br. J. Psychiatry,* 149, 365, 1987.

78. **Lipper, S., Davidson, J. R. T., Grady, T. A., Edinger, J. D., Hammett, E., Mahorney, S. L., and Cavenar, J. O.,** Efficacy of carbamazepine in post-traumatic stress disorder: a preliminary evaluation, *Psychosomatics,* 27, 249, 1986.

79. **Davidson, J., Glover, V., Clow, A., Kudler, H., Meador, K., and Sandler, M.,** unpublished results.

Chapter 7

EPILEPSY AND NEUROSIS

Peter F. Liddle and Thomas R. E. Barnes

TABLE OF CONTENTS

I. INTRODUCTION

Hippocrates wrote "melancholics ordinarily become epileptics and epileptics melancholics: of these two states what determines the preference is the direction the malady takes; if it bears upon body, epilepsy, if upon the intelligence, melancholy".[1] Hippocrates' clinical observation of a relationship between melancholy and epilepsy, vindicated by several studies in this century, is of value because it antedates the confounding influences of anticonvulsant medication by more than two thousand years. A contribution from social stigma to the genesis of melancholy in epileptics in Hippocrates time cannot be excluded, though it is probable that many of his contemporaries regarded epilepsy as a sign of divine rather than evil influence, and therefore a cause for reverence rather than contempt. Hippocrates himself believed that epilepsy was due to a disorder of the brain, and his added proposition that epilepsy and psychological disturbance can be alternative manifestations of a single underlying process, receives at least partial support from the extensive evidence that brain lesions can cause both epilepsy and psychiatric disturbance. In this chapter we will examine some of this evidence, and in addition consider the extent to which the evidence supports the more speculative hypothesis that the specific pathological processes underlying epileptic seizures are closely related to the pathological processes that generate psychiatric symptoms.

In the latter part of the 19th century, psychiatrists whose clinical experience was gained in large asylums, regarded epilepsy as a disease associated with inevitable intellectual, moral, and social degeneration. In his influential book, *Body and Mind*, published in 1883, Maudsley employed the term "epileptic neurosis", encompassing concepts of moral perversion and willful viciousness, and he emphasized the similarities between epileptic neurosis and insanity.[2] Contemporary neurologists, who treated many more patients not living in large institutions, formed a different impression. For example, in 1861 Reynolds reported that 40% of epileptics were mentally well.[2] Reynolds appreciated that those in whom seizures were symptomatic of an identifiable brain disease were more likely to have psychiatric disturbance than those suffering from idiopathic epilepsy.

The foundations for an enlightened era in the understanding of epilepsy were laid in the 1930s when electroencephalography eventually vindicated Hughlings Jackson's assertion that epilepsy consists of "occasional, sudden, excessive, rapid, and local discharges in grey matter".[2] In many cases, the origin of the seizure could be traced to a focal lesion in the brain, and by implication, psychopathology in epileptic patients could be attributed to the location or the extent of pathological lesions in the brain, rather than being an intrinsic attribute of the epilepsy itself.[3] In parallel with the rejection of the view that epilepsy is intrinsically linked to a process of mental degeneration, increased emphasis was placed upon the role of drug treatment and of psychological stresses such as social rejection, in the genesis or exacerbation of psychological problems in epileptic patients.

While it is now widely accepted that there is not an inevitable association between epilepsy and psychopathology, the possibility of an intrinsic relationship between temporal lobe epilepsy and psychopathology remains a subject of active debate,[4,5] and much of the evidence we shall consider in this chapter is relevant to this debate. After a brief review of the nature of epilepsy, we shall examine the findings of epidemiological studies of psychiatric disturbance in epileptic patients, paying particular attention to the evidence concerning risk factors predicting psychiatric disturbance in epileptic patients. We will then turn our attention to more specialized studies which provide evidence of the relationship between epilepsy and specific aspects of psychopathology. Finally, we will address the question of the features common to the pathological processes underlying epilepsy and psychopathology.

Guided by Maudsley, who used the term "epileptic neurosis" to describe a wide range of psychopathology occurring in epileptic patients,[2] we will consider not only transient neurotic illnesses, such as depression and anxiety, but also disorders of personality, sexual dysfunction, and aggressive tendencies. We will exclude consideration of the relationship between schizophrenia and epilepsy, and only touch on manic-depressive illness insofar as it arises out of considering the relationship between depression and epilepsy.

II. THE NATURE OF EPILEPSY

A. Clinical Aspects

Brain[6] defines epilepsy as "a paroxysmal and transitory disturbance of the function of the brain which develops suddenly, ceases spontaneously and exhibits a conspicuous tendency to recurrence." Despite the role of electroencephalography in clarifying the nature of epilepsy, diagnosis in clinical practice must be based mainly on an account of episodes of altered consciousness, perception and/or behavior. In the case of generalized convulsive seizures, diagnosis usually presents few difficulties, because convulsive limb movements and loss of consciousness are relatively easy to ascertain. In other forms of epilepsy, especially in cases in which the altered behavior during the seizure consists of psychomotor, psychosensory, or ideational symptoms, the diagnosis is often more difficult. Alteration of the level of consciousness occurs in many types of seizures, and is therefore a sign of special importance in making a diagnosis of epilepsy.

A distinction of major importance with regard to the likelihood of associated psychiatric disturbance is that between seizures which are generalized and cause loss of consciousness from the onset (such as primary generalized convulsive seizures and absence seizures) and partial seizures, which have a focal onset, although they might become generalized subsequently. Partial seizures have either elementary symptomatology, such as clonic limb movement, or complex symptomatology, such as emotional experiences, cognitive symptoms, and impaired consciousness. Partial seizures are often associated with identifiable brain lesions. In the case of complex partial seizures, the identified lesion usually lies within the temporal lobe, consistent with the psychomotor role of the limbic system, much of which lies in the medial part of the temporal lobe. In adult patients, the majority of seizures are associated with a focus in the temporal lobe.

Some seizures which begin with a loss of consciousness and would be classified on clinical grounds as generalized, might arise from a focus in the limbic system. This is important in assessing the results of all studies which compare psychopathology in patients suffering complex partial seizures with patients having generalized epilepsy. Recording from electrodes placed deep in the brains of patients with "grand mal" seizures, Heath[7] observed a consistent pattern of progression of paroxysmal electrical activity, beginning in deep temporal nuclei with bursts which became continuous before involving the septal region and then spreading to involve the whole brain. Unfortunately, it is not clear whether these seizures would have been classified on clinical grounds as primary generalized seizures or partial seizures with secondary generalization. Another process by which limbic foci can be associated with generalized seizures is the development of secondary foci in the limbic system as a consequence of epileptic seizures arising elsewhere.[8] Therefore, clinical characteristics reflecting limbic disorder might be expected to occur in patients with either complex partial or generalized seizures, especially those with chronic disorder. Bear[5] proposes that the most important distinction with regard to psychopathology is between partial seizures arising in the temporal lobe and focal seizures confined to cortex outside of the temporal lobe.

The essential component of epilepsy is the seizure itself, which is characteristically episodic. In considering the relationship between epilepsy and psychiatric disturbance, the state prior to, during, and after the seizure are all relevant. It is possible that the abnormality creating the epileptic diathesis gives rise to different, possibly even antithetical, pathophysiological processes, during ictal, peri-ictal and inter-ictal phases. In many cases, there is a prodromal phase during which the patient experiences dysphoria, especially irritability. In some patients the seizure appears to relieve the preceding dysphoria. The seizure might be a compensatory response to the preceding phase; alternatively, the prodromal dysphoria might arise from a compensatory process mounted in an attempt to stave off the impending seizure. This rather simplistic notion of antagonism between psychiatric disturbance and seizures, which is similar to the notion that prompted von Meduna to introduce convulsive therapy for schizophrenia, receives circumstantial support not only from the proven efficacy of electroconvulsive therapy for treating depression, but also from the findings of pharmacological and metabolic studies which we shall consider later.

Study of the relationships between ictal and inter-ictal psychological symptoms might be expected to yield useful insights into the relationship the pathological processes underlying epilepsy and psychological symptoms. In complex partial epilepsy, experiences such as depersonalization, hallucinations, and passivity phenomena, or emotions such as fear, sadness, elation, or anger, occur as a component of the seizure itself. There have been very few studies of the relationships between ictal and interictal phenomena. Hermann et al.[9] found that ictal fear is a predictor of interictal psychiatric disturbance, while Stark-Adamec et al.[10] found that several types of aura, including visual and auditory illusions, sudden unexplained hatred and experience of the mind becoming stuck on a single idea, were associated with interictal psychological disturbance.

B. Electroencephalographic Aspects

Electroencephalographic records of paroxysmal fluctuations in electrical potential between electrodes attached to the scalp during a seizure justify the pathophysiological description of an epileptic seizure as an abnormal electrical discharge of neurons which spreads through the brain. This pathophysiological description provides a basis, at least in principle, for distinguishing epilepsy from other episodic disturbances of experience and behavior, such as migraine and transient ischemic attacks.

Thus, electroencephalography (EEG) has helped to refine the clinical definition of epilepsy, and it might seem logical to define epilepsy in terms of abnormal electrical activity detected either in the scalp or in the substance of the brain. Such an approach has major practical limitations. For example, the episodic nature of epileptic seizures makes it difficult to obtain even scalp recordings during seizures. Recording from electrodes in the substance of the brain is even more difficult and invasive, and is only justified in exceptional circumstances. Nonetheless, a theoretical issue of great relevance to the subject of this chapter, is the relationship of paroxysmal electrical activity, such as EEG spikes, to epilepsy.

An EEG spike consists of a rapid rise in electrical potential over a period of about 30 msec, to a sharp peak, followed by a rapid fall of similar duration. It probably reflects the synchronous activation of numerous synapses on the same neuron or group of neurons.[11] In some circumstances, spikes recorded from scalp electrodes during the inter-ictal period provide evidence suggesting a diathesis towards epilepsy, and might help locate the site of a causative lesion, but are not adequate evidence for a diagnosis of epilepsy. Spikes are often recorded in individuals without a history of clinical seizures. In particular, spikes have been recorded in subjects prone to violence,[12] in children with behavior problems, and in patients with schizophrenia.[13] EEG records similar to those of epileptic patients can occur in normal children. Cavazzuti et al.[14] followed up 100 normal children with EEG records typical of epilepsy, for a period of 9 years. In the majority, the EEG abnormality disappeared, and only seven children developed epileptic attacks.

Recording from electrodes placed in or on the brain itself provides further evidence that paroxysmal electrical abnormalities are not necessarily accompanied by clinical seizures. Consideration of two individual cases illustrates the complexity of the relationships between paroxysmal electrical activity, clinical seizures, and psychological processes. The first case is that of a 23-year-old man described by Stevens.[15] He suffered from seizures triggered by a picture of a ship, or imagining a ship sailing on water. These seizures were accompanied by genital paraesthesiae. Interictal recordings from intracranial subdural electrodes beneath the right temporal lobe revealed continuous epileptiform activity, at a time when the patient was reported to be entirely free of symptoms. The epileptiform activity was not discernable in simultaneous recordings from the scalp.

The second case is one of three similar cases of episodic rage reported by Smith.[16] A 35-year-old man from a very violent family, who had worked as a side-show boxer and had frequently been knocked unconscious in his teens, had exhibited unpremeditated and unprovoked violence since age 19. Sometimes he inflicted very severe injuries on his victims, one of whom died. His rage attacks followed a stereotyped pattern. For several days he would become progressively more irritable. Observers reported that immediately prior to an attack he became vague, and then launched into an episode of violent aggression lasting about 1 min. Afterwards he felt calm and usually had little memory for the events of the attack. Repeated scalp EEGs failed to demonstrate any abnormality. EEG recordings from electrodes implanted in the brain revealed paroxysms of spike and wave activity lasting several seconds, in the vicinity of the right amygdala, when the patient was resting. When he became spontaneously enraged the paroxysms became more frequent and more prolonged, lasting up to 25 sec. When he was asked to work himself up into a rage by thinking of some provocative situation, he would suffer a full psychomotor seizure, beginning with a rising epigastric sensation, followed by violent struggling and clouding of consciousness. Furthermore, he learned that by giving vent to his rage in a controlled environment, he could obtain relief for several hours from his chronic feelings of irritability.

These cases demonstrate several points relevant to our present concerns. First of all, it is possible to record epileptiform EEG traces from deeply placed electrodes at times when scalp electrodes reveal no evidence of seizure activity. On such occasions, the subject might be experiencing no symptoms whatever, or psychological disturbance ranging from irritability to rage. Secondly, psychological stimuli can play a prominent role in generating abnormal electrical discharges. Thirdly, a seizure can be followed by a period of relief from preceding psychological symptoms.

These observations raise the possibility that pathological processes of the type underlying epilepsy might occur without any clinical evidence of seizures, and that such processes might influence the mental state. Heath[17] has reported cases similar to those reported by Smith, in which episodic rage in patients with clinically established epilepsy, was associated with spike and wave activity interspersed with bursts of 12 to 18 spikes per second (spindles), in the hippocampus and amygdala. In addition, he has recorded traces showing some similar features in the amygdala and hippocampus of patients without epilepsy during periods of emotional dyscontrol. However, in the patients without epilepsy there were only fast spindles, without spike and wave activity. With our present limited knowledge of the nature of the pathological process underlying epilepsy, and also our lack of knowledge about the relationship between EEG traces and the electrophysiological processes occuring within neurons, it would be unwise to extend the diagnosis of epilepsy to any conditions other than those in which clinically identifiable seizures occur.

III. EPIDEMIOLOGICAL STUDIES

The discrepancies between the observations of psychiatrists responsible for the care of epileptic patients in asylums in the 19th century and the observations of contemporary neurologists[2] emphasize the importance of examining the association between epilepsy and psychiatric disturbance in samples representative of the general population.

Pond and Bidwell[18] studied 254 epileptic patients from 14 general practices chosen to be representative of the population of England and Wales, and found that 29% had psychiatric problems, about half of which were classified as neurotic. Unfortunately, the psychiatric assessment was relatively unstructured, and therefore comparison of the observed prevalence of psychiatric disorder in the sample of epileptic patients with the expected frequency of disorder in the general population is of limited value. With regard to risk factors, Pond and Bidwell found a higher prevalence of psychiatric problems in those with temporal lobe epilepsy.

In a survey of all identified cases of epilepsy in Iceland, Gudmundsson[19] found that disorder of the personality occured in 54% of the 883 patients whom he examined. The most prevalent abnormality was "ixoid personality", a term denoting a type of abnormal personality derived from the concept of "ixophrenia" introduced by Sjobring to describe the cluster of mental changes such as perservation, adhesiveness, and explosive temperament, presumed to be characteristic of epilepsy. Ixoid personality occurred in 28.3% of the cases, and was more common in males (33.3%) than females (22.4%). On the other hand, "neurotic personality" which occurred in 18.5% of the total patient group, was more common in females (27.4%) than in males (11%). Unfortunately, Gudmundsson does not provide a detailed account of the criteria he used to ascertain personality traits.

It is difficult to compare Gudmundsson's estimate of the prevalence of ixoid personality in epileptics with the expected prevalence in the general population. In the case of neurotic personality, Gudmundsson's estimates can be compared with the findings of a survey of psychiatric morbidity in the Icelandic general population by Helgason.[93] Helgason found a prevalence of neurosis of 17% in women and 9% in men. Thus, Gudmundsson's data suggest a higher prevalence of neurotic disorder in epileptic women than in the women in the general population. Gudmundsson found a history of psychosis in 5.5% of the males with epilepsy and 9.1% of the females, in comparison with Helgason's finding in the general population of a probability of becoming psychotic before age 61 years, of 4.7% for males and 6.9% for females.

Turning to the question of risk factors, Gudmundsson found increased psychiatric disorder in those with epilepsy of known etiology, compared with those in whom no cause for their epilepsy could be identified, and concluded that extent of brain injury was the most important risk factor. He also found that psychiatric disorder was more prevalent in those in whom seizures began in childhood. In regard to the relevance of the site of origin of the seizures, he found that the prevalence of psychological disturbance in the 68 cases with EEG evidence of a temporal lobe focus was 69%, which is substantially higher than the figure of 54% for the total group. Analysis of Gudmundsson's data for the prevalence of abnormal personality types in patient groups classified according to clinical diagnosis of seizure type, reveals that the highest prevalence of abnormality was in those with "grand mal and focal seizures", of whom 11 out of 16 cases (69%) were judged to be abnormal. Of the 81 cases, 43 (53%) with a clinical diagnosis of psychomotor epilepsy were abnormal, a prevalence nearly identical to that in the total patient group. On the other hand, none of the 18 patients with a clinical diagnosis of petit mal (absence seizures) showed evidence of psychiatric disturbance.

The interpretation of the findings of Pond and Bidwell and of Gudmundsson is limited because they did not use well-defined criteria of established reliability to ascertain psychopathology. One of the most thorough attempts to apply rigorous criteria in a psychiatric epidemiological study was the study of children, aged 5 to 14, on the Isle of Wight by Rutter and his colleagues.[20] Among those children attending school, they found a prevalence of psychiatric disorder of 28.6% in the children with uncomplicated epilepsy, compared with a prevalence of 11.6% in those with chronic disability not directly involving the brain, and 6.6% in a representative group comprising all children aged 10 to 11 attending school.

In contrast to the prevalence of 28.6% in the 63 children with uncomplicated epilepsy, the prevalence of psychiatric disorder in the 12 children with epilepsy and an identified lesion above the brain-stem was 58.3%. In the 24 children with a lesion above the brain stem, but no seizures, the prevalence was 37.5%. These prevalence estimates are based on children attending school and therefore do not take account of those excluded from the educational system because of mental subnormality.

Although the prevalence of psychiatric disorder was higher in the group with uncomplicated epilepsy than in the control group without neuroepileptic disorders, the relative frequencies of different types of psychiatric disorder in these two groups of children were similar. The two most prevalent classes of disorder were neurotic disorder and conduct disorder. In the group with uncomplicated epilepsy, 44% of the psychiatric disorders were neurotic, 33% were conduct disorders, and 17% were mixed neurotic and conduct disorders. In the children without neuro-epileptic disorder, 38% of the disorders were neurotic disorders, 37% were conduct disorders and 23% were mixed. In those children with identified lesions above the brain-stem, the pattern of relative prevalence was similar to that occurring in uncomplicated epilepsy, except that there was a trend towards a higher prevalence of hyperkinetic syndrome, but the numbers of cases with this diagnosis were too small to justify any conclusion.

The higher prevalence of psychiatric disorder in children with uncomplicated epilepsy and in those with epilepsy and overt brain disorder, compared with the prevalence in those with chronic disability not involving the brain, suggests that the very high rate of psychiatric disorders in neuro-epileptic children is associated specifically with a lesion of the brain rather than merely reflecting the consequence of handicap in itself. It is unlikely that differences in the severity of the handicap between the different groups could account for the findings, because psychiatric disorder was considerably more common among the least handicapped neuroepileptic children than in the group of children with noncerebral disorders. In addition, the rate of psychiatric disorder in those epileptic children whose condition was made public by a restriction on physical activities was only slightly higher (by a statistically insignificant amount) than in those whose epilepsy was not publicly identified by the imposition of restrictions, suggesting that the influence of social prejudice was small in comparison with the contribution of other factors.

In the group of children with uncomplicated epilepsy, there was a statistically significant association between having psychomotor seizures, and psychiatric disorder. Of the 18 cases with uncomplicated epilepsy and psychiatric disorder, 33% had psychomotor epilepsy, while of the 45 with uncomplicated epilepsy but no psychiatric disorder, only 4% had psychomotor epilepsy. Other types of focal attacks were not significantly associated with psychiatric disturbance. Rutter and his colleagues concluded that psychiatric disorder is considerably more common in children with psychomotor seizures than in children with other varieties of epilepsy.

Recently, Edeh and Toone have assessed psychiatric morbidity using the Clinical Interview Schedule in a sample of adult epileptic patients attending general practitioners in a part of south London.[21] In this study they examined the relationship of psychiatric morbidity to anticonvulsant treatment, and hematological and neurological variables. The aspect of their study of relevance to our present concern is their use of a standardized scale to assess psychiatric morbidity in a group representative of epileptic patients living in the general community. They found that 37 out of 82 patients had significant psychiatric morbidity. Nineteen patients were depressed, 11 suffered from anxiety, 2 had a personality disorder, and there was one case each of "other neurosis", schizophrenia, affective psychosis, organic psychosis, and mental subnormality.

In conclusion, the epidemiological studies reviewed demonstrate that between a quarter and a half of those with epilepsy have substantial psychiatric disorder. A range of different psychiatric disorders occur. In Gudmundsson's study there is equivocal evidence that both psychotic and neurotic disorders are more common than in the general population, and there is some evidence of a specific personality disorder, for which Gudmundsson uses the Scandinavian term "ixoid." In the study by Edeh and Toone, depression was the most prevalent disorder. Epileptic children exhibit an increased prevalence of psychiatric disorders but the relative proportions of emotional and conduct disorders are similar to those occurring in children without epilepsy. The most clearly established factor predicting the likelihood of psychiatric disturbance in epileptic patients is the presence of an identified brain lesion. In addition, there is some evidence supporting the hypothesis that a diagnosis of psychomotor seizures or EEG evidence of a temporal lobe focus is associated with a higher risk of psychiatric disturbance.

IV. SPECIFIC PSYCHIATRIC DISORDERS

A. Depression and Anxiety

Clinicians have long recognized that depression is common in epileptic patients in the interictal period.[1] In a study of 666 patients with temporal lobe epilepsy, Currie et al.[22] found depression, which occurred in 19%, and anxiety, which occurred in 11%, were the two commonest psychological complaints. Betts[23] examined the main psychiatric diagnoses in 72 epileptic patients admitted to a psychiatric hospital and found that a diagnosis of depressive illness, made in 31% of cases, was the commonest diagnosis. Standage and Fenton[24] assessed 37 patients with epilepsy and 27 control patients suffering from chronic locomotor disorders, using the Present State Examination. They found somatic symptoms of depression in 60% of the epileptic patients and 30% of the controls. Autonomic symptoms of anxiety occured in 50% of those with epilepsy, and 25% of the controls. Trimble and Perez[25] administered the Middlesex Hospital Questionnaire to 281 patients admitted to a neurological hospital for assessment and treatment of their epilepsy, and found significantly higher scores for anxiety, depression, and obsessionality than occur in the normal population. Kogeorgos et al.[26] compared epileptic patients attending a neurological clinic with 50 consecutive nonepileptic outpatients, using the Middlesex Hospital Questionnaire and the General Health Questionnaire. Psychiatric disorder was more prevalent and more severe in the epileptic group, of whom 45.5% were classified as probable psychiatric cases, and in these cases depression, anxiety, and hysterical symptoms were the most common symptoms.

The demonstration by Standage and Fenton[24] that depressive symptoms and autonomic anxiety are twice as common in epileptic subjects as in patients with chronic locomotor disorders, suggests that these symptoms in epileptic patients are not merely a psychological response to chronic illness. However, the unpredictability of seizures and the loss of control over bodily function during seizures can be sources of inescapable stress for some patients with epilepsy. Hermann[27] has drawn attention to the possible parallel between the psychological problems of epileptic patients and the behavioral responses that Seligman[28] induced in dogs by exposing them to unpredictable, uncontrollable aversive events.

The relationship between anticonvulsant medication and depression has been the subject of many studies. Reynolds et al.[29] suggested that the decrease in folic acid levels produced by many anticonvulsants might cause depression, and several studies[21,30,31] have provided evidence supporting this hypothesis, although Robertson[32] found that low serum and red blood cell folic acid were not associated with severity of depression, nor with type, dose, or serum level of anticonvulsant. Phenobarbitone has long been regarded by clinicians as especially likely to cause dysphoria, and Robertson[32] confirmed this association. The simultaneous use of several different anticonvulsants tends to increase the risk of dysphoria. Shorvon and Reynolds[33] have shown that changing from polytherapy to monotherapy, especially by withdrawal of either phenobarbitone or phenytoin, has beneficial effects on alertness, mood, and sociability, and rarely compromises seizure control.

One of the reasons for the rise in the usage of the anticonvulsant carbamazepine has been the lower frequency of mood disturbance in patients treated with it. This is possibly due to a specific antidepressant action of this drug, which has a tricyclic structure similar to that of antidepressants such as imipramine. Sillanpaa[34] and Post[35] have reviewed the many studies which demonstrate that carbamazepine has beneficial psychotrophic effects. Robertson[32] found that patients on carbamazepine were less depressed and had a lower trait anxiety score than those on other anticonvulsants, and that there was a significant negative correlation between trait anxiety score and both dose and serum level of carbamazepine.

The question of an association between interictal affective disturbance and site of the seizure focus has been an issue of special interest since Flor-Henry[36] proposed that affective psychosis in patients with epilepsy is associated with an epileptic focus in the right temporal lobe, while schizophrenia is associated with a left temporal lobe focus. Flor-Henry's hypothesis was based on the findings in a sample of 50 patients with temporal lobe epilepsy of whom 9 suffered from manic-depressive illness. In 4 of these cases there was a right unilateral temporal focus, while 2 had a left unilateral temporal focus and in the remaining 3 there were bilateral foci.

Unfortunately, there have been few studies of the site of the epileptic focus in patients with affective disorder, but the available evidence does not provide strong support for Flor-Henry's proposal of an association between right temporal focus and affective psychosis. In a recent study of 66 patients with epilepsy and depression satisfying Research Diagnostic Criteria for major depressive disorder and a score of 13 or more on the Hamilton depression rating scale, Robertson[32] found only 6 with unilateral right temporal EEG abnormalities and 3 with bilateral temporal abnormality, while 17 had left temporal abnormality and 18 had generalized diffuse abnormality. Perini and Mendius[37] administered the Beck depression scale and the State and Trait Anxiety Inventory to 9 epileptic patients with left temporal lobe foci and 8 with right temporal foci, and to 19 normal controls. Those with a left temporal focus had significantly higher scores for depression and anxiety than both the group with a right temporal focus and the normal control group, whereas the right temporal focus group did not differ significantly from the normal control group. Thus, the evidence suggests that depression is associated with a left temporal lobe focus. The situation with regard to manic-depressive psychosis remains uncertain.

With regard to the possibility of antagonism between psychiatric disturbance and seizures, it is of interest that both Flor-Henry,[36] Betts,[23] and Robertson[32] all found evidence that depression is more likely to occur during a period of reduced seizure frequency.

B. Suicide and Deliberate Self-Harm

Barraclough[38] has reviewed the evidence from 11 studies which have yielded data concerning the frequency of suicide in epileptic patients. Pooling the numbers of suicides in epileptic patients and the numbers of suicides expected in the corresponding general population yielded a total of 86 suicides in epileptic patients compared with an expected total of 16. Thus, Barraclough concludes that the risk is increased approximately five-fold. One of the soundest studies methodologically was the enquiry into the mortality of patients treated at the Chalfont Centre for Epilepsy.[39] Of 425 recorded deaths, 21 were from suicide. The expected number, computed from English life tables, was 3.9. Since the patients in this study had all been treated at a special referral center for epilepsy, they are not representative of all epileptic patients. However, even in a cohort of patients accepted for life insurance, who are likely to represent cases with less severe epilepsy, the mortality from suicide was 2.5 times that expected (Society of Actuaries, 1954, cited by Barraclough[38]).

Barraclough[38] found 4 studies dealing with temporal lobe epilepsy from which it was possible to estimate the risk of suicide in that condition. All 4 studies indicated an increased risk: the total number of suicides recorded was 15, compared with an expected number of 0.6, giving an estimated 25-fold increase in risk of suicide among patients suffering from temporal lobe epilepsy. However, 9 of the total of 15 suicides were from Taylor and Marsh's[40] study of outcome in cases treated surgically. These cases cannot be regarded as typical cases of temporal lobe epilepsy. In particular, it is likely that behavioral problems were one factor contributing to their selection.

Deliberate self-harm is also more common in epileptic patients than in the general population.[41,42] In many cases the self-harming behavior consists of an over-dose of anticonvulsant medication,[42] and it is possible that availability of these drugs contributes to the increased risk.

C. Sexual Dysfunction

Hyposexuality is the most common and the best documented of the many forms of sexual dysfunction and deviation reported in epileptic patients. Guerrant et al.[43] compared the sexual function of epileptic patients with that of a control group with other chronic medical illness, mainly diabetes. They found that 53% of patients with temporal lobe epilepsy reported impotence or frigidity, compared with 33% of patients with other forms of epilepsy, and 32% of patients with other chronic medical illness. Shukla, Srivastava, and Katiyar[44] compared the sexual function of 70 patients with temporal lobe epilepsy with that of 70 patients with generalized epilepsy, diagnosed on the basis of EEG criteria. The groups were matched for age, sex, marital status, duration of epilepsy, and degree of control of seizures. In the case of male patients, 63% of those with a temporal lobe focus and 12% of those with generalized epilepsy reported less than one sexual outlet per month. Similarly, 64% of females with a temporal lobe focus and 8% of the females with generalized epilepsy reported less than one sexual outlet per month. Patients who had had a temporal lobe focus since childhood had never developed any interest in sexual activity.

The evidence from series of patients subjected to surgery for temporal lobe epilepsy[45,46] also indicates that lack of sexual drive occurs in more than 50% of cases, and confirms that onset of seizures in childhood is associated with an especially low level of sexual interest. Thus, the evidence for a specific relationship between temporal lobe epilepsy and hyposexuality is strong. While all of the relevant studies confirm the presence of hyposexuality in temporal lobe epilepsy, there is some evidence that hyposexuality also occurs in patients with generalized epilepsy. For example, Toone et al.[47] found hyposexuality was equally common in generalized epilepsy and temporal lobe epilepsy. Furthermore, they found an association between hyposexuality and use of anticonvulsant medication.

Although hyposexuality is the most common sexual dysfunction in patients with temporal lobe epilepsy, hypersexuality is reported in some cases. For example, Blumer[48] reported hypersexuality in 15% of cases prior to temporal lobectomy, in comparison with hyposexuality in 58%. This observation is consistent with the observations from animal studies that limbic lesions can produce either hyper- or hyposexuality.[49]

D. Aggression

The relationship between epilepsy and aggression poses some especially challenging questions. The first question is whether or not aggression occurs during undisputed seizures, and if so, what form does the aggression take. Study of selected individual cases leaves no doubt that ictal aggression does occur. However, it is probably rare, and usually takes the form of disorganized, unsustained violence.

In 1980, an international panel of experts on epilepsy convened in Bethesda, Maryland, to address the question of the nature of ictal aggression.[50] They examined video recordings and concurrent EEG records of 33 epileptic seizures suffered by 19 patients selected on the grounds of apparent ictal aggressive behavior from a collection of video records of seizures in a total of 54,000 patients. In 13 of these 19 cases, the panel agreed that a clinical diagnosis of epilepsy was beyond doubt. In 6 of the 13 cases with undoubted epilepsy the panel judged that the video record showed either no aggression, or only nondirected aggressive motion without damage to person or property. In 4 cases the consensus assessment was of violence to property only, in 3 cases there was threatened violence to a person, and in 1 case the ictal behavior was judged to include actual violence of mild degree to a person. In all 7 cases in which there was either violence to property, or threatened or actual violence to persons, the seizures were classified as complex partial seizures. In 3 cases the ictal EEG abnormality was diffuse, and in 4 it was localized in either temporal cortex, hippocampus or amygdala. The relevance of the findings from retrospective examination of video-recordings is limited by the possibility that seizures which have been video-recorded are not representative of all seizures.

The second question is whether aggression occurs more commonly in the interictal period in epileptic patients than in the general population. In principle, the answer to this question should be sought by comparing the frequency of aggression, assessed according to a standardized procedure, in a sample of patients representative of unselected epileptic patients, with an age- and sex-matched sample of the general population. No study addressing the question of aggression in epilepsy has studied a sample of truly unselected epileptic patients. Therefore, one of the most important factors to consider in evaluating the results of the studies addressing this question, is the criteria by which patients were selected.

Studies of patient samples biased in favor of cases of severe epilepsy with concomitant psychiatric symptoms and/or persistent neurological disorder, report a relatively high proportion of cases with a history of aggression. For example, Serafetinides[51] studied the first 100 cases submitted to temporal lobectomy at the Maudsley Hospital, on account of intractible complex partial seizures. Of these, 36 had a history of recurring, unprovoked physical aggression towards persons or property. The majority of these patients were males whose seizures had commenced before age 10. In the series of 100 children with temporal lobe epilepsy studied by Ounstead[52] 36 had outbursts of catastrophic rage. The majority of this series had severe, intractable epilepsy, with a high proportion having suffered severe brain trauma or prolonged febrile convulsions. The children with rage outbursts were more likely to have a history of brain insult and to have had an onset of seizures in early life. All of the children with rage outbursts exhibited other types of seizures in addition to temporal lobe attacks.

In an analysis of the records of 700 cases seen at the Epilepsy Centre of Michigan, Rodin[53] found that 34 (4.8%) had a history of destructive-assaultive behavior. When these 34 cases were compared with a sample of 34 nonaggressive patients matched for age, sex, and IQ, the aggressive group were found to have a higher proportion of males, to be less religious, to have more evidence of behavioral disturbance since childhood, and to show more signs of diffuse cerebral disease. There was no excess of cases of temporal lobe epilepsy in the aggressive group.

Studies which compare the scores obtained by patients with different seizure types on various standardized scales purporting to measure aggressive tendency fail to demonstrate an association between temporal lobe epilepsy and aggression. For example, Hermann et al.[54] compared 153 temporal lobe epilepsy patients with 79 patients suffering from generalized seizures on an index of aggression derived from the MMPI, and found no significant difference between the groups after controlling for potentially confounding variables such as age of onset of seizures, sex, and age. Although they did not make a direct comparison with healthy controls, the mean aggression score for the epileptic patients was significantly greater than that obtained by Huesmann et al.[55] in a group of nondelinquent subjects matched with the epileptic subjects for age. Standage and Fenton[24] found no difference between those with temporal lobe epilepsy and those with other seizure types, and no difference between patients with epilepsy and a control group with chronic illness, on the irritability item of the Present State Examination. The score for this item reflects contributions from subjective experience of irritability, verbal aggression, and violent or destructive behavior.

In a small but careful study of 14 patients with uncomplicated temporal lobe epilepsy matched for age, sex, social class, seizure frequency, duration of epilepsy, and current anticonvulsant medication, with 14 patients having idiopathic tonic-clonic seizures, Brown et al.[56] found that complaints of irritability were significantly more common in the temporal lobe group.

Overall, the evidence suggests that the major part of the excess of aggression reported in epileptic patients can be accounted for by variables such as history of brain trauma and prolonged febrile convulsions, the presence of signs of cerebral damage, the occurrence of multiple seizure types, and early onset of seizures. Location of the epileptic focus in the temporal lobe does not appear to make a major contribution to the risk of aggression. In general, it appears that variables reflecting the extent of the cerebral lesion rather than its site are more important in determining the likelihood of aggression.

However, aggression is not a homogenous phenomenon, and a failure to discriminate between various types of aggression, or various components of aggressive behavior, will favor the finding of a nonspecific association with brain damage. If there is an association between a particular type of epilepsy and a specific aspect of aggression, this is unlikely to be revealed unless the relevant aspect of aggression is examined explicitly. Mungas[57] performed a cluster analysis based on ratings of various behavioral attributes of violence recorded in 138 neuropsychiatric outpatients. He found that the patients could be divided into five different groups on the basis of the pattern of violence they exhibited. Two of the groups, which differed only in the severity of violence, exhibited frequent episodes of poorly organized violence arising from minor provocation. Mungas described these patients as having poor impulse control. The pattern of violence exhibited by these two groups of patients resembles that controversially designated the episodic dyscontrol syndrome. These two groups had a significantly higher likelihood of suffering from seizures, in comparison with the other groups.

However, even within each of the relatively homogeneous groups identified by Mungas, there was evidence suggesting that a multiplicity of biological and psychosocial factors contributed to the violence. The presence of a temporal lobe abnormality did not discriminate significantly between the groups, but there was a tendency for the two groups with poor impulse control to exhibit a higher prevalence of temporal lobe abnormality than the other groups of violent patients, so Mungas' study neither confirms nor refutes the hypothesis that frequent poorly organized unprovoked violence is associated with temporal lobe abnormality.

What are the implications of cases such as those reported by Smith,[16] one of which we described above when considering the nature of epilepsy? In that case, aggressive behavior occurred during an episode which commenced with the experience of an abnormal epigastric sensation followed by rage and clouding of consciousness. These clinical characteristics are strongly suggestive of a complex partial seizure. The same patient also suffered episodes of rage, without adequate clinical indices to justify describing these episodes as seizures. During these episodes, spike and wave activity was recorded via electrodes implanted in the amygdala. In addition, Smith reports that electrical stimulation applied via the implanted electrodes produced experiences such as the abnormal epigastric sensation. This suggests that the spike and wave activity recorded via these electrodes reflected electrical activity of the type occurring during seizures. Other investigators have also reported the occurrence of spike and wave activity recorded via electrodes in the amygdala during ictal rage or aggression.[17,50] Therefore, there is indirect evidence supporting the proposal that in Smith's patient, electrical activity of the type occurring during ictal aggression accompanied rage which was not ictal according to clinical criteria. Whether or not there is a causal relationship between the rage and the epileptiform electrical activity, and if so whether the rage caused the epileptiform electrical activity, or the epileptiform electrical activity caused the rage, remains a subject for speculation.

B. The Epileptic Personality

Perhaps the most controversial of the long-standing debates concerning the mental state of patients with epilepsy concerns the existence of the epileptic personality. Issues concerning criteria for the selection of patients, and the reliability and relevance of the method used to ascertain the putative personality abnormalities, confound the interpretation of studies of personality in epileptic patients. Nonetheless, the evidence, including the epidemiological studies reviewed above, provides strong grounds for asserting that a large proportion of epileptic patients do not have a significant disorder of personality. Therefore, the principal issue at stake is whether a subgroup of epileptic patients have a characteristic type of personality. Erna and Frederick Gibbs[58] studied the relationships between clinical characteristics and EEG features in over 10,000 patients and found psychological abnormalities in 40% of those with psychomotor seizures only, 50% of those with psychomotor and grand mal seizures, and less than 10% of those with other types of epilepsy. Following this work, much effort has been devoted to study of the personality of patients with temporal lobe epilepsy.

A large number of such studies, reviewed by Hermann and Stevens,[59] have used the Minnesota Multiphasic Personality Inventory (MMPI), and in general, these studies have not found significant differences between patients with temporal lobe epilepsy and those with other forms of epilepsy. Nonetheless, some of the studies have identified clinical variables that are associated with abnormal MMPI scores. Meier and French[60] found that patients with independent bitemporal EEG spikes had more abnormal MMPI scores than patients with unilateral foci. Hermann et al.[61] found that patients with temporal lobe seizures that commenced in adolescence scored higher on the MMPI Pd (psychopathic deviate); Pa (paranoia); Pt (psychasthenia); and Sc (schizophrenia) scales than those with temporal lobe seizures commencing at other ages. Age of onset did not affect MMPI scores in patients with generalized epilepsy. In a later study, Hermann et al.[9] found that those with temporal lobe epilepsy who experienced ictal fear scored significantly higher than a control group with temporal lobe epilepsy but without ictal fear, on the Pa, Pt, Sc and Si (social introversion) scales. On the other hand, the scores of the group without ictal fear were not significantly different from those of a group with generalized epilepsy.

Hughes and Hermann[62] administered the MMPI to 25 patients with the controversial EEG pattern of rhythmic mid-temporal discharges. This EEG pattern, which is also known by the term ''psychomotor variant'', is characterized by bursts of 6/sec waves, often accompanied by a harmonic of 12/sec, maximal in the mid-temporal regions. The clinical significance of this EEG pattern remains a subject of debate. Hughes and Hermann found that the MMPI scores for the group of patients with rhythmic mid-temporal discharges were approximately two standard deviations above the normal mean for the hypochondriasis, schizophrenia, depression, and hysteria scales. The MMPI profile of these patients was similar to that for patients with definite temporal lobe epilepsy, but different from that for a group of patients with general medical disorders.

Bear and Fedio[63] argued that the most appropriate way to investigate the existence of a specific type of personality disorder in patients with temporal lobe epilepsy is to devise a rating scale for the assessment of various personality traits which clinicians have claimed are more common in these patients. They constructed two scales, one intended for self-rating and the other for rating by an observer, each containing a total of 90 items designed to elicit evidence of 18 personality traits which had previously been reported to be characteristic of patients with temporal lobe epilepsy. These traits, listed in Table 1, embrace characteristics similar to those identified as typical of epileptic patients by Sjobring,[19] such as circumstantiality, adhesiveness, and explosive rage, and in addition, include traits such as religiosity, hypergraphia, and altered sexual interest, which Waxman and Geschwind[64] had claimed were characteristic of the interictal behavior of patients with temporal lobe epilepsy.

Table 1
PERSONALITY TRAITS ARRANGED IN ORDER OF DECREASING DISCRIMINATION BETWEEN THE SELF-REPORTED BEHAVIORS OF EPILEPTIC AND NONEPILEPTIC PATIENTS [63]

Humorlessness, sobriety
Dependence, passivity
Circumstantiality
Sense of personal destiny
Obsessionalism
Viscosity
Emotionality
Guilt
Philosophical interest
Anger
Religiosity
Hypermoralism
Paranoia
Sadness
Hypergraphia
Elation
Aggression
Altered sexual interest

Note: Traits assessed by the Bear and Fedio Inventory in patients studied by Bear and Fedio.[63]

Bear and Fedio[63] compared the scores of 27 patients with unilateral temporal lobe spikes with the scores obtained by 9 patients with other neurological disease and 12 normal adults. The epileptic patients scored significantly higher than both control groups on all 18 items of the self-reported scale. Step-wise discriminant analysis demonstrated that humorless sobriety, dependence, and obsessionalism were the self-reported traits which discriminated best between epileptic patients and controls. The scores assigned by observers for 14 of the 18 traits distinguished epileptic patients from controls at a significant level. Discriminant analysis demonstrated that circumstantiality, philosophical interests, and anger were the most discriminating observer rated traits. Patients with a right temporal lobe focus displayed more emotional tendencies, such as elation, while those with a left-sided focus displayed more ideational traits, such as paranoia. In addition, comparison of self-report and observer ratings demonstrated that those with a right-sided focus tended to deny abnormalities while those with a left-sided focus exaggerated their abnormalities.

Hermann and Reil[65] attempted to replicate the findings of Bear and Fedio, using only the self-report scale. They compared 15 temporal lobe patients with 15 matched patients suffering from generalized epilepsy. The temporal lobe patients had significantly higher scores for the items measuring dependence, personal destiny, philosophical interest, and paranoia. However, for several of the items of the scale there was a nonsignificant trend for the temporal lobe patients to have lower scores than the patients with generalized epilepsy.

Mungas[66] used the Bear and Fedio scales in a comparison of patients having temporal lobe epilepsy and behavioral-psychiatric disorders with a group of patients having concomitant neurological and behavioral-psychiatric disorders, and concluded that the personality traits assessed by these scales reflect nonspecific psychopathology and do not necessarily indicate the presence of a specific behavioral syndrome in temporal lobe epilepsy. Bear[67] responded by emphasizing that individual traits from his inventory might be expected to occur in a variety of psychiatric conditions, and the composite score of a control group suffering from a mixture of psychiatric illnesses might not differ significantly from the composite score of a temporal lobe epilepsy group. The important question is whether or not a combination of traits occurs in patients with temporal lobe epilepsy but not in any other single psychiatric syndrome. Mungas[68] addressed this question by a further analysis of his data and found that the coincidence of at least 3 of 5 selected cardinal traits, was not significantly more common in the temporal lobe group.

However, in the meantime, Bear and his colleagues[69] compared the behavioral profile of temporal lobe epileptic patients with the profiles of patients suffering from bipolar affective disorder, schizophrenia, and aggressive character disorder. In this study they used a clinical interview which addressed aspects of behavior similar to those covered in the Bear and Fedio inventory. They found that the temporal lobe group could be distinguished from each of the individual psychiatric syndrome. Circumstantiality, interpersonal cohesiveness, deepened affect, and interest in religion and philosophy were particularly characteristic of the patients with temporal lobe epilepsy.

Subsequently, Master et al.[70] compared 55 patients with temporal lobe epilepsy, 16 with primary generalized epilepsy, 27 psychiatric controls, and 40 normal controls, and found that the 3 patient groups scored higher than the normal controls on all items of the self-rated Bear and Fedio inventory (except anger), but the differences between scores of the three patient groups were not statistically significant. They concluded that temporal lobe epilepsy is a sufficient but not a necessary condition for elevation of Bear and Fedio inventory scores, and support Mungas' finding that nonepileptic patients with psychiatric disorder obtain high scores.

Stark-Adamec and colleagues[10] administered a modified form of the Bear and Fedio inventory to patients with temporal lobe epilepsy, psychiatric patients, dialysis patients, and normal controls. They found substantial similarities between the profiles of the epileptic patient group and the psychiatric group, but the epileptic group was markedly heterogeneous, with about one third having profiles resembling normal controls and one third resembling psychiatric patients. Brandt, Seidman, and Kohl[71] compared the Bear and Fedio self-report inventory scores of patients with left temporal lobe epilepsy, right temporal lobe epilepsy, generalized epilepsy, and normal controls, and found that the patients with left temporal epilepsy and those with generalized epilepsy had elevated trait scores, while the right temporal epileptics were not significantly different from the normal controls. This finding might merely confirm Bear and Fedio's conclusion that patients with a right-sided focus tend to deny problems, but is also consistent with the hypothesis that left-sided brain lesions cause greater psychological disturbance than right-sided lesions.

Rodin and Schmaltz[72] used the Bear and Fedio self-rating scale to compare treated epilepsy patients, patients with psychiatric disorders, and normal controls. They found that epileptic patients scored higher than normal controls for all of the traits, but the group of patients with a variety of psychiatric diagnoses obtained even higher scores than the epileptic patients. Within the epilepsy group, those with temporal lobe epilepsy scored higher than those with diffuse spike and wave discharges for 13 of the 18 traits, but the difference was statistically significant only for one trait, hypergraphia. While these results provide only very limited support for Bear and Fedio's hypothesis that the traits in their inventory are specific to temporal lobe epilepsy, analysis of correlations of scores with anticonvulsant levels provided a potentially important finding. Total score and scores for elation, philosophical interests, sense of destiny, altered sexuality, and hypergraphia were correlated inversely with serum carbamazepine levels. This finding warrants an attempt at replication as it raises the possibility the traits measured by the Bear and Fedio scale are produced by a pathological process which can be reversed by carbamazepine.

On balance, the evidence from the studies employing the Bear and Fedio inventory suggests that the traits assessed using these scales are more common in patients with temporal lobe epilepsy than in normal individuals, but there is still debate over whether or not a combination of these traits constitutes a syndrome which is specific to temporal lobe epilepsy.

V. NEUROPHYSIOLOGICAL ABNORMALITIES IN EPILEPSY AND PSYCHIATRIC ILLNESS

A. Paroxysmal Electrical Activity and Mono-amine Neurotransmission: A Hypothesis

A clinical epileptic seizure occurs when a neuronal discharge at an epileptogenic focus spreads through the surrounding brain. The first question to consider is whether the essential abnormality lies in the excitability of individual neurons, or in the transmission between neurons. Recording from within individual neurons in some animal models of epilepsy reveals giant depolarization of the neuronal membrane, known as paroxysmal depolarization shifts.[73] Apparently normal discharges can also be recorded from the same neurons. It is possible to produce electrical activity similar to paroxysmal depolarization shifts in normal nerve cells, in vitro, by hyperpolarizing the cell such that the transmembrane potential is more negative than -65 m.[74] The resulting giant depolarization involves an inward flux of calcium ions, in contrast to the sodium influx occurring in normal depolarization.

EEG spike activity can be recorded in limbic structures in nonepileptic individuals during periods of intense emotion, including sexual orgasm.[75] It is not known whether or not the electrical events generating these spikes are paroxysmal depolarization shifts, or some other form of paroxysmal activity. Nonetheless, the evidence suggests that paroxysmal electrical activity reflects a normal mode of operation of some systems of the brain, and is manifest during periods of strong emotion. In circumstances where transmission between neurons is not adequately regulated, such paroxysmal activity might lead to epileptic seizures.

The evidence suggests that this is the case. Stevens[76] has reviewed the substantial evidence that blockade of catecholamine neurotransmitters or destruction of catecholamine terminals promotes the propagation of spike activity in the brain. For example, local instillation of 6-hydroxydopamine, which destroys catecholamine terminals, in the mid-brain tegmentum promotes the propagation of spike activity in animals previously subjected to repeated electrical stimulation in the ventral tegmental area. In primates, dopamine blocking agents such as chlorpromazine, lower the seizure threshold, while agents such as amphetamine, which augment catecholamine function, inhibit spike and wave discharges. Thus, monoamine activity appears to inhibit the propagation of epileptiform electrical activity in the brain. Conversely, seizures can produce a reduction in the numbers of various kinds of neurotransmitter receptors, including monoamine receptors.[77]

Stevens[77] addresses the issue of the possible physiological role of epileptiform discharges in the normal brain, and proposes that such discharges are involved in the maintenance of the equilibrium between excitatory and inhibitory systems. Furthermore, she explores the implications of these ideas for the association between epilepsy and psychosis, a topic outside the remit of this chapter. Post and Uhde[78] also speculate on the implications of the psychotropic effects of seizures. For example, faced with the paradox that both electroconvulsive therapy and anticonvulsant drugs, especially carbamazepine which is the treatment of choice for temporal lobe epilepsy, are effective treatments for depression, they propose that the antidepressant effect of electroconvulsive therapy might be closely related to its limbic anticonvulsant effect.

In view of the evidence, on the one hand, that electroconvulsive therapy has an antidepressant effect, and furthermore, that paroxysmal electrical activity can occur in normal individuals in association with strong emotion, while on the other hand, monoamine activity can suppress the spread of seizures, we are led to the hypothesis that the opposing interaction between paroxysmal electrical activity and monoamine activity plays a role in brain homeostasis, especially is the regulation of emotion. We will use this hypothesis as a framework to assemble the diverse evidence relevant to the underlying neurophysiological mechanism in epilepsy and neurotic disorder.

B. Pharmacological Evidence

Pharmacological evidence supports the notion that monoamine activity and paroxysmal electrical activity are intimately related, and the interaction is relevant to the manifestation of psychiatric symptoms. Drugs affecting monoamine transmission influence both mood and seizure threshold. For example, tricyclic antidepressants are the main treatment for depression and act via the 5-hydroxytryptamine and noradrenergic systems. They can precipitate EEG abnormalities in the temporal lobe and elsewhere, and can provoke generalized convulsions.[79]

However, the evidence does not support the simplistic proposal that antidepressants and anticonvulsants are inevitably antagonistic. As pointed out above, carbamazepine is both antidepressant and anticonvulsant. Amphetamine inhibits spike and wave discharges and augments monoamine function, but tends to be euphoriant rather than depressant. The relationship between monoamine function and mood is complex, and present evidence allows only the conclusion that affective disorder is associated with monoamine imbalance.

Agents such as bicuculline, which block the action of the neurotransmitter gamma-amino butyric acid (GABA), lower the seizure threshold, while agents which potentiate the action of GABA, such as benzodiazepines, are effective anticonvulsants. Benzodiazepines are also effective against anxiety, and despite the synthesis of many drugs of this class, it has not yet proved possible to segregate the anxiolytic, anticonvulsant, and hypnotic properties of these drugs.

C. Sleep and Sleep Deprivation

The neurophysiological processes associated with sleep are closely related to the manifestation of paroxysmal electrical activity and to disturbances of affect. Sleep and also sleep deprivation are potent means of eliciting or enhancing focal epileptiform discharges.[80] Sleep deprivation has a marked, but transient, antidepressant effect[81] but exacerbates panic attacks.[80] These observations suggest that sleep deprivation might produce episodes of panic accompanied by epileptiform EEG activity in patients with a liability to panic attacks. The current evidence is equivocal. Brodsky et al.[82] reported that 24 hr of sleep deprivation revealed temporal lobe epileptiform foci in 10 patients suffering from attacks of anxiety. In contrast, Roy-Byrne and his colleagues[80] found no evidence of abnormal EEG activity recorded via naso-pharyngeal leads after sleep deprivation in 12 patients with panic disorder.

D. Psychopathological Features Common to Affective Disorder and Epilepsy

The hypothesis that the interaction between paroxysmal electrical activity and monoamine activity plays a role in brain homeostasis predicts that patients with affective disorder and the patients with epilepsy will be prone to instability of the balance between monoamine activity and paroxysmal electrical activity, and hence patients with either disorder might suffer from symptoms reflecting this instability.

Those cases in which the source of the instability resulted in an excessive production or spread of paroxysmal depolarization, would present with epileptic seizures, but in addition would be expected to suffer affective symptoms such as depression or anxiety as a result of compensatory adjustments by monoamine neurons. Those in whom the lesion affected the monoamine system directly would be more likely to present with affective illness, but nonetheless, would be expected to suffer episodic phenomena reflecting inadequately controlled paroxysmal electrical activity. We would expect these phenomena to take the form of abrupt but transient alterations in the content or level of consciousness, of the type which occur in seizures.

Ulett and colleagues[83] found that anxiety-prone individuals respond to repetitive photic stimulation with EEG waveforms richer in higher harmonics, and are particularly liable to report strange or unpleasant experiences such as dizziness, depersonalization, and illusions during photic stimulation.

Silberman et al.[84] investigated the frequency of occurrence of 42 different transient cognitive, perceptual, and affective phenomena in 44 patients with affective illness, 37 patients, with temporal lobe epilepsy, and a control group of 30 hypertensive patients. The group with affective illness and the group with temporal lobe epilepsy both reported experiencing significantly more of the phenomena than did the hypertensive controls. The patients with affective illness reported having suffered a mean of 7 out of the 42 phenomena during episodes of illness, and a mean of 3 of the phenomena during intervening periods. The epileptic patients reported a mean of 8 of the phenomena during seizures and 3 of the phenomena during intervening periods. In contrast, the mean number of phenomena reported by a group of 30 hypertensive controls was 2. There were 12 phenomena which were common in both epileptic and affective patients. These were: illusions of significance, jumbled thoughts, altered sound intensity, altered odor intensity, formed auditory hallucinations, altered color intensity, olfactory hallucinations, derealization, metamorphosia, altered distance, epigastric hallucinations, and amnestic episodes.

When Silberman and his colleagues analyzed the correlations between phenomena, they found no significant correlations between phenomena reported by the control group, but for both the affective group and the epileptic group there were strong correlations between reported phenomena. These correlations between phenomena suggest the possibility that specific groups of phenomena arise from definable underlying mechanisms. Factor analysis of the scores for the 12 phenomena which were common in both affective and epileptic patients revealed a factor structure that was similar for the two groups. The similarity of the factor structure in the affective and epileptic patient groups suggests similarities of the underlying mechanisms.

On the other hand, there were substantial differences between affective and epileptic patients in frequency of reporting some phenomena. For example, speeded thoughts, slow thoughts, and sudden, intense, unexplained sexual sensations, as well as various sudden, intense, unexplained emotions such as fear, depression, and pleasure, were more common in affective patients than in patients with temporal lobe epilepsy. On the other hand, time disorientation, vestibular hallucinations, and gustatory hallucinations were more common in the epileptic patients than the affective patients.

Silberman's data are consistent with the hypothesis that epileptic patients and patients with affective disorder are prone to instability of an equilibrium between monoamine activity and paroxysmal electrical activity. While the similarities in the patterns of experiences reported by epileptic and affective patients support the proposal that the phenomena arise from instability in the same regulatory network, differences between the reported frequencies of some of the phenomena would be expected on the grounds that the presumed causes of the instability differ between the different cases.

Furthermore, the hypothesis that monoamine activity and paroxysmal electrical activity play opposing roles in the regulation of emotion leads to the prediction of a sequential relationship between dysphoria and the phenomena reflecting paroxysmal electrical activity. A possible example of such a sequential relationship is the anxiety-depersonalization syndrome described by Roth,[85] in which an episode of intense anxiety culminates with an experience of depersonalization. There is no direct evidence that depersonalization occurring in association with anxiety is a manifestation of paroxysmal electrical activity, but the fact that depersonalization can occur in psychomotor seizures is consistent with the proposal that, at least under some circumstances, it is a manifestation of paroxysmal electrical activity.

E. EEG Concomitants of Psychiatric Disturbance

The most direct way in which to investigate the hypothesis that paroxysmal electrical activity plays a role in the regulation of emotion is to study the temporal relationship between mood and EEG activity, in both epileptic and nonepileptic subjects. Recording from electrodes placed within limbic structures would offer the best prospect of recording the relevant paroxysmal electrical activity. Heath[17] has reported that spike activity occurs in the amygdala of nonepileptic individuals in association with irritability. Could this reflect a compensatory mechanism acting to regulate the irritability? In the case of the patient reported by Smith,[16] rage appeared to generate a seizure which alleviated the patient's former irritability. However, an overt seizure was not necessary for this effect, because Smith reports that the patient learned to alleviate his chronic irritability by controlled release of anger. Conversely, it is possible that the rare paradoxical increase in irritability produced by benzodiazepines, which are anticonvulsants, arises from the inhibition of paroxysmal electrical activity.

Dongier's[86] compilation of clinical and electroencephalographic manifestations of 536 psychotic episodes occurring between clinical seizures, in 516 epileptic patients (presented at a colloquium in Marseilles in 1956) provides a major source of information about EEG abnormalities recorded concurrently with the experience of psychiatric symptoms. In 39% of the patients, depression occurred during the psychotic episode, but only 14% of those with bisynchronous spike and wave discharges during the episode were depressed. In contrast, patients who lost their pre-existing focal (particularly temporal) discharges during the psychotic episode were frequently depressed. This phenomenon is analogous to the phenomenon of forced normalization described by Landholt, and defined by Wolf[87] as a condition where disappearance or quasi-disappearance of an epileptic EEG discharge is accompanied by the development of psychiatric symptomatology. Anxiety was reported during the psychotic episode in 46% of the patients, but 28 out of 29 cases in which the episode was accompanied by bisynchronous spike and wave discharges showed no anxiety. On the other hand, there was a trend for anxiety to be more common in those in whom previously recorded discharges disappeared during the episode. Anxiety was reported in 60% of such cases.

Dongier's inclusion of episodes in which bisynchronous spike and wave discharges accompanied an alteration in the content of consciousness, in a study of "psychotic episodes occurring between seizures" is potentially misleading, because such states are generally regarded to be seizures, and could be classified as absence status. However, the observation that bisynchronous spike and wave activity tends to exclude simultaneous depression or anxiety remains relevant to our present hypothesis irrespective of whether such episodes are classified as seizures or not. Perhaps a more important reason for caution in the interpretation of Dongier's findings in the cases with bisynchronous spike and wave activity is the fact that many of these 29 cases were children.

F. Regional Cerebral Metabolism

Positron Emission Tomography (PET) studies of regional cerebral energy metabolism provide indirect evidence supporting the hypothesis that paroxysmal electrical activity is one component of a regulatory system involving the interplay of mutually opposing processes. Engel et al.[88] have demonstrated local regions of decreased glucose metabolism, between seizures, in the brains of patients with focal seizures, and Mazziotta and Engel[89] have presented evidence that regions which show interictal hypometabolism become hypermetabolic during seizures. Furthermore, Gur et al.[90] demonstrated that focal interictal hypometabolism resolved postoperatively in a patient with the Lennox Gastaut syndrome whose seizures were relieved by resection of the corpus callosum. Thus, the interictal hypometabolism observed in patients with focal seizures appears to be due to a reversible process. If the PET observations are relevant to the hypothesis that paroxysmal electrical activity plays a part in the regulation of emotion, we would predict that patients with depression would also have reversible regional cerebral hypometabolism. Several studies[91,92] have reported regional hypometabolism in patients with affective disorder.

Thus, current evidence provides at least some support for the hypothesis that the opposing interaction between paroxysmal electrical activity and monoamine activity plays a role in the regulation of emotion. However, many details of the proposed regulatory system remain speculative. In particular, there is little evidence to suggest which neurons exhibit paroxysmal electrical activity, or with which particular monoamine neurons they interact. In principle, longitudinal studies during which mental state, EEG activity and regional cerebral metabolism are monitored might clarify these issues.

VI. CONCLUSIONS

While many epileptic patients show no sign of significant psychiatric disturbance between seizures, a substantial proportion do exhibit psychiatric symptoms. Depression, anxiety, and suicide are all more common in epileptic patients than in the general population. Sexual dysfunction, especially hyposexuality, is more common in those with temporal lobe epilepsy than in the general population. Aggressive behavior is relatively common in series of epileptic patients biased towards severe, intractable illness with overt evidence of brain damage, but much less common in samples representative of cases referred to an epilepsy service. The balance of evidence suggests that the clinical concept of a specific epileptic personality type does reflect some real peculiarity occurring is a proportion of epileptic patients, but the exact specification of the abnormalities of personality, and the characteristics of the epileptic illness in which the abnormalities occur, remain uncertain.

With regard to the comparison between patients with temporal lobe epilepsy and those with other forms of epilepsy, much inconclusive evidence points toward a higher prevalence of psychopathology of various kinds in those with temporal lobe epilepsy, but hyposexuality is the only abnormality for which the majority of the evidence suggests a specific association.

The question of whether the pathological process underlying epilepsy itself plays a causal role in interictal psychopathology remains unanswered. It is clear that the presence of a readily identifiable brain lesion is one of the strongest risk factors for psychiatric disorder in epileptic patients, implying that a substantial part of the disorder arises from the presence of the lesion which causes the epilepsy rather from the epilepsy itself. There is evidence that the occurrence of multiple types of seizures, or the presence of multiple independent spikes on EEG, and age of onset at or before puberty, increase the likelihood of psychiatric disturbance.

Despite the paucity of evidence for a specific relationship between the pathological processes underlying epilepsy and psychiatric symptoms, there is a consistent thread of evidence hinting at a homeostatic balance between paroxysmal electrical activity and activity is monoamine system subserving emotion. In epileptic patients, it might be predicted that this equilibrium would be unstable, making them prone to alternating mood disturbance and seizures.

REFERENCES

1. **Robertson, M. M. and Trimble, M. R.,** Depressive illness in patients with epilepsy: a review, *Epilepsia,* 24 (suppl. 2), S109, 1983.
2. **Hill, D.,** Historical review, in *Epilepsy and Psychiatry,* Reynolds, E. H. and Trimble, M. R., Eds., Churchill Livingstone, Edinburgh, 1981, chap. 1.
3. **Lennox, W. G.,** Brain injury, drugs and environment as causes of decay in epilepsy, *Am. J. Psychiatry,* 99, 174, 1942.
4. **Stevens, J. R.,** Epilepsy and psychosis: neuropathological studies of six cases, in *Aspects of Epilepsy and Psychiatry,* Trimble, M. R. and Bolwig, T. G., Eds., Wiley, Chichester, 1986, 117.
5. **Bear, D. M.,** Behavioural changes in temoral lobe epilepsy, in *Aspects of Epilepsy and Psychiatry,* Trimble, M. R. and Bolwig, T. G., Eds., Wiley, Chichester, 1986, 19.

6. **Walton, J.**, *Brain's Diseases of the Nervous System*, 9th ed., Oxford University Press, Oxford, 1985, 609.
7. **Heath, R. G.**, Sub-cortical brain function correlates of psychopathology and epilepsy, in *Psychopathology and Brain Dysfunction*, Shagass, C., Gershon, S. and Friedhoff, A. J., Eds., Raven Press, New York, 1977, 51.
8. **Falconer, M. A.**, Reversibility by temporal lobe resection of the behavioural abnormalities of temporal lobe epilepsy, *N. Engl. J. Med.*, 289, 451, 1973.
9. **Hermann, B. P., Dickman, S., Schwartz, M. S., and Karnes, W. E.**, Interictal psychopathology in patients with ictal fear: a quantitative investigation, *Neurology*, 32, 7, 1982.
10. **Stark-Adamec, C., Adamec, R. E., Graham, J. M., Hicks, R. C., and Bruun-Meyer, R. C.**, Complexities in the complex partial seizures personality controversy, *Psychiatr. J. Univ. Ottawa*, 10, 231, 1985.
11. **Matsumoto, H., Ayala, F. F., and Gumnit, R. J.**, Neuronal behaviour and triggering mechanisms in cortical epileptic focus, *J. Neurophysiol.*, 32, 688, 1969.
12. **Riley, T. and Niedermeyer, E.**, Rage attacks and episodic violent behaviour: encephalographic findings and general consideration, *Clin. Encephalogr.*, 9, 113, 1978.
13. **Stevens, J. R., Bigalow, L., Denney, D., Lipkin, J., Livermore, A., Rauscher, F., and Wyatt, R. J.**, Telemetered EEG-EOG during psychotic behaviours of schizophrenia, *Arch. Gen. Psychiatry*, 36, 251, 1979.
14. **Cavazzuti, G. B., Capella, L., and Nalin, A.**, Longitudinal study of epileptiform EEG patterns in normal children, *Epilepsia*, 21, 43, 1980.
15. **Stevens, J. R.**, Risk factors for psychopathology in individuals with epilepsy, *Adv. Biol. Psychiatry*, 8, 56, 1982.
16. **Smith, J. S.**, Episodic rage, in *Limbic Epilepsy and the Dyscontrol Syndrome*, Girgis, M. and Kiloh, L. G., Eds., Elsevier, Amsterdam, 1980, 255.
17. **Heath, R. G.**, Psychosis and epilepsy: similarities and differences in the anatomic-physiologic substrate, *Adv. Biol. Psychiatry*, 8, 106, 1982.
18. **Pond, D. A. and Bidwell, B. H.**, A survey of epilepsy in 14 general practices. II. social and psychological aspects, *Epilepsia*, 1, 285, 1959.
19. **Gudmundsson, G.**, Epilepsy in Iceland. A clinical and epidemiological investigation, *Acta Neurol. Scand.*, 43, suppl. 25, 1966.
20. **Rutter, M., Graham, P., and Yule, W.**, *A Neuropsychiatric Study in Childhood*, Heinemann, London, 1970.
21. **Edeh, J. and Toone, B. J.**, Anti-epileptic therapy, folate deficiency and psychiatric morbidity: a general practice survey, *Epilepsia*, 26, 434, 1985.
22. **Currie, S., Heathfield, K. W. G., Henson, R. A., and Scott, D. F.**, Clinical course and prognosis of temporal lobe epilepsy. A survey of 666 patients, *Brain*, 94, 173, 1971.
23. **Betts, T. A.**, A follow-up study of a cohort of patients with epilepsy admitted to psychiatric care in an English city, in *Epilepsy: Proceedings of the Hans Berger Centenary Symposium*, Harris, P. and Mawdsley, C., Eds., Churchill Livingstone, Edinburgh, 1974, 326.
24. **Standage, K. F. and Fenton, G. W.**, Psychiatric symptom profiles of patients with epilepsy. A controlled investigation, *Psychol. Med.*, 5, 152, 1975.
25. **Trimble, M. R. and Perez, M. M.**, Quantification of psychopathology in adult patients with epilepsy, in *Epilepsy and Behaviour, 1979*, Kulig, B. M., Meinardi, H., and Stores, G., Eds., Swets and Zeitlinger, Lisse, 1980, 118.
26. **Kogeorgos, J., Fonagy, P., and Scott, D. F.**, Psychiatric symptom patterns of chronic epileptics attending a neurological clinic: A controlled investigation, *Br. J. Psychiatry*, 140, 236, 1982.
27. **Hermann, B. P.**, Psychopathology in epilepsy and learned helplessness, *Med. Hypothesis*, 5, 723, 1979.
28. **Seligman, M. E. P.**, *Helplessness*, Freeman, San Francisco, 1975.
29. **Reynolds, E. H., Milner, G., Matthews, D. M., and Chanarin, I.**, Anticonvulsant therapy, megaloblastic haemopoiesis and folic acid metabolism, *Q. J. Med.*, 35, 521, 1966.
30. **Snaith, R. P., Mehta, S., and Raby, A. H.**, Serum folate and vitamin B_{12} in epileptics with and without mental illness, *Br. J. Psychiatry*, 116, 179, 1970.
31. **Trimble, M. R., Corbett, J. A., and Donaldson, D.**, Folic acid and mental symptoms in children with epilepsy, *J. Neurol. Neurosurg. Psychiatry*, 43, 1030, 1980.
32. **Robertson, M. M.**, Ictal and interictal depression in patients with epilepsy, in *Aspects of Epilepsy and Psychiatry*, Trimble, M. R. and Bolwig, T. G., Eds., Wiley, Chichester, 1986, 213.
33. **Shorvon, S. D. and Reynolds, E. H.**, Reduction of polypharmacy for epilepsy, *Br. Med. J.*, 2, 1023, 1979.
34. **Sillanpaa, M.**, Carbamazepine, pharmacology and clinical uses, *Acta Neurol. Scand.*, 64, suppl. 88, 1981.
35. **Post, R. M.**, The use of the anticonvulsant carbamazepine in primary and secondary affective illness: clinical and theoretical implications, *Psychol. Med.*, 12, 701, 1982.
36. **Flor-Henry, P.**, Psychosis and temporal lobe epilepsy. A controlled investigation, *Epilepsia*, 10, 363, 1969.

37. **Perini, G. and Mendius, R.,** Depression and anxiety in complex partial seizures, *J. Nerv. Ment. Dis.,* 172, 287, 1984.

38. **Barraclough, B.,** Suicide and epilepsy, in *Epilepsy and Psychiatry,* Reynolds, E. H. and Trimble, M. R., Eds., Churchill Livingstone, Edinburgh, 1981, 76.

39. **White, S. J., McClean, A. E. M., and Howland, C.,** Anti-convulsant drugs and cancer (a cohort study in patients with severe epilepsy), *Lancet,* 2, 458, 1979.

40. **Taylor, D. C. and Marsh, S. M.,** Implications of long term follow-up studies in epilepsy: with a note on the cause of death, in *Epilepsy: the Eighth International Symposium,* Penry, J. K., Ed., Raven Press, New York, 1977, 27.

41. **Mackay, A.,** Self-poisoning — a complication of epilepsy, *Br. J. Psychiatry,* 134, 277, 1979.

42. **Hawton, K., Fagg, J., and Marsack, P.,** Association between epilepsy and attempted suicide, *J. Neurol. Neurosurg. Psychiatry,* 43, 168, 1980.

43. **Guerrant, J., Anderson, W., Fischer, A., Weinstein, M., and Jaros, R. M.,** *Personality in Epilepsy,* Charles C Thomas, Springfield, Ill., 1962.

44. **Shukla, G. D., Srivastava, O. N., and Katiyar, B. C.,** Sexual disturbances in temporal lobe epilepsy, *Br. J. Psychiatry,* 109, 95, 1979.

45. **Taylor, D. C.,** Sexual behaviour and temporal lobe epilepsy, *Arch. Neurol.,* 21, 510, 1969.

46. **Blumer, D. and Walker, A. E.,** Sexual behaviour in temporal lobe epilepsy, *Arch. Neurol.,* 16, 37, 1967.

47. **Toone, B. K., Wheeler, M., Nanjee, M., Fenwick, P. B. C., and Grant, R.,** Sex hormones, sexual drive and plasma anticonvulsant levels in male epileptics, *J. Neurol. Neurosurg. Psychiatry,* 46, 824, 1983.

48. **Blumer, D.,** Hypersexual episodes in temporal lobe epilepsy, *Am. J. Psychiatry,* 126, 1099, 1970.

49. **Koella, W. P.,** The functions of the limbic System — evidence from animal experimentation, *Adv. Biol. Psychiatry,* 8, 12, 1982.

50. **Delgado-Escueta, A. V., Mattson, R. H., King, L., Goldensohn, E. S., Spiegel, H., Masden, J., Crandall, P., Dreifuss, F., and Porter, R. J.,** Special report. The nature of aggression during epileptic seizures, *N. Engl. J. Med.,* 305, 711, 1981.

51. **Serafetinides, E. A.,** Aggressiveness in temporal lobe epileptics and its relation to cerebral dysfunction and environmental factors, *Epilepsia,* 6, 33, 1965.

52. **Ounstead, C.,** Aggression and epilepsy rage in children with temporal lobe epilepsy, *J. Psychom. Res.,* 13, 237, 1959.

53. **Rodin, E.,** Psychomotor epilepsy and aggressive behaviour, *Arch. Gen. Psychiatry,* 28, 210, 1973.

54. **Hermann, B. P., Schwartz, M. S., Whitman, S., and Karnes, W.,** Aggression in epilepsy: seizure type comparisons and high risk variables, *Epilepsia,* 22, 691, 1980.

55. **Huesmann, L. R., Lefkowitz, M. M., and Eron, L. D.,** Sum of MMPI scales F, 4, and 9 as a measure of aggression, *J. Consult. Clin. Psychology,* 46, 1071, 1978.

56. **Brown, S. W., McGowan, M. E. L., and Reynolds, E. H.,** The influence of seizure type and medication on psychiatric symptoms in epileptic patients, *Br. J. Psychiatry,* 148, 300, 1986.

57. **Mungas, D.,** An empirical analysis of specific syndromes of violent behaviour, *J. Nerv. Ment. Dis.,* 171, 354, 1983.

58. **Gibbs, F. A. and Gibbs, E. L.,** *Atlas of Encephalography,* Vol. 2 and 3, Addison-Wesley, Cambridge, Mass., 1984.

59. **Hermann, B. P. and Stevens, J. R.,** Interictal behavioural correlates of the epilepsies, in *Multidisciplinary Handbook of Epilepsy,* Charles C Thomas, Springfield, Ill., 1980, 272.

60. **Meier, M. J. and French, L. A.,** Some personality correlates of unilateral and bilateral EEG abnormalities in psychomotor epilepsy, *J. Clin. Psychology,* 21, 3, 1965.

61. **Herman, B. P., Schwartz, M. S., Karnes, W., and Vadhat, P.,** Psychopathology in epilepsy: relationship of seizure type to age at onset, *Epilepsia,* 21, 15, 1980.

62. **Hughes, J. R. and Hermann, B. P.,** Evidence for psychopathology in patients with rhythmic midtemporal discharges, *Biol. Psychiatry,* 19, 1623, 1984.

63. **Bear, D. M. and Fedio, P.,** Quantitative analysis of interictal behaviour in temporal lobe epilepsy, *Arch. Neurol.,* 34, 454, 1977.

64. **Waxman, S. G. and Geschwind, N.,** The interictal behaviour syndrome of temporal lobe epilepsy, *Arch. Gen. Psychiatry,* 32, 1580, 1975.

65. **Hermann, B. P. and Reil, P.,** Interictal personality and behavioural traits in temporal lobe and generalised epilepsy, *Cortex,* 17, 125, 1981.

66. **Mungas, D.,** Interictal behaviour abnormality in temporal lobe epilepsy, *Arch. Gen. Psychiatry,* 39, 108, 1982.

67. **Bear, D. M.,** Behavioural symptoms in temporal lobe epilepsy (letter), *Arch. Gen. Psychiatry,* 40, 467, 1983.

68. **Mungas, D.,** Behavioural symptoms in temporal lobe epilepsy (reply to letter), *Arch. Gen. Psychiatry,* 40, 468, 1983.

69. **Bear, D. M., Levin, K., Blumer, D., Chetham, D., and Reider, J.,** Interictal behaviour in hospitalized temporal lobe epileptics: relationship to idiopathic psychiatric syndromes, *J. Neurol. Neurosurg. Psychiatry*, 45, 481, 1982.

70. **Master, D. R., Toone, B. K., and Scott, D. F.,** Interictal behaviour in temporal lobe epilepsy, in *Advances in Epileptology: XVth Epilepsy International Symposium*, Porter, R. J., Ed., Raven Press, New York, 1984, 557.

71. **Brandt, J., Seidman, L. J., and Kohl, D.,** Personality characteristics of epileptic patients: a controlled study of generalized and temporal lobe cases, *J. Clin. Exp. Neuropsychol.*, 7, 25, 1985.

72. **Rodin, E. and Schmaltz, S.,** The Bear-Fedio personality inventory and temporal lobe epilepsy, *Neurology*, 34, 591, 1984.

73. **Matsumoto, H. and Ajmone-Marsan, C.,** Cortical cellular phenomena in experimental epilepsy: ictal manifestations, *Exp. Neurol.*, 9, 305, 1984.

74. **Llinas, R. and Jahnsen, H.,** Electrophysiology of mammalian thalamic neurones in vitro, *Nature*, 297, 406, 1982.

75. **Heath, R. G.,** Pleasure and brain activity in man: deep, and surface electroencaphalograms during orgasm, *J. Nerv. Ment. Dis.*, 154, 3, 1972.

76. **Stevens, J. R.,** All that spikes is not fits, in *Psychopathology and Brain Dysfunction*, Shagass, C., Gershon, S., and Friedman, A. H., Eds., Raven Press, New York, 1977, 183.

77. **Stevens, J. R.,** Epilepsy, personality, behaviour and psychopathology — the state of the evidence and directions for future research and treatment, *Folia Psychiatr. Neurol. Japon.*, 37, 203, 1983.

78. **Post, R. M. and Udhe, T. W.,** Anticonvlsants in non-epileptic psychosis, in *Aspects of Epilepsy and Psychiatry*, Trimble, M. R. and Bolwig, T. G., Eds., Wiley, Chichester, 1986, 177.

79. **Trimble, M. R.,** Non-MAOI antidepressants and epilepsy, *Epilepsia*, 19, 241, 1978.

80. **Roy-Byrne, P. P., Uhde, T. W., and Post, R. M.,** Effects of one night's sleep deprivation on mood and behaviour in panic disorder, *Arch. Gen. Psychiatry*, 43, 895, 1986.

81. **Barnes, T. R. E.,** Miscellaneous treatments, in *Treatment and Management in Adult Psychiatry*, Berrios, G. E. and Dowson, J. H., Bailliere Tindall, London, 1983, 243.

82. **Brodsky, L., Zuniga, J. S., Casenas, E. R., Ernstoff, R., and Sachdev, H. S.,** Refractory anxiety: a masked epileptiform disorder? *Psychiatr. J. Univ. Ottawa*, 8, 42, 1983.

83. **Ulett, G. A., Gleser, G., Winokur, G., and Lawler, A.,** The EEG and reaction to photic stimulation as an index of anxiety-proneness, *Electroencephalogr. Clin. Neurophsyiol.*, 5, 23, 1953.

84. **Silberman, E. K., Post, R. M., Nurnberger, J., Theodore, W., and Boulenger, J-P.,** Transient sensory, cognitive and affective phenomena in affective illness, *Br. J. Psychiatry*, 146, 81, 1985.

85. **Roth, M.,** The phobic anxiety-depersonalization syndrome, *Proc. R. Soc. Med.*, 52, 587, 1959.

86. **Dongier, S.,** Statistical study of clinical and electroencephalographic manifestations of 536 psychotic episodes occurring in 516 epileptics between clinical seizures, *Epilepsia*, 1, 117, 1959.

87. **Wolf, P.,** Forced normalization, in *Aspects of Epilepsy and Psychiatry*, Trimble, M. R. and Bolwig, T. G., Eds., Wiley, Chichester, 1986, 101.

88. **Engel, J., Brown, W. J., Kuhl, D. E., Phelps, M. E., Mazziotta, J. C., and Crandall, P. H.,** Pathological findings of focal temporal hypometabolism in partial epilepsy, *Ann. Neurol.*, 12, 518, 1982.

89. **Mazziotta, J. C. and Engel, J.,** The use and impact of positron computed tomography scanning in epilepsy, *Epilepsia*, 25 (suppl 2), s86, 1984.

90. **Gur, R. C., Sussman, N. M., Alavi, A., et al.,** Positron emission tomography in two cases of childhood epileptic encephalopathy, *Neurology*, 32, 1191, 1982.

91. **Baxter, L. R., Phelps, M. E., Mazziotta, J. C., Schwartz, J. M., Gerner, R. H., Selin, C. E., and Sumida, R. M.,** Cerebral Metabolic Rates for Glucose in Mood Disorders, *Arch. Gen. Psychiatry*, 42, 441, 1985.

92. **Buchsbaum, M. S., Cappelletti, J., Ball, R., Hazlett, E., King, A. C., Johnson, J., Wu, J., and DeLisi, L. E.,** Positron emission tomographic image measurement in schizophrenia and affective disorders, *Ann. Neurol.*, 15 (suppl), S157, 1984.

93. **Helgason, T.,** Epidemiology of mental disorders in Iceland, *Acta Psychiatr. Scand.*, Suppl. 173, 1964.

Chapter 8

NEUROSIS AND NEUROLOGICAL DISORDER

Daniel Rogers

TABLE OF CONTENTS

I. INTRODUCTION

In 1936, Comroe[1] carried out a follow-up study of 100 patients who had received a diagnosis of neurosis. He found that within an average follow-up period of eight months, 24% now had definite evidence of physical disorder. He suggested that this was not a chance association, and that neurotic symptoms could either precede or follow the appearance of physical symptoms of different physical disorders. The 24 patients who showed evidence of physical disorder on follow-up, had between them 21 different physical disorders. None of these were primary neurological disorders. Following Comroe, different investigators[2-11] found a prevalence of physical disorder among different populations of psychiatric patients ranging from 9 to 80% (median 43%). No particular psychiatric diagnosis was especially associated with physical disorder and no particular group of physical disorders was particularly represented, the number of different physical disorders found in different studies often approaching the total number of patients found with physical disorder. Of the different physical disorders reported, only 18% were neurological disorders. The commonest of these were different manifestations of cerebrovascular disease, epilepsy, and intracranial tumors, these three categories accounting for more than 50% of the cases reported.

Such studies, while emphasizing the association of psychiatric disorder and physical disorder, do not suggest any particular association of psychiatric disorder with neurological disorder, or that any such association is more than nonspecific. When a different approach is taken, however, and the psychiatric disorder associated with specific neurological disorders is considered, a different picture emerges. Davison and Bagley,[12] by means of a comprehensive literature reivew, were able to show that there was an association of schizophreniform psychosis and a wide range of neurological disorders. They concluded that this was not a chance association and that in most cases the neurological disorder appeared to be a necessary cause of the psychiatric disorder. Although not as well established, the same is probably true of the association of neurotic disorder with many neurological disorders, and whereas the prevalence of schizophrenia in those neurological disorders where it occurs is relatively rare, that of neurotic disorder is relatively common.

II. HUNTINGTON'S CHOREA

Whittier,[13] reviewing the literature on the psychopathology of Huntington's disease, concluded that it is almost as characteristic for the disease as the chorea. Huntington, in his original report,[14] included a tendency to insanity and suicide as one of the three marked peculiarities of the disease. Hughes,[15] in one of the first comprehensive studies of the associated psychiatric disorder in 1925, found that early temperamental traits, including excitability, restlessness, and discontent, had been present in 79% of 218 cases later developing Huntington's disease, and that after the onset of neurological features, 70% showed behavioral disturbance, including seclusiveness, suspiciousness, stealing, assaultiveness, suicidal attempts, alcoholism, and drug addiction. Parker,[16] in a later study, maintained that every variety of psychiatric syndrome could be observed in association with Huntington's disease. Five studies,[13,17-20] which included altogether 755 cases of Huntington's disease, showed that the earliest and most numerous symptoms of the condition were psychiatric. The commonest of these were personality change, and depressive and anxiety symptoms; in contrast, phobic, hysterical, and obsessive-compulsive disorders were rare.

Mindham and colleagues[21] recently studied matched groups of patients suffering from Huntington's disease and Alzheimer's disease, and compared them for psychiatric morbidity. The patients with Alzheimer's disease were randomly selected from an outpatient clinic. They were matched by sex with patients with Huntington's disease and dementia, taken from a register covering a defined geographical area, with only one affected member, the eldest, from any affected family being chosen. For the purposes of the study, only a history of psychiatric disorder before the onset of dementia in either condition was considered. Diagnoses were made according to the categories of DSM III. There were 27 patients in each group. In the Huntington's group, 16 patients (59%) received psychiatric diagnoses: 12 had a diagnosis of major affective disorder, 5 of anxiety disorder of childhood, 4 of substance or alcohol abuse, 1 of dysthymic disorder, 1 of phobic disorder, and 1 of antisocial personality disorder. In the Alzheimer's group 8 patients (30%) received psychiatric diagnoses: 6 of major affective disorder, 1 of dysthymic disorder, and 1 of alcohol abuse. The occurence of psychiatric illness in both neurological disorders was high, and a specific association of Huntington's disorder with affective disorder was suggested.

III. TOURETTE SYNDROME

This rare syndrome is defined by the DSM III criteria of an age of onset between 2 and 15 years, the presence of recurrent, involuntary repetitive, rapid, purposeless movements affecting multiple muscle groups, multiple vocal tics, ability to suppress the movements voluntarily for minutes to hours, variations in the intensity of the symptoms over weeks or months, and duration of more than one year. Associated features are coprolalia and echo phenomena.

Different authors have examined the associated psychiatric disorder of patients with Tourette syndrome, and these studies have been comprehensively reviewed by Robertson.[22] The first report[23] of associated psychiatric disorder appeared within a year of Gilles de la Tourette's original description of the disorder, and he himself[24] noted that the family history of affected patients was almost invariably "loaded for nervous disorder". This was confirmed in a modern study[25] which showed that 70% of 30 first-degree relatives of 15 patients with Tourette syndrome satisfied Feighner criteria for psychiatric illness, the most common diagnoses being unipolar depression, obsessive-compulsive illness, and panic disorder. Some modern investigators, e.g., Shapiro and Shapiro[26] with a personal experience of over 1000 patients, have not found any association with specific psychiatric syndromes. Morphew and Sim,[27] however, using a literature search involving reports on 43 patients, showed that a significant number of patients with Tourette syndrome had evidence of childhood neurotic traits, phobic features, and personality disorders, including hysterical, psychopathic, paranoid, and hysterical varieties. Robertson and her colleagues[28] have used a battery of standardized rating scales to assess the psychiatric disorder of 54 adult patients from a study group of 90 consecutively referred adults and children with Tourette syndrome. The mean depression scores of the patients were higher than those of normal populations, and this was unrelated to medication. Other features of the Tourette patients were high scores for anxiety and hostility, the hostility being associated with coprolalia, one of the major features of the disorder.

Many authors have reported an association with obsessional disorder. Obsessional symptoms, traits, or illness have been reported with a prevalence of from 11 to 80% in different studies.[25,29-34,27,35,36,37] Robertson found that 37% of her group of 90 patients reported obsessive-compulsive behavior and that rating scales for obsessionality showed much higher scores in the Tourette patients compared to normals.[28] These scores were associated with coprolalia and echo phenomena. Frankel and his colleagues[38] compared patients with Tourette syndrome, patients with obsessive-compulsive disorder diagnosed according to DSM III criteria, and normal controls with a specially designed inventory. Of the Tourette patients, 51% achieved a score indicative of obsessive-compulsive disorder, as did 12% of the normal controls. The highest score of any patient was from a Tourette patient. Cummings and Frankel[39] have examined the similarities between Tourette syndrome and obsessive-compulsive disorder, including age of onset, life-long course, waxing and waning of symptoms, involuntary and intrusive, "ego-alien" behavior and experience, occurrence in the same families, and worsening with depression and anxiety. There is also an increased prevalence of obsessive-compulsive disorder among families of patients with Tourette syndrome. Some authors[40] go as far as to regard Tourette syndrome as an obsessive-compulsive disorder.

IV. MULTIPLE SCLEROSIS

The psychiatric aspects of Multiple Sclerosis (MS) have been well reviewed by Trimble and Grant.[41] An associated psychopathology was included in the first definitive description of Multiple Sclerosis by Charcot in the 19th century.[42] One of the first comprehensive accounts of the psychiatric associations of the condition was that of Brown and Davis,[43] who in 1922 noted mental alterations in 90% of their cases. In a later study of 75 patients, Braceland and Giffin[44] described a significant prevalence of mood disorder, including depression, lability of mood, euphoria, and irritability. Pratt[45] compared the psychiatric disorder of 100 patients with MS with that of 100 patients with other central nervous system disorders, and confirmed the increased prevalence of mood disorder, with inappropriate cheerfulness and lability of mood being prominent features; hysteria and anxiety, by contrast, were not a major manifestation in MS patients. Studies with the Minnesota Multiphasic Personality Inventory showed elevations of the so-called "neurotic triad" — hypochondriasis, depression, and hysteria — in patients with MS, although this is also found in other central nervous system disorders.[46,47]

More recently, Surridge[48] compared 108 patients with MS with 39 patients with muscular dystrophy. Depression was present in 18% of the MS patients, being severe in 4%, and in 13% of the control group; euphoria was present in 26% of the MS patients and none of the controls. Whitlock and Siskind[49] compared 30 MS patients without dementia and a control group of patients with other chronic or progressive neurologic syndromes, matched for sex and age. The two groups were matched for general disability, and there was a correlation between the degree of depression and disability in both groups, but the MS patients had significantly higher depression scores. In this and other studies, depressive disorder could precede the onset of neurologic disabilities, suggesting that, in some cases, depression could in fact be one of the first signs of MS.

Schiffer and Babigian[50] used a computer search technique to identify all in-patients of one hospital who had received diagnoses of Multiple Sclerosis, Temporal Lobe Epilepsy (TLE) and Amyotrophic Lateral Sclerosis (ALS) over a 13-year period. These patients were matched against a psychiatric register of patients from the same geographical area. The prevalence for psychiatric contact for the 368 MS patients was 71 (19%), and for the 402 TLE patients it was 92 (23%). On the other hand, for the 124 ALS patients, who suffered from a disease with a particularly ominous prognosis, it was only 6 (5%), and in 5 of these the psychiatric contact had been long before the onset of the ALS. In the MS patients, 62% of the register matches were for depressed affective disorder, 42% for personality disorder, 34% for anxiety disorder, 25% for transient situational disturbance, and 4% for alcoholism. In the TLE patients, 48% of the register matches were for personality disorder, 40% for depressed affective disorder, 32% for transient situational disorder, 21% for anxiety disorder, and 11% for alcoholism. Temporal lobe epilepsy like Multiple sclerosis, seems especially associated with depressive disorder. This is not just of theoretical interest. Temporal lobe epilepsy has an increased risk of suicide by about 25 times,[51] and the suicide rate in one study of Multiple sclerosis was 14 times higher than that of the local general population.[52]

V. CEREBROVASCULAR DISEASE

Robinson and colleagues, in a series of studies[53-57] have elegantly studied the relations of stroke and psychiatric disorder. Of a consecutive series of 103 stroke patients capable of undergoing a psychiatric interview, nearly 50% had clinically significant depression in the acute stroke period. Using DSM III criteria, 27% had major depression, and 20% dysthymic (minor) depression. In addition, 9% were unduly cheerful. Functional physical impairment, intellectual impairment, quality of social support, and age all contributed to, or modified, the depression but the most important factor in determining frequency and severity of depression was the localization of the lesion in the brain. On follow-up, 3 months and 6 months after the stroke, the association between severity of depression and functional impairment increased, but the association with lesion site persisted, suggesting an interaction between neurological and psychosocial factors in the determination of post-stroke depression.

Storey[58] studied the psychiatric sequelae of 261 cases of subarachnoid hemorrhage; symptoms restricted to the first 3 months convalescence were excluded. Psychiatric symptoms were almost entirely depressive states and anxiety symptoms, often occurring together; in patients with pre-existing anxiety or depression only a worsening was counted. Of the 231 with aneurysms demonstrated on investigation, 55% had psychiatric morbidity. In 31% it was moderate or severe, and in 24% mild. Most of this morbidity was associated with brain damage. It was also associated with physical disability, and intellectual and personality impairment, but not more definitely than with brain damage — implying that the depression was not mainly a reaction to disability, although this element contributed in many cases.

VI. PARKINSON'S DISEASE AND DEPRESSION

In 1922, Patrick and Levy[59] described 140 cases of Parkinson's disease, excluding post-encephalitic cases, in which 34% had mental symptoms, largely in the form of depressive reaction, occurring as frequently before as after the onset of the neurological features; others at this time reported similar findings.[60] Economo[61] pointed out that, as well as parkinsonism, postencephalitic patients developed states of depression which frequently, as in genuine melancholia, remitted a little in the evening. Steck[62] made an extensive study of the psychiatric sequelae of encephalitis lethargica. He suggested that postencephalitic depressive syndromes were diverse in nature — he contrasted neurasthenia-like states which could precede parkinsonism and melancholic states with tearfulness; ideas of misery, guilt, and suicide; and hypochondriasis. With the subsequent predominance of a psychological approach to understanding depression, there was a relative lack of interest in depression in parkinsonism. A typical view was that of Wilson[63] who suggested that symptoms of depression or hypochondriasis in parkinsonism were understandable since it was doubtless natural for the parkinsonian, "finding life a burden and pondering a hopeless future, to react by loss of spirit and chilling moodiness". The revival of interest in the 1960s in the psychiatric aspects of parkinsonism prompted a series of studies which found an incidence of depression in different groups of patients with parkinsonism higher than in matched control groups and ranging from 29 to 90%,[64-73] which appeared to be independent of degree of disability, duration of illness, or medication.[74] In a study comparing Parkinson's, Huntington's, and Alzheimer's diseases,[75] depression, absent in patients with Alzheimer's, was present in half the patients with Huntington's and Parkinson's disease.

With the appreciation of the high incidence of depression in parkinsonism came one of the degree of overlap of these two clinical syndromes, up till then considered within two completely distinct frames of reference. An overlap in response to treatments considered specific for either condition was the subject of reports, such as the mild but definite beneficial effect of tricyclics on the motor state of parkinsonism untreated with L-dopa[76-78] but not when already treated with L-dopa,[79] the effective intervention with electro-convulsive therapy in patients with both conditions,[80-82] and the equal effectiveness of bromocriptine as an antiparkinsonian and antidepressant.[83]

The most significant overlap in treatment was that with L-dopa, since this focused attention more precisely on the areas of overlap. Ajuriaguerra,[66] in his study of 204 patients with parkinsonism, had found depressive states to be a feature in 70%. Like the authors in the 1920s he divided these depressive states into two different types. Of his cases, 60% had "simple" depressive states characterized by asthenia and inertia with a certain sadness. All of these had a significant akinesia, and L-dopa produced a parallel improvement in motor and affective disorder. A further 10% had more severe depressive symptoms or "melancholia" with ideas of guilt, self-reproach, suicide, and hypochondriasis. In these, L-dopa could produce a significant improvement in motor disorder with a less marked effect on the depressive symptoms, needing the addition of a tricyclic. In Anderson and colleagues' placebo-controlled trial of the effect of nortriptyline in 19 patients with L-dopa-treated parkinsonism and depressive symptoms,[79] the tricyclic had a significant clinical effect on the depressive symptoms, but scores for inhibition of thought and motor retardation in these L-dopa-treated patients were not improved to the same degree as other items. The authors suggested that the inhibition of thought and motor retardation of both parkinsonism and depression might be related.

Psychomotor retardation, the most consistent feature characterizing major depressive illness,[84] may also be the result of dopaminergic dysfunction. Dopamine turnover, as measured by homovanillic acid concentration in the cerebrospinal fluid, is reduced in depressed patients with retardation but not in those without,[85,86] and L-dopa has a therapeutic effect in depressive illness but usually only when motor retardation is a prominent feature.[87-89] Dysfunction of the meso-cortico-limbic dopaminergic system has been demonstrated in Parkinson's disease, with the suggestion that, in this condition, loss of dopaminergic innervation in the prefrontal cortex could be the basis of akinesia and slowing of cognitive processing speed, or bradyphrenia, and in certain parts of the limbic forebrain, of affective disturbance.[90-92] Innervation of different structures by a single neuro-transmitter system would explain the association between akinesia, cognitive impairment, and depression in Parkinson's disease.[66,93,94]

In a recent study,[95] 30 consecutively referred, newly diagnosed patients with Parkinson's disease had a significantly higher mean Hamilton depression rating[96] than a group of 30 age- and education-matched normal controls, their Hamilton scores overlapping with both the normal controls and 30 age- and education-matched patients with primary depressive illness. All the depressed patients satisfied the RDC diagnostic criteria[97] for major depressive disorder. Of the parkinsonian patients, 15 had at least 5 positive items on the RDC symptom check-list, a necessary, though not sufficient, criterion for a diagnosis of major depressive disorder. Furthermore, the 30 patients with primary depressive illness in this study had a Webster rating for motor impairment[98] intermediate between that of the Parkinson's patients and the normal controls, and significantly higher than that of the normal controls. Each of the ten individual items on this scale contributed to the rating in the depressed patients, with the highest ratings for arm swing, facial expression, and speech impairment. The only items with a mean rating per patient in the parkinsonian group of over twice that in the depressed group were tremor, rigidity, self-care, and bradykinesia.

After 3 to 12 months, 12 of the parkinsonian patients, who had all been treated with a dopaminergic agonist, did not show significant change in mean depression rating, as others have found,[99-101] nor in motor impairment or cognitive processing time. Individual change in depression rating, however, correlated significantly with change in the measure of cognitive processing time. Twelve depressed patients retested after a comparable period of time on conventional antidepressant medication, whose mean depression rating had significantly improved, also showed a significant improvement in cognitive processing speed and motor impairment rating. These findings support the involvement of a common cerebral system, such as the meso-cortico-limbic dopaminergic system, in the affective and cognitive disorder of Parkinson's disease and psycho-motor retardation in primary depressive illness.

VII. DISCUSSION

Already in 1930, in an editorial on the neuroses,[102] it was suggested that "old distinctions" between functional and organic types of symptoms had proved valueless except for clinical diagnoses and that, even in this field, their usefulness was over. Theoretically and etymologically, it was argued, the term neurosis indicated a disturbance of function of nervous tissue; in fact any neurological syndrome was a neurosis. Some, however, had restricted its use to cases where no recognizable physiological disability could be detected, and others, taking this one stage further, had equated the absence of any recognizable physiological etiology with a psychological origin. It was pointed out that chorea and Parkinson's disease had originally been classified as neuroses, and that these conditions pointed the way to an understanding of neurosis as "physiological disarray". The alternative viewpoint was represented, for example, by Sief,[103] who at this time, was arguing that neurosis was a social disturbance — a "half-and-half adjustment to the tasks of life". The neurotic person was cooperative so far as his superiority was approximately guaranteed, but the moment this superiority was endangered in the face of blame, defeat, decision, or responsibility, he tended to secure his self-esteem with the "alibi" of neurotic symptoms; neurosis was acquired as the result of a mistaken estimate of self and life during childhood.

Janet, in his book on obsessional disorder and psychasthenia,[104] saw the psychological viewpoint as a necessary precursor to "anatomical" thinking, since if "one should always think anatomically then one would have to resign oneself to not thinking at all when it comes to psychiatry". Our understanding of the neurophysiological basis of psychiatric disorder has advanced immeasureably since this was written, but until recently the psychological viewpoint has been preferred to the neurophysiological. In the 1930s and 40s, the psychological approach was applied to neurological disorder; Jelliffe[105,106] suggested that the tremor of parkinsonism represented repressed sadism and that drooling of saliva represented displaced seminal emission. In the 1950s, when ACTH was successfully introduced in the treatment of such disorders as rheumatoid arthritis, it precipitated psychiatric disturbance in some patients; this, it was argued, was due to the too-rapid removal of physical symptoms which had been the mask and means of expression of unexpressed psychological tensions and conflicts.[107] When Capstick and Seldrup[108] in 1977 showed a relationship between a history of abnormal birth and the later development of obsessional states, they favored an explanation in terms of psychological factors at the time of the birth.

These two approaches, physiological and psychological, are alternative ways of explaining any particular psychiatric disorder. The two approaches are not mutually exclusive, and therefore should not be used as a basis for classification; this is easily done, however. Olin and Weisman,[109] for example, examining the relationship of psychiatric and neurological disorder, distinguished "mental" symptoms produced by impaired neurological functional capacities from "psychiatric" symptoms brought about by conflict and incompatible motivations. They found no instance of psychiatric misdiagnosis of neurologic disease in the records of 2200 patients admitted to a general hospital psychiatric service and 475 patients with brain tumors. In contrast, Tissenbaum and colleagues[110] had earlier reviewed 395 neurological cases under treatment in a Veterans Administration clinic, and found that 53 (13%) had been incorrectly diagnosed for a mean of 4 1/2 years, most having been considered "functional". The most common misdiagnoses were of parkinsonism and multiple sclerosis, and they considered that functional signs of one type or another were extremely frequent as precursors of organic disease in certain neurological disorders, and might very well be part and parcel of the total disease process.

In an important study, Brown and colleagues[94] found that the fluctuations in motor function of the on/off phenomenon in parkinsonian patients was associated with similar fluctuations in cognitive function and a measure of affect-arousal. When a regression analysis was carried out, only the impairment of affect-arousal made a significant contribution to the cognitive impairment when considered independently, the correlation with motor impairment being mediated through the impairment of affect-arousal. The authors, when discussing this finding, felt that a crucial question was to what extent the changes in affect-arousal were the result of neurochemical fluctuations, possibly in the meso-cortico-limbic dopaminergic system, and to what extent they were "reactive" to the fluctuations in motor disability. Quite rightly they pointed out that it was difficult to see how this question might be answered with currently available techniques. It is, in fact, a philosophical issue; the choice is between two different approaches to understanding psychiatric disorder. The decision as to which approach is most appropriate in a particular case cannot be resolved by investigative techniques. Historically, however, explanation in terms of psychological reaction to physical disorder represents an approach which has become increasingly incapable of incorporating the results of studies, such as those reviewed in this chapter, of the neurotic disorder associated with different neurologic disorders.

The modern position of neurosis is discussed by Tyrer.[111] Over the years, neurosis has been increasingly subdivided. The International Classification of Diseases now has nine categories of neurosis, and the third edition of the Diagnosic and Statistical Manual of Mental Disorders (DSM III) has 21 now included in affective, anxiety, somatoform, and associative disorders. Tyrer asks if this represents a real advance. If the subdivisions are fundamentally distinct from each other they should help to predict prognosis and decide treatment; on both these ground he finds the categorization of neurosis defective. In one study of patients, who had been admitted twice to a psychiatric hospital within a 5-year period, less than 40% of those with neurotic disorders retained their diagnosis;[112] in other studies, using different diagnostic instruments supposed to define independent diagnoses, the presence of one psychiatric diagnosis increased the patient's chance of having a second, with the greatest diagnostic overlap being for neurotic syndromes.[113,114] With regard to treatment, diagnoses of primary depression or anxiety, for example, do not predict response to treatment, and antidepressants are equally effective for both anxiety and depressive symptoms.[115,116]

The studies reviewed in this chapter support the clinical overlap between different neurotic syndromes. By and large, a particular neurologic disorder can be associated with a wide spectrum of different neurotic disorders. However, the association of particular neurotic syndromes and particular neurologic disorders, such as obsessive-compulsive disorder with Tourette syndrome and depressive disorder with multiple sclerosis, cerebrovascular disease, and Parkinson's disease, suggest a certain specificity in the association. In the case of schizophreniform psychosis, Davison and Bagley[12] were able to show a particular association with neurological disorders affecting particular brain areas, namely, the temporal lobes and diencephalon. Some neurological disorders, such as Huntington's disease, have an association with both schizophreniform psychosis and neurotic disorder, while others, such as stroke and Parkinson's disease, have a marked association with affective disorder but very little association with schizophreniform psychosis, and in the case of Parkinson's disease possibly even a negative association. It is possible that different neurotic syndromes, too, are particularly associated with disorder of specific brain areas.

Concepts of cerebral localization have developed since the time of Janet's rejection of anatomical models for psychiatry, and localization is now attempted in terms of discrete cerebral systems, commonly defined by the predominant neurotransmitter involved, and which are not necessarily confined to one "geographical" location in the brain. Interestingly, the seeds of this new conception of cerebral localization were already present at the time Janet was despairing of a brain-based psychiatry.[117] Examples of such possible cerebral localization are the dopaminergic meso-cortico-limbic projection pathways as a basis for the cognitive and affective disorder of Parkinson's disease, and similar catecholamine subcortico-cortical projection systems as a basis for the cognitive and affective disorder following stroke, and the cognitive impairment of primary affective disorders.[93,118,119] The concept of cerebral localization by brain system would explain why such apparently heterogeneous cerebral disorders as frontal meningiomas, temporal/hippocampal gliomas, and neurosarcoidosis, with periventricular white matter attenuation on CT scan, can present as depressive disorders.[120-122]

VIII. TREATMENT ASPECTS

Few studies have addressed themselves directly to the treatment of neurotic disorder associated with different neurological disorders, but the general view is that their treatment should be as for idiopathic neurotic conditions.[41,123] It is helpful to have psychiatrists familiar with neurological disorder involved in the treatment, for the neurological and psychiatric components of different disorders can be intimately related. The study of these relationships can also contribute to the understanding and treatment of primary neurotic disorders. It leads, for example, to the "dissection" of psychiatric syndromes into their component parts, a principle advocated by such investigators as Barbeau[124] and van Praag.[85] The study of affective disorder in Parkinson's disease suggests that different components of the disorder respond differently to different pharmacological treatments. What some authors have called "simple" depression, with lack of drive and energy, responds to dopaminergic medication, whereas "melancholia" with tearfulness, guilt, suicidal ideas, and hypochondriasis does not, but can respond to tricyclic medication. Van Praag and his colleagues[85] reported the use of the probenicid technique to study central dopamine metabolism in patients with depressions and Parkinson's disease. They concluded that decreased dopamine turnover was associated with, and probably related etiologically to, hypomotility, rather than being specific for any particular disease or syndrome. They suggested that hypokinesia was a central symptom, not only in Parkinson's disease, but also in depressive illness with motor retardation, where the retardation could be reversed by treatment with L-dopa without significant change in mood score. This view has been supported by other studies, where only depressive illnesses with a marked retardation "component" responded to dopaminergic treatment.[87,88] Such treatment might be considered for affective disorder where retardation is the main, only, or a residual component. This should be done carefully, because of the theoretical possibility of precipitation of psychosis with dopaminergic treatment.

IX. CONCLUSION

The change in viewpoint from a psychological to a neurophysiological one has only recently started gathering momentum. Once a physiological framework of explanation for psychiatric disorder is adopted, as Schwab maintained in his presidential address to the American Neurological Association in 1921,[125] there is no such thing as a psychiatric method of thinking or approach that can be divorced from the neurological one. With such an approach, emotional and physical traumas are seen as having identical effects on cerebral mechanisms,[126,127] and neurotic symptoms become the most sensitive indicators of brain disorder.[128] Neurological examination and investigation is performed on patients with psychiatric disorder for the information it provides on the disorder, and not only to define those cases needing referral to a neurologist. Neurological abnormalities detected are no longer described as incidental or noncontributory, neurological dysfunction is incorporated as a component of psychiatric syndromes, and concepts such as the "neurology of depression" become respectable.[129,130] Needless to say, a neurological approach to the disorder in no way precludes a caring or supportive approach to the patient.

The "neurological" approach to the neurotic disorders associated with the specific neurologic disorders reviewed in this chapter show that such an approach is both satisfying conceptually and practically relevant. There is no need for it to be restricted to psychiatric disorder associated with known cerebral disorder. The neurologic disorders discussed vary in prevalence from 6 per 100,000 of the population for Huntington's disease to 6 per 1000 for cerebrovascular disease; most are uncommon disorders. Taken together, however, they point the way to a biological understanding and therapy of the more common idiopathic neurotic disorders.

REFERENCES

1. **Comroe, B. I.,** Follow-up study of 100 patients diagnosed as "neurosis", *J. Nerv. Ment. Dis.*, 83, 679, 1936.
2. **Marshall, H. E. S.,** Incidence of Physical disorders among psychiatric in-patients, a study of 175 cases, *Brit. Med. J.*, 2, 468, 1949.
3. **Herridge, C. D.,** Physical disorders in psychiatric illness, a study of 209 consecutive admissions, *Lancet*, 2, 949, 1960.
4. **Davies, D. W.,** Physical illness in psychiatric out-patients, *Brit. J. Psychiat.*, 111, 27, 1965.
5. **Johnson, D. A. W.,** The evaluation of routine physical examination in psychiatric cases, *Practitioner*, 200, 686, 1968.
6. **Maguire, G. P. and Granville-Grossman, K. L.,** Physical illness in psychiatric patients, *Brit. J. Psychiat.*, 115, 1365, 1968.
7. **Eastwood, M. R., Mindham, R. H. S., and Tennant, T. G.,** The physical status of psychiatric emergencies, *Brit. J. Psychiat.*, 116, 545, 1970.
8. **Hall, R. C. W., Popkin, M. K., Devaul, R. A., Faillace, L. A., and Stickney, S. K.,** Physical illness presenting as psychiatric disease, *Arch. Gen. Psychiatry*, 35, 1315, 1978.
9. **Burke, A. W.,** Physical disorder among day hospital patients, *Brit. J. Psychiat.*, 133, 22, 1978.
10. **Koranyi, E. K.,** Morbidity and rate of undiagnosed physical illnesses in a psychiatric clinic population, *Arch. Gen. Psychiatry*, 36, 414, 1979.
11. **Hall, R. C. W., Gardner, E. R., Stickney, S. K., LeCann, A. F., and Popkin, M. K.,** Physical illness manifesting as psychiatric disease, analysis of a state hospital inpatient population, *Arch. Gen. Psychiat.*, 37, 989, 1980.
12. **Davison, K. and Bagley, C. R.,** Schizophrenia-like psychoses associated with organic disorders of the central nervous system, a review of the literature, in *Current Problems in Neuropsychiatry, British Journal of Psychiatry special publication no. 4,* Herrington, R. N., Ed., Headley Brothers, Ashford, Kent, 1969, 113.
13. **Whittier, J. R.,** Hereditary chorea (Huntington's chorea), a paradigm of brain dysfunction with psychopathology, in *Psychopathology and Brain Dysfunction,* Shagass, C., Gershon, S., and Friedhoff, A. J., Eds., Raven Press, New York, 1977, 267.
14. **Huntington, G.,** On chorea, *Med. Surg. Reporter,* 26, 317, 1872.
15. **Hughes, E. M.,** Social significance of Huntington's chorea, *Am. J. Psychiatry*, 4, 537, 1925.
16. **Parker, N.,** Observations on Huntington's chorea based on a Queensland survey, *Med. J. Aust.*, 1, 359, 1958.
17. **Brothers, C. R. D.,** Huntington's chorea in Victoria and Tasmania, *J. Neurol. Sci.*, 1, 405, 1964.
18. **Heathfield, K. W. G.,** Huntington's chorea, investigation into the prevalence of this disease in the area covered by the North East Metropolitan Regional Hospital Board, *Brain*, 90, 203, 1967.
19. **Oliver, J. E.,** Huntington's chorea in Northamptonshire, *Brit. J. Psychiat.*, 116, 241, 1970.
20. **Mattsson, B.,** Clinical, genetic and pharmacological studies in Huntington's chorea; Huntington's chorea in Sweden: social and clinical data, *Umea Univ. Med. Dissertations*, 7, 33, 1974.
21. **Mindham, R. H. S., Steele, C., Folstein, M. F., and Lucas, J.,** A comparison of the frequency of major affective disorder in Huntington's disease and Alzheimer's disease, *J. Neurol. Neurosurg. Psychiat.*, 48, 1172, 1985.
22. **Robertson, M. M.,** Personal communication, 1986.
23. **Guinon, G.,** Sur la maladie des tics convulsifs, *Rev. Med.*, 6, 50, 1886.
24. **Tourette, G. de la,** La maladie des tics convulsifs, *Semaine Med.*, 19, 153, 1899.
25. **Montgomery, M. A., Clayton, P. J., and Friedhoff, A. J.,** Psychiatric illness in Tourette syndrome patients and first degree relatives, in *Gilles de la Tourette Syndrome,* Friedhoff, A. J. and Chase, T. N., Eds., Raven Press, New York, 1982, 335.

26. **Shapiro, A. K. and Shapiro, E.**, Tourette syndrome: history and present status, in *Gilles de la Tourette Syndrome*, Friedhoff, A. J. and Chase, T. N., Eds., Raven Press, New York, 1982, 17.

27. **Morphew, J. A. and Sim, M.**, Gilles de la Tourette's syndrome: a clinical and psychopathological study, *Brit. J. Med. Psychol.*, 42, 293, 1969.

28. **Robertson, M. M., Trimble, M. R., and Lees, A. J.**, A clinical and psychopathological study of Gilles de la Tourette syndrome in the United Kingdom, Abstract 531.10, Proc. IVth World Congress of Biological Psychiatry, Philadelphia, 1985.

29. **Kelman, D. H.**, Gilles de la Tourette's disease in children: a review of the literature, *J. Child Psychol. Psychiat.*, 6, 219, 1965.

30. **Fernando, S. J. M.**, Gilles de la Tourette's syndrome, a report on four cases and a review of published case reports, *Brit. J. Psychiat.*, 113, 607, 1967.

31. **Comings, D. E. and Comings, B. G.**, Tourette syndrome: clinical and psychological aspects of 250 cases, *Am. J. Hum. Gen.*, 37, 435, 1985.

32. **Abuzzahab, F. A. and Anderson, F. O.**, Gilles de la Tourette's syndrome: international registry, *Minnesota Medicine*, p. 492, June 1973.

33. **Asam, V.**, A follow-up study of Tourette Syndrome, in *Gilles de la Tourette Syndrome*, Friedhoff, A. J. and Chase, T. N., Eds., Raven Press, New York, 1982, 285.

34. **Hagin, R. A., Beecher, R., Pagano, G., and Kreeger, M.**, Effects of Tourette syndrome on learning, in *Gilles de la Tourette Syndrome*, Friedhoff, A. J. and Chase, T. N., Eds., Raven Press, New York, 1982, 323.

35. **Nee, L. E., Caine, E. D., Polinsky, R. J., Eldridge, R., and Ebert, M. H.**, Gilles de la Tourette syndrome: clinical and family study of 50 cases, *Ann. Neurol.*, 7, 41, 1980.

36. **Stefl, M. E.**, Mental health needs associated with Tourette syndrome, *Am. J. Publ. Hlth.*, 74, 1310, 1984.

37. **Yaryura-Tobias, J. A., Neziroglu, F., Howard, S., and Fuller, B.**, Clinical aspects of Gilles de la Tourette Syndrome, *J. Orthomol. Psychiat.*, 10, 263, 1981.

38. **Frankel, M., Cummings, J. L., Robertson, M. M., Trimble, M. R., Hill, M. A., and Benson, D. F.**, Obsessions and compulsions in Gilles de la Tourette's syndrome, *Neurology*, 36, 378, 1986.

39. **Cummings, J. L. and Frankel, M.**, Gilles de la Tourette syndrome and the neurological basis of obsessions and compulsions, *Biol. Psychiatry*, 20, 1117, 1985.

40. **Yaryura-Tobias, J. A.**, *Obsessive-compulsive Disorders; Pathogenesis — Diagnosis — Treatment*, Marcel Dekker, New York, 1983.

41. **Trimble, M. R. and Grant, I.**, Psychiatric aspects of multiple sclerosis, *Psychiatric Aspects of Neurologic Disease*, Vol. 2, Benson, D. F. and Blumer, D., Eds., Grune and Stratton, New York, 1982, 279.

42. **Charcot, J. M.**, *Lectures on the Diseases of the Nervous System Delivered at the Salpetriere*, New Sydenham Society, London, 1877, 194.

43. **Brown, S. and Davis, T. K.**, The mental symptoms of multiple sclerosis, *Arch. Neurol. Psychiatry*, 7, 629, 1922.

44. **Braceland, F. J. and Giffin, M. E.**, The mental changes associated with multiple sclerosis, *Res. Publ. Assoc. Res. Nerv. Ment. Dis.*, 28, 450, 1950.

45. **Pratt, R. T. C.**, An investigation of the psychiatric aspects of disseminated sclerosis, *J. Neurol. Neurosurg. Psychiat.*, 14, 326, 1951.

46. **Baldwin, M. V.**, A clinic-experimental investigation in to the psychologic aspects of multiple sclerosis, *J. Nerv. Ment. Dis.*, 115, 299, 1952.

47. **Ross, A. T. and Reitan, R. M.**, Intellectual and affective functions in multiple sclerosis, *Arch. Neurol. Psychiatry*, 73, 663, 1955.

48. **Surridge, D.**, An investigation into some psychiatric aspects of multiple sclerosis, *Brit. J. Psychiat.*, 115, 749, 1969.

49. **Whitlock, F. A. and Siskind, M. M.**, Depression as a major symptom of multiple sclerosis, *J. Neurol. Neurosurg. Psychiat.*, 43, 861, 1980.

50. **Schiffer, R. B. and Babigian, H. M.**, Behavioral disorders in multiple sclerosis, temporal lobe epilepsy, and amyotrophic lateral sclerosis, An epidemiologic study, *Arch. Neurol.*, 41, 1067, 1984.

51. **Barraclough, B.**, Suicide and epilepsy, in *Epilepsy and Psychiatry*, Reynolds, E. H. and Trimble, M. R., Eds. Churchill Livingstone, Edinburgh, 1981, 72.

52. **Liebowitz, U., Kahana, E., Jacobson, S. G., and Alter, M.**, Cause of death in multiple sclerosis, in *Proc. Int. Symp. Progress in Multiple Sclerosis, Research and Treatment*, Liebowitz, U., Ed., Academic Press, New York, 1970.

53. **Robinson, R. G., Starr, L. B., Kubos, K. L., and Price, T. R.**, A two-year longitudinal study of post-stroke mood disorders: findings during the initial evaluation, *Stroke*, 14, 736, 1983.

54. **Robinson, R. G., Kubos, K. L., Starr, L. B., Rao, K., and Price, T. R.**, Mood disorders in stroke patients, Importance of location of lesion, *Brain*, 107, 81, 1984.

55. **Robinson, R. G., Starr, L. B., and Price, T. R.**, A two year longitudinal study of mood disorders following stroke; Prevalance and duration at six months follow-up, *Brit. J. Psychiat.*, 144, 256, 1984.

56. **Robinson, R. G., Starr, L. B., Lipsey, J. R., Rao, K., and Price, T. R.,** A two year longitudinal study of post stroke mood disorders: dynamic changes in associated variables over the first six months of follow-up, *Stroke*, 15, 510, 1984.

57. **Robinson, R. G., Starr, L. B., Lipsey, J. R., Rao, K., and Price, T. R.,** A two year longitudinal study of poststroke mood disorders, In-hosptital prognostic factors associated with six-month outcome, *J. Nerv. Ment. Dis.*, 173, 221, 1985.

58. **Storey, P. B.,** Psychiatric sequelae of subarachnoid haemorrhage, *Brit. Med. J.*, 2, 261, 1967.

59. **Patrick, H. T. and Levy, D. M.,** Parkinson's disease: a clinical study of 146 cases, *Arch. Neurol. Psychiatry*, 7, 711, 1922.

60. **Jackson, J. A., Free, G. B. M., and Pike, H. V.,** The psychic manifestations in paralysis agitans, *Arch. Neurol. Psychiatry*, 10, 680, 1923.

61. **Economo, C. V.,** *Encephalitis lethargica, its sequelae and treatment*, Newman, K. O., Transl., Oxford University Press, London, 1931.

62. **Steck, H.,** Les syndromes mentaux postencephalitiques, *Schweizer Archiv. Neurol. Psychiatr.*, 27, 137, 1931.

63. **Wilson, S. A. K.,** *Neurology*, vol. 1., Arnold, London, 1940.

64. **Warburton, J. W.,** Depressive symptoms in Parkinson patients referred for thalamotomy, *J. Neurol. Neurosurg. Psychiat.*, 30, 368, 1967.

65. **Mindham, R. H.,** Psychiatric symptoms in Parkinsonism. *J. Neurol. Neurosurg. Psychiat.*, 33, 188, 1970.

66. **Ajuriaguerra, J. de,** Etude psychopathologique des Parkinsoniens, in *Monoamines, Noyaux Gris Centraux et Syndrome de Parkinson*, Ajuriaguerra, J. de and Gauthier, G., Eds., Georg, Geneva, 1971, 327.

67. **Brown, G. L. and Wilson, W. P.,** Parkinsonism and depression, *South. Med. J.*, 65, 540, 1972.

68. **Celesia, G. G. and Wanamaker, W. M.,** Psychiatric disturbances in Parkinson's disease, *Dis. Nerv. System*, 33, 577, 1972.

69. **Marsh, G. G. and Markham, C. H.,** Does levodopa alter depression and psychopathology in parkinsonian patients?, *J. Neurol. Neurosurg. Psychiat.*, 36, 925, 1973.

70. **Horn, S.,** Some psychological factors in parkinsonism, *J. Neurol. Neurosurg. Psychiat.*, 37, 27, 1974.

71. **Robins, A. H.,** Depression in patients with parkinsonism, *Brit. J. Psychiat.*, 128, 141, 1976.

72. **Lieberman, A., Dziatolowski, M., Kupersmith, M., Serby, M., Goodchild, A., Korein, J., and Goldstein, M.,** Dementia in Parkinson's disease, *Ann. Neurol.*, 6, 355, 1979.

73. **Mayeux, R., Stern, Y., Rosen, J., and Levanthal, J.,** Depression, intellectual impairment and Parkinson's disease, *Neurology*, 31, 645, 1981.

74. **Mayeux, R., Williams, J. B. W., Stern, Y., and Cote, L.,** Depression and Parkinson's disease, in *Advances in Neurology*, vol. 40, Hassler, R. G. and Christ, J. F., Raven Press, New York, 1984, 241.

75. **Mayeux, R., Stern, Y., Rosen, J., and Benson, D. F.,** Is "subcortical dementia" a recognisable clinical entity?, *Ann. Neurol.*, 14, 278, 1983.

76. **Strang, R. R.,** Imipramine in treatment of parkinsonism: a blind placebo study, *Brit. Med. J.*, 2, 33, 1965.

77. **Laitinen, L.,** Desimipramine in the treatment of Parkinson's disease, *Acta Neurol. Scand.*, 45, 109, 1969.

78. **Sigwald, J.,** Therapeutique des syndromes extrapyramidaux par les traitements classiques, in *Monoamines, Noyaux Gris Centraux et Syndrome de Parkinson*, Ajuriaguerra, J. de and Gauthier, G., Eds., Georg, Geneva, 1971, 369.

79. **Anderson, J., Aabro, E., Gulmann, N., Hjelmsted, A., and Pederson, H. E.,** Antidepressive treatment in Parkinson's disease, A controlled trial of the effect of Nortriptyline in patients with Parkinson's disease treated with L-dopa, *Acta Neurol. Scand.*, 62, 210, 1980.

80. **Lebensohn, Z. M. and Jenkins, R. B.,** Improvement of parkinsonism in depressed patients treated with ECT, *Am. J. Psychiatry*, 132, 283, 1975.

81. **Asnis, C.,** Parkinson's disease, depression and ECT: A review and case study, *Am. J. Psychiatry*, 134, 191, 1977.

82. **Yudofsky, S. C.,** Parkinson's disease, depression and convulsive therapy: a clinical and neurobiologic synthesis, *Comp. Psychiatry*, 20, 579, 1979.

83. **Jouvent, R., Abensour, P., Bonnet, A. M., Widlocher, D., Agid, Y., and Lhermitte, F.,** Antiparkinsonian and antidepressant effects of high doses of bromocryptine, *J. Affect. Disorders*, 5, 141, 1982.

84. **Nelson, J. C. and Charney, D. S.,** The symptoms of major depression, *Am. J. Psychiatry*, 138, 1, 1981.

85. **Praag, H. M. V., Korf, J., Lakke, J. P. W. F., and Schut, T.,** Dopamine metabolism in depressions, psychoses, and Parkinson's Disease: the problem of the specificity of biological variables in behaviour disorders, *Psychol. Med.*, 5, 138, 1975.

86. **Banki, C. M.,** Correlation between CSF metabolites and psychomotor activity in affective disorders, *J. Neurochem.*, 28, 255, 1977.

87. **Goodwin, F. K., Brodie, H. K., Murphy, D. L., and Bunney, W. E.,** Administration of a peripheral decarboxylase inhibitor with L-dopa to depressed patients, *Lancet*, 1, 908, 1970.

88. **Mattussek, N., Benkert, O., Schneider, K., Otten, H., and Pohlmeir, H.,** L-dopa plus decarboxlylase inhibitor in depression, *Lancet*, 2, 660, 1970.

89. **Kanzler, M. and Malitz, S.,** L-dopa for the treatment of depression, in L-*dopa and Behaviour*, Malitz, S., Ed., Raven Press, New York, 1972, 103.

90. **Price, K. S., Farley, I. J., and Hornykiewicz, O.,** Neurochemistry of Parkinson's disease: relation between striatal and limbic dopamine, in *Advances in Biochemical Psychopharmacology*, vol. 19, Roberts, P. J., Woodruff, G. N., and Iverson, L. L., Eds., Raven Press, New York, 1978, 293.

91. **Javoy-Agid, F. and Agid, Y.,** Is the mesocortical dopaminergic system involved in Parkinson's disease?, *Neurology*, 30, 1326, 1980.

92. **Scatton, B., Rouquier, L., Javoy-Agid, F., and Agid, Y.,** Dopamine deficiency in the cerebral cortex in Parkinson's disease, *Neurology*, 32, 1039, 1982.

93. **Agid, Y., Ruberg, M., Dubois, B., and Javoy-Agid, F.,** Biochemical substrates of mental disturbances in Parkinson's disease, in *Advances in Neurology*, vol. 40, Hassler, R. G. and Christ, J. F., Raven Press, New York, 1984, 211.

94. **Brown, R. G., Marsden, C. D., Quinn, N., and Wyke, M. A.,** Alterations in cognitive performance and affect-arousal state during fluctuations in motor function in Parkinson's disease, *J. Neurol. Neurosurg. Psychiat.*, 47, 454, 1984.

95. **Rogers, D., Lees, A., Smith, E., Trimble, M., and Stern,** Bradyphrenia in Parkinson's Disease and psychomotor retardation in depressive illness: an experimental study, *Brain*, in press.

96. **Mamilton, M.,** Development of a rating scale for primary depressive illness, *Brit. J. Soc. Clin. Psychol.*, 6, 278, 1967.

97. **Spitzer, R. L., Endicott, J., and Robins, E.,** Research Diagnostic Criteria: rationale and reliability, *Arch. Gen. Psychiatry*, 35, 773, 1978.

98. **Webster, D. D.,** Clinical analysis of the disability in Parkinson's disease, *Mod. Treatment*, 5, 257, 1968.

99. **Beardsley, J. J. and Puletti, F.,** Personality (MMPI) and cognitive (WAIS) changes after levodopa treatment, *Arch. Neurol.*, 25, 145, 1971.

100. **Marsh, G. G. and Markham, C. H.,** Does levodopa alter depression and psychopathology in Parkinsonism patients? *J. Neurol. Neurosurg. Psychiat.*, 36, 925, 1973.

101. **Portin, R. and Rinne, U. K.,** Neuropsychological responses of Parkinsonian patients to long-term levodopa treatment, in *Parkinson's Disease — Current Progress, Problems and Management*, Rinne, U. K., Klinger, M., and Stamm, G., Eds., Elsevier, Amsterdam, 1980, 271.

102. Editorial, The neuroses, *J. Neurol. Psychopath.*, 11, 163, 1930.

103. **Seif, L.,** Individual psychology and the treatment of neurosis, *Arch. Neurol. Psychiat.*, 19, 190, 1928.

104. **Janet, P.,** *Les Obsessions et la Psychasthenie*, Alcan, Paris, 1903.

105. **Jelliffe, S. E.,** The parkinsonian body posture, some considerations of unconscious hostility, *Psychoanaly. Rev.*, 27, 467, 1940.

106. **Jelliffe, S. E.,** The mental pictures in schizophrenia and in epidemic encephalitis, *Am. J. Psychiatry*, 6, 413, 1927.

107. **Doust, J. W. L.,** Psychiatric aspects of somatic immunity; differential incidence of physical disease in the histories of psychiatric patients, *Brit. J. Soc. Med.*, 6, 49, 1952.

108. **Capstick, N. and Seldrup, J.,** Obsessional states; a study in the relationship between abnormalities occurring at the time of birth and the subsequent development of obsessional symptoms, *Acta Psychiat. Scand.*, 56, 427, 1977.

109. **Olin, H. S. and Weisman, A. D.,** Psychiatric misdiagnosis in early neurologic disease, *JAMA*, 189, 533, 1964.

110. **Tissenbaum, M. J., Harter, H. M., and Friedman, A. P.,** Organic neurological syndromes diagnosed as functional disorders, *JAMA*, 147, 1519, 1951.

111. **Tyrer, P.,** Neurosis divisible? *Lancet*, 1, 685, 1985.

112. **Kendall, R. E.,** The stability of psychiatric diagnoses, *Br. J. Psychiat.*, 124, 352, 1974.

113. **Boyd, J. H., Burke, J. D., Jr., Gruenberg, E., Holzer, C. E., Rae, D. S., George, L. K., Karno, M., Stoltzman, R., McEvoy, L., and Nestadt, G.,** Exclusion criteria of DSM III: a study of co-occurrence of hierarchy-free syndromes, *Arch. Gen. Psychiat.*, 41, 983, 1984.

114. **Sturt, E.,** Hierarchical patterns in the distribution of psychiatric symptoms, *Psychol. Med.*, 11, 783, 1981.

115. **Johnstone, E. C., Owens, D. G. C., Frith, C. D., McPherson, K., Dowie, C., Riley, G., and Gold, A.,** Neurotic illness and its response to anxiolytic and antidepressant treatment, *Psychol. Med.*, 10, 321, 1980.

116. **Ancill, R. J., Poyser, J., Davey, A., and Kennerson, A.,** Management of mixed affective symptoms in primary care: a critical experiment, *Acta Psychiat. Scand.*, 70, 463, 1984.

117. **Bastian, H. C.,** *The Brain as an Organ of Mind*, Kegan Paul, London, 1880, 522.

118. **Robinson, R. G., Bolla-Wilson, K., Kaplan, E., Lipsey, J. R., and Price, T. R.,** Depression influences intellectual impairment in stroke patients, *Brit. J. Psychiat.*, 148, 541, 1986.

119. **McHugh, P. R. and Robinson, R. G.,** The two-way trade — psychiatry and neuroscience, *Brit. J. Psychiat.*, 143, 303, 1983.

120. **Hunter, R., Blackwood, W., and Bull, J.,** Three cases of frontal meningiomas presenting psychiatrically, *Brit. Med. J.,* 3, 9, 1968.

121. **Malamud, N.,** Psychiatric disorder with intracranial tumors of limbic system, *Arch. Neurol.,* 17, 113, 1967.

122. **Stiller, J., Goodman, A., Kahmi, L. M., Sacher, M., and Bender, M. B.,** Neurosarcoidosis presenting as major depression, *J. Neurol. Neurosurg. Psychiat.,* 47, 1050, 1984.

123. **Lipsey, J. R., Robinson, R. G., Pearlson, G. D., Rao, K., and Price, T. R.,** Nortriptyline treatment of post-stroke depression: a double blind treatment trial, *Lancet,* 1, 297, 1984.

124. **Barbeau, A.,** Dopamine and mental function, in L-dopa and Behavior, Malitz, S., Ed., Raven Press, New York, 1970, 9.

125. **Schwab, S. I.,** The neurologic dilemma, *Arch. Neurol. Psychiat.,* 6, 255, 1921.

126. **Vujic, V.,** Larvate encephalitis and psychoneurosis, *J. Nerv. Ment. Dis.,* 116, 1051, 1952.

127. **Whitty, C. W. M.,** Mental changes as a presenting feature in subcortical cerebral lesions, *J. Ment. Sci.,* 102, 719, 1956.

128. **Hermann, K.,** Atrophia cerebri: some remarks on the clinic of atrophy of the brain, *Acta Psychiat. Scand.,* (suppl.) 74, 165, 1951.

129. **Ross, E. D. and Rush, A. J.,** Diagnosis and neuroanatomical correlates of depression in brain-damaged patients; implications for a neurology of depression, *Arch. Gen. Psychiat.,* 38, 1344, 1981.

130. **Freeman, R. L., Galaburda, A. M., Cabal, R. D., and Geschwind, N.,** The neurology of depression; cognitive and behavioral deficits with focal findings in depression and resolution after electroconvulsive therapy, *Arch. Neurol.,* 42, 289, 1985.

Chapter 9

NEUROTIC DISORDERS ASSOCIATED WITH MENSTRUATION, PREGNANCY, AND THE POSTPARTUM PERIOD

Gary Bell and Cornelius Katona

TABLE OF CONTENTS

I. INTRODUCTION

The physiological changes that occur in association with menstruation, during pregnancy, and after childbirth, although normal, are nonetheless of a profound and dramatic nature. Psychological symptoms and behavioral changes have long been recognized as common accompaniments of the menstrual cycle and the early postpartum period. The prevalence of these symptoms is such that many regard them as ''normal'' accompaniments of the physiological changes.[1,2] Although there is substantial evidence that such mood changes follow predictable patterns and are disabling, there has been considerable reluctance to recognize specific neurotic illnesses as being related to reproductive processes in women, or to implicate concurrent biological changes as causal — particularly when personality disturbance and social stresses are present and appear to influence the clinical picture.

Research to date has also been bedevilled by varied and poorly formulated definitions of the disorders in question, poor subject selection, and lack of adequate comparison groups. This chapter presents a critical review of the etiology of neurotic disorders specifically associated with menstruation, pregnancy, and the postpartum period, focusing on the evidence for a biological contribution and on the relevance of such a contribution to the development of management strategies. The well-recognized but relatively rare postpartum psychoses will not form part of this review.

II. MENSTRUATION

The Premenstrual Syndrome, Dysmenorrhea, and the Climacteric Syndrome will be considered here.

A. The Premenstrual Syndrome

The Premenstrual Syndrome (PMS) has been the subject of extensive investigation, yet continues to be a confusing and controversial diagnostic entity. The lack of a generally accepted definition has resulted in many apparently contradictory research findings and a wide variety of treatment approaches, most of which remain illogical and unproven.

1. Definition

Since Frank's[3] first description of PMS in 1931, little progress has been made in establishing a generally accepted definition. Although a cyclical pattern of symptoms is clearly necessary for a diagnosis of PMS, there is considerable disagreement over the nature and duration of symptoms in the premenstrual and menstrual phases and over their relationship to any symptoms persisting after menstruation. Kramp[4] defined the premenstrual phase as the last 6 days of the luteal phase and the first 2 days of menstruation, whereas others[5] have stated that premenstrual symptoms must, by definition, vanish with the onset of menstruation. Dalton[6] advocates the use of a strict definition of PMS, yet her own "the recurrence of symptoms in the premenstruum with absence of symptoms in the postmenstruum" is vague and imprecise. Moos[7] observed that psychological symptoms were more distressing premenstrually, whereas physical symptoms predominated after the onset of menstruation. A review of 24 prospective studies has shown that the majority report negative mood disturbances as occurring more frequently during premenstrual and menstrual weeks.[8] In contrast to this, Slade[9] recently found that whereas physical symptoms were cyclical, peaking premenstrually and menstrually, psychological symptoms occurred randomly throughout the cycle.

The psychological symptoms most frequently reported are depression, anxiety, tension, irritability, impaired concentration, and sleep disturbance, while commonly encountered physical symptoms are breast swelling and tenderness, abdominal swelling, peripheral edema, weight gain, and headaches.[10,11] Some do not consider physical symptoms necessary for the diagnosis to be made,[6,12,13] whereas others see them as an integral part of the syndrome.[7,11] The persistence of symptoms after menstruation is another point of disagreement. Dalton[6] considers such patients to be suffering from "menstrual distress" rather than true premenstrual syndrome, whereas Magos and Studd[11] feel that a significant reduction in symptoms during the postmenstrual phase is acceptable for a diagnosis of PMS.

In addition to the physical and psychological symptoms encountered in PMS, a number of behavioral changes have been reported. Clare[10] has reviewed these under the headings of (1) aggressive behavior, (2) illness behavior and accidents, (3) examinations and other test performances, and (4) sporting performance. The majority of such studies suffer from either poor methodology or have produced inconclusive results. Of particular interest in recent years has been the reported relationship of PMS to criminal behavior. A study by d'Orban and Dalton[14] of 50 women charged with crimes of violence found that 44% had committed their offences in the "paramenstruum". Although this is consistent with the findings of an earlier study,[15] neither study showed women who complained of premenstrual symptoms to be more violent in the premenstruum when compared to noncomplainers. A cyclical pattern of violent or criminal behavior needs to be identified before a clear link between PMS and criminal behavior can be established. Dalton[16] reported three cases, all with long histories of repeated misdemeanors, in which severe PMS was accepted as a contributory factor in crimes of manslaughter, arson, and assault. Bizarre behavior recurring in the premenstruum was demonstrated in all three; however, as Clare[10] points out, there are immense difficulties in establishing retrospectively that there exists a clear temporal association between the premenstrual phase and episodes of criminal behavior.

Our own conceptual definition of PMS, modified after Magos and Studd[11] is "Distressing physical and psychological symptoms, not caused by organic disease, which regularly recur during the luteal phase of each menstrual cycle, and which significantly regress or disappear at or soon after the onset of menstruation".

Operational definitions of PMS have also been developed. Steiner et al.[1] have developed research diagnostic criteria that use both an interview and a daily rating scale, although their criteria fail to include physical symptoms and exclude patients with a past psychiatric history. The Premenstrual Assessment Form by Halbreich et al.[17] has 95 items, and allows for the development of inclusion and exclusion criteria for a number of subsyndromes primarily concerned with changes in mood and behavior; however its length severely limits its usefulness on a daily basis.

2. Prevalence

Estimates of the prevalence of PMS vary widely and are dependent on the retrospective or prospective nature of the studies, as well as on different definitions and methods of subject selection. In retrospective studies, the proportion of women reporting cyclical changes varies between 29% and 97%.[18] Andersch[19] found 2-3% of women to report severe premenstrual symptoms in a retrospective epidemiological study. Cross-cultural studies have shown no difference in prevalence in varying groups of women, although there are substantial differences in group symptom profiles.[20,21]

3. Measurement

Moos,[7] who developed the retrospective 47-item Menstrual Distress Questionnaire (MDQ), focused largely on physical complaints at the expense of psychological symptoms and behavioral changes. The MDQ was later modified by Clare and Wiggins[22] to enable prospective daily ratings to be made over a number of cycles. More recently, Sanders[23] in Edinburgh has devised a daily rating scale of core physical and psychological symptoms with the option of recording additional symptoms peculiar to the individual. Scoring is on visual analogue scales. A similar scale has been developed by Rubinow et al.[24]

The development of operational definitions of PMS together with valid and reliable measures of severity and change, much enhance recent studies in this field. Although prospective studies over a number of cycles are to be preferred, Logue and Moos[25] have shown that retrospective ratings are nevertheless accurate, and we have therefore not dismissed retrospective studies from our discussion.

4. Etiology

Numerous hypotheses, both psychosocial and biochemical, have been advanced as possible explanations for PMS.

The relationship between PMS and psychiatric illness has received considerable attention. Many studies, reviewed by Clare,[10] have looked at a variety of psychiatric disorders, including affective disorders, neurotic illness, personality disorder, and suicide and parasuicide. Despite problems of methodology, there is increasing evidence for an association between subtypes of PMS and psychiatric disorder, particularly subtypes of affective illness.[26] Clare has suggested that the severity of psychological, but not physical, symptoms in PMS sufferers can be used to discriminate between patients with and without psychiatric histories.

Many biological hypotheses have been proposed as explanations for the etiology of PMS. Circulating levels of gonadal steroids, prolactin, and aldosterone have been studied most intensively, although several other biochemical variables have also received attention, and will be reviewed here.

a. Gonadal Steroids

A recent study by Sanders et al.[27] demonstrated a clear relationship between mood, physical states, and hormonal phases of the menstrual cycle, thus strongly supporting a hormonal basis for PMS. Ruble,[28] however, showed that women report an increase in premenstrual symptoms prior to expected menstruation, yet others have shown a continuation of symptoms after hysterectomy. Bäckstrom et al.[29] reported that women with severe PMS who had a corpus lute-ectomy performed at the time of hysterectomy to disrupt their normal ovarian function, experienced a resumption of PMS-type symptoms when ovarian function recovered, although symptoms were of a less severe degree.

Frank[3] originally postulated that excess estrogen was responsible for symptoms, but research has produced conflicting results. A number of studies[30-33] have demonstrated significantly elevated estrogen levels and elevated estrogen-progesterone ratios in the premenstruum, whereas others have found no difference.[34-36] Bäckstrom and Mattsson[37] found correlations between estrogen levels, anxiety, and irritability, but Rubinow and Roy-Byrne[38] point out that symptoms and blood samples were often obtained in different menstrual cycles and there was wide variation in blood sampling times. Some studies have reported lower progesterone levels during the luteal phase in PMS, although it does not appear to be related to either the onset or the severity of symptoms.[30,32] Normal[35,36] and elevated[39] luteal levels of progesterone levels following ovulation have been reported in PMS sufferers. Bäckstrom et al.[31] further reported an increase in the concentration of follicle stimulating hormone (FSH) in the mid-luteal phase in PMS sufferers, suggesting that this might mediate the subsequent rise in estrogen levels found in these women. Surprisingly, interactions between pituitary and gonadal hormone secretion has otherwise received scant attention.

Most of the above studies have used clinical or partially operationalized entry criteria and a variety of undefined levels of severity of PMS. This may, in part, explain the inconsistent findings reported, although Magos and Studd have suggested that these very inconsistencies coupled with the observation that symptoms are usually worst immediately before menstruation — a time when estrogen and progesterone levels are falling — implies that fluctation in hormones and their metabolites, rather than absolute hormone levels, may be critical to symptom production.

b. Prolactin

Studies of prolactin (PRL), which causes retention of sodium, water and potassium, have similarly reported contradictory findings.[38] Halbreich et al.[40] found PRL to be increased during the premenstrual phase of the cycle more in PMS sufferers than in controls, while other workers[41,42] have found PRL levels to be higher in PMS sufferers with low progesterone levels. Andersch et al.[35] found PRL levels to be lower in the follicular phase in PMS sufferers with no difference between them and controls during the luteal phase.

c. Aldosterone

Aldosterone has been implicated in the etiology of PMS because of its salt- and water-retaining properties. Dalton[43] has suggested that excessive aldosterone action, related to low progesterone levels, is responsible for fluid accumulation premenstrually, and a number of earlier studies reviewed by Rubinow and Roy-Byrne[38] have supported this. However, a recent well-conducted prospective study[32] failed to show any significant difference in aldosterone levels between PMS sufferers and controls.

d. Other

Other variables which have been putatively linked with PMS, but in which investigations have generated negative results are pyridoxine,[18] serotonin uptake,[44] prostaglandins,[45] and androgens.[46] Ried and Yen[47] have postulated that luteal sensitivity to, and subsequent withdrawal from, the central effects of β-endorphin and α-melanocyte stimulating hormone (MSH) and variable gonadal steroid modulation of response to these neuropeptides, may provide an explanation for the clinical manifestations of PMS.

5. Treatment

Studies of treatment response have been lucidly reviewed by Rubinow and Roy-Byrne.[38] Many of these have methodological flaws similar to the studies of etiology cited above. In addition, their interpretation is confounded by the variable, but often high, placebo response rate. Magos et al.,[48] for example, reported a 94% placebo response which waned over successive cycles. The main treatment modality that has received attention is hormonal, although several other drugs have been tried. Despite the lack of evidence implicating low levels in PMS sufferers, progesterone has been widely used. Studies of its therapeutic use have, however, been inadequate and two double blind placebo-controlled cross-over trials have both been negative.[36,49] The synthetic progestogen, dydrogesterone, has shown more encouraging results.[50]

Specific attempts to treat PMS by suppression of cyclical ovarian activity have produced more encouraging results. Muse et al.[51] administered a gonadotropin-releasing hormone (GnRH) analogue to eight subjects in a double-blind cross-over trial and succeeded in abolishing the cyclical occurrence of premenstrual symptoms. Magos et al.,[48] in a study combining good patient selection criteria with adequate measures of changing menstrual symptoms, suppressed ovulation with subcutaneous estradiol implants in 68 women treated in a double-blind longitudinal study. Norethistrone was also administered to produce regular withdrawal bleeds. The initial very high placebo response waned, but the response to active treatment was maintained for the 10 months of the study, being significantly superior to placebo at and after 2 months.

Bromocryptine has been shown to be significantly more effective than placebo, in a number of double-blind studies reviewed by Dennerstein et al.,[52] in relieving premenstrual symptoms, particularly cyclical mastalgia. It is not known, however, whether this reflects suppression of PRL levels or other effects of bromocryptine.

Diuretics have been widely advocated in the treatment of PMS, both because of the possible involvement of vasopressin in the etiology of the condition and because of the commonly prominent symptom of bloating. A number of studies, reviewed by Rubinow and Roy-Byrne[38] have produced generally inconclusive results. The possible relationship of affective illness to PMS[26] suggests that lithium may be a rational treatment. Some response to lithium has been reported[53] in PMS sufferers with subsyndromal affective disorder, but lithium appeared no better than placebo in two double-blind cross-over studies.[54,55]

Despite the paucity of evidence suggesting a role for pyridoxine in PMS, it remains widely used as an over-the-counter preparation. Stokes and Mendels[56] failed to show pyridoxine to be more effective than placebo, though Abraham and Hargrove[57] have more recently reported it to be significantly superior. In contrast, Wood and Jakubowicz[58] found the prostaglandin (PG) synthetase inhibitor, mefenamic acid, to be significantly superior to placebo, despite a lack of theoretical justification for its use.

The nature of the relationship between subtypes of PMS and depression, and the potential to treat PMS by inhibiting ovarian function, have recently been and should continue to be fruitful areas of inquiry. For the present, however, although the biological basis for PMS and the disabling nature of its physical and psychological symptoms are relatively clear, our understanding of its precise etiology remains insufficient to allow rational and consistently effective treatments.

B. Dysmenorrhea

Dysmenorrhea, perhaps even more than PMS, is a condition which, because no underlying organic mechanisms have been unequivocally identified, has been labeled as "functional" or "psychological". Recent research, however, has made significant steps toward a more precise understanding of its etiology. In dysmenorrhea, unlike PMS, there appears to be little evidence for a direct connection between physiological symptoms and neurotic disturbance.

1. Definition and Prevalence

Dysmenorrhea is defined as "painful menstruation" and can be classified as primary (idiopathic) or secondary to organic pathology such as uterine fibroids, endometriosis, or pelvic inflammatory disease. Only primary dysmenorrhea will be considered here. Typically, the patient experiences cramping lower abdominal pain, radiating into the back and lower limbs, usually worst at the onset of menstruation. Other symptoms which commonly occur are nausea, vomiting, diarrhea, headaches, and urinary frequency.

The prevalence of primary dysmenorrhea, which most commonly affects teenage girls and diminishes with increasing parity but not with advancing age alone, has been estimated to occur with sufficient severity to require time off work or studies in 7% to 39%.[59] This disparity may be a reflection not only of the variation in definition and diagnostic criteria found in the literature, but also of the samples studied which have included specific groups of women, such as students, outpatient attenders, and industrial workers.[60] Andersch and Milsom[60] investigated the prevalence and severity of dysmenorrhea in a representative sample of 19-year-old women from Gothenburg, and found the overall prevalence of dysmenorrhea to be 72%, whereas that of severe dysmenorrhea was 15%.

2. Measurement

In a review of 51 clinical trials of the use of PG synthetase inhibitors in the treatment of primary dysmenorrhea, Owen[61] commented that the criteria and methods used to assess and analyze drug response were often imprecise and inadequate. The majority of trials relied on three- to six-point unidimensional rating scales and evaluated the severity of the dysmenorrhea solely on the subjective and often retrospective ratings of the patient. Ylikorkala and Dawood[62] had previously commented that such unidimensional rating scales reporting pain intensity or relief of pain neglect pain-associated effects such as interference with daily activities, thus preventing valid comparisons with other studies. In the study of prevalence by Andersch and Milsom,[60] a multidimensional scoring system was used. Pain was graded on a four-point scale, and an assessment of the effect of symptoms on daily activity, systemic symptoms, and analgesic requirements was also made. Studies of both prevalence and treatment need to adopt similar standard forms of assessment, giving both subjective and objective measurements of the degree of severity of dysmenorrhea.

3. Etiology

The interaction of psychological and physical factors is important in understanding the severity of any painful condition. It is inevitable that personality and social learning will influence the way symptoms, particularly pain, are perceived by an individual and will modify their behavior.[63] Research into the significance of personality as an etiological factor has produced inconclusive results. Coppen[64] found no relationship between dysmenorrhea and neuroticism, although Spero[65] reported that women with a menstrual disorder were more hypochondriacal. Bloom et al.[66] found women with dysmenorrhea to be more anxious and depressed than controls, but did not conclude that they were "neurotic". Cox and Santirocco[67] have noted life stresses and social supports to be significantly related to symptoms reported. The authors consider that the individual's personality and her social circumstances are important in determining such a subjective experience as pain, but their review offers little support for a primary psychological etiology for dysmenorrhea.

In contrast, a number of biological factors have been identified as playing an important part in the etiology of dysmenorrhea. The main areas of recent investigation have been (1) prostaglandins, (2) steroid hormones, and (3) vasopressin.

a. Prostagladins

$PGF_{2\alpha}$ is present in the endometrium and rises rapidly during the last part of the menstrual cycle, reaching its peak levels at the onset of menstruation.[68] Lundstrom[69] has shown that $PGF_{2\alpha}$, a potent myometrial stimulant in vitro, causes an increase in contractility when administered in vivo, and is consistently accompanied by dysmenorrhea-like pain. $PGF_{2\alpha}$ has now been shown to be present both in the endometrium and in the menstrual fluid in significantly greater amounts in those suffering from dysmenorrhea.[70] $PGF_{2\alpha}$ is also a powerful vasoconstrictor and prostacyclin (PGI_2) is one of the most potent vasodilators known. $PGE_{2\alpha}$ can also increase the sensitivity of nerve endings.[70] Further supporting evidence for the involvement of PGs in dysmenorrhea comes from the findings that PG synthetase inhibitors not only decrease menstrual fluid PG concentrations and decrease uterine contractility, but also relieve dysmenorrhea.[71]

b. Steroid Hormones

The role of steroid hormones in dysmenorrhea is less clear, although it is well recognized that, unlike PMS,[18] primary dysmenorrhea occurs almost exclusively in ovulatory menstrual cycles.[18,59,71] Supression of ovulation by any method results in relief of the pain of primary dysmenorrhea.[59] The rise in progesterone during the secretory phase of the cycle is accompanied by higher concentrations of PG in the endometrium,[72] although the exact nature of their relationship remains unknown.[68] In addition, uterine contractility is at its maximum when steroid hormone levels drop at the onset of menstruation.[71] In anovulatory cycles, there is no rise in endometrial PG, the pattern of myometrial contractions is similar to that seen during the follicular phase of the cycle, and the subsequent menstruation is painless.[59]

c. Vasopressin

Vasopressin has a powerful stimulant effect on the uterus, particularly at the onset of menstruation,[73] causes a decrease in myometrial blood flow, and is present in higher concentration at the onset of menstruation in those with dysmenorrhea.[74]

An elucidation of the role of PGs, particularly $PGF_{2\alpha}$ and the finding that PG synthetase inhibitors are capable of relieving menstrual pain, have given support to the early theory of uterine ischemia as the most likely explanation for dysmenorrhea.[68] Lumsden[71] has suggested a vicious cycle in which $PGF_{2\alpha}$ is absorbed from the degenerating endometrium on day 1 of menstruation, resulting in increased myometrial contractility and vasoconstriction, which in turn worsens ischemia and further increases the production of PG. Dysmenorrhea can be regarded as a function of a complex interaction of PGs, gonadal hormones, vasopressin, and pain perception, and it may be postulated that the biological abnormalities act as a stress on the background of a psychological diathesis.

4. Treatment

In view of the etiological evidence, rational attempts at treatment have included both PG synthetase inhibitors and the oral contraceptive pill. PG synthetase inhibitors have an analgesic effect, as well as inhibiting the synthesis of PGs, while the oral contraceptive pill exerts its effect by inhibiting ovulation, and thereby eliminating the luteal phase rise in endometrial PG concentration.

Both oral contraceptives and PG synthetase inhibitors have been subjected to controlled trials. The efficacy of oral contraceptives in reducing the frequency and severity of dysmenorrhea has been clearly shown in two large open studies[75,76] and in a double-blind controlled study.[77] Owen[61] reviewed 51 clinical trials of PG synthetase inhibitors and found that significant pain relief occurred in 72% of women, with a 15% placebo response rate. The methodologically sounder trials showed a somewhat more modest, but nevertheless significantly superior, response over placebo. The fenamate subgroup of PG synthetase inhibitors (flufenamic acid and mefenamic acid) was shown to be particularly effective in providing pain relief.

Symptom relief is often dramatic, with complete resolution of psychological, as well as physical, symptoms. The choice of treatment is usually determined by the contraceptive needs of the patient, although if one treatment is unsuccessful the other is usually tried. In view of the very high acceptability and effectiveness of the combined oral contraceptive pill in relieving primary dysmenorrhea, it is not necessarily inappropriate to administer this treatment to young girls who do not require contraceptive protection. If both treatments fail, calcium antagonists (nifedipine) are sometimes successful in low doses (15 to 30 mg/day) during menstruation, though the rationale for their use is unclear. During treatment, psychological factors cannot be ignored, and supportive psychotherapy should form an important part of the overall management. Surgical approaches to the treatment of primary dysmenorrhea have included dilatation and curettage, and presacral neurectomy. The theoretical justification for such hazardous procedures is doubtful, and they should not be used today.[71]

C. The Climacteric Syndrome

Decline in gonadal function in women has long been associated with psychological disturbance. Is there a physiological basis to such disturbance, or is it related to the major life events which often occur at this time in a woman's life?

1. Definition

A workshop, convened during the First International Congress on the Menopause, produced a composite definition of climacteric, menopause, and climacteric syndrome to facilitate comparability in research.[78] These definitions are summarized below.

1. The climacteric is that phase in the aging process of women, marking the transition from the reproductive to the nonreproductive period of life.
2. The menopause indicates the final menstrual period and occurs during the climacteric.
3. Specific symptomatology associated with the climacteric is termed the "climacteric syndrome" and consists of three main components:
   a. Decreased ovarian activity with subsequent hormonal deficiency results in early symptoms such as hot flushes, perspiration, and atrophic vaginitis; and late symptoms related to metabolic change in the end organ affected.
   b. Socio-cultural factors.
   c. Psychological factors.

The variety in symptomatology in the climacteric syndrome results from the interaction between these three components. The menopause is merely an event occurring within the climacteric, indicating that estrogen deficiency has reached a critical point.

In addition to the above conceptual definitions, operational definitions[79,80] have been developed. McKinlay and Jeffreys define a woman's menstrual status in terms of her last menstrual bleed, whereas Jaszmann et al. base their classification on changes in the pattern of the menstrual cycle. The disadvantage of McKinlay's relatively simple and more objective method, which only requires a woman to recall when her last bleed occurred, is that a woman might be classified as premenstrual when she is really perimenstrual.

2. Prevalence

Greene[81] reviewed 14 studies undertaken to determine which of the many symptoms attributed to the climacteric, actually increased in prevalence around the time of the menopause. Despite the wide variation in methods employed, there was remarkable consistency in identifying vasomotor symptoms (hot flushes, perspiration) as being closely associated with the menopause, occurring in 61 to 75% of subjects. There was little agreement between studies, however, as to the prevalence of other symptoms and their relationship to the menopause. Although symptoms such as depression, anxiety, fatigue, poor concentration, lability, headaches, and insomnia do increase significantly around the time of the menopause, the increase is not to the same degree as for vasomotor symptoms.

3. Measurement

Most studies of the climacteric syndrome to date have used symptom checklists.[81] The best of these have rated the severity of each symptom as well as its presence or absence. A number of factor analytic studies, reviewed by Greene,[81] have analyzed the relationship between these symptoms, attempting to identify clusters that might each have a single, but separate etiology. These studies have generated remarkably similar findings, the main clusters identified being vasomotor, psychological, and somatic symptoms. Greene and Cooke[82] have devised a climacteric symptom rating scale consisting of 18 items, each rated on a four-point scale of severity, allowing total scores for each of the above clusters to be calculated.

This scale offers the best available synthesis of earlier rating instruments. It provides the opportunity for increased comparability in subsequent studies of prevalence and severity of the climacteric syndrome, and the standardized evaluation of treatment interventions.

4. Etiology

There is a general consensus among both clinicians and researchers that symptoms of vasomotor instability and atrophic vaginitis are among the earliest and most characteristic of symptoms associated with the menopause. Vasomotor, and to a lesser extent psychological, symptoms have been correlated with declining levels of estrogen.[83] There is also general agreement that these symptoms are a consequence of estrogen deficiency.[84] Support for this view comes both from prevalence studies described above, and from the evidence discussed below, that considerable relief of symptoms is obtained with estrogen replacement therapy.

Whether estrogen deficiency is directly related to the increase in other symptoms, both physical and psychological, reported around the time of the menopause, still remains to be established. Earlier studies suffer from various methodological flaws, particularly of sampling. Recent, better-designed studies have attempted to overcome some of these earlier shortcomings, but have still produced conflicting results. Greene and Cooke[82] examined the relationship between different types of climacteric symptoms and menopausal status in a sample of urban Scottish women, and showed a significant peaking for both psychological and somatic symptoms within the early climacteric period, which was a few years prior to when the majority experienced vasomotor symptoms. In contrast to this, Bungay et al.[85] conducted a community survey in Oxfordshire of both men and women (aged 30 to 64 years) to investigate ''general health'', the subjects being unaware of the menopausal focus of the study. The main psychological symptoms reported by menopausal women were difficulty in making decisions, loss of confidence, anxiety, forgetfulness, difficulty in concentration, and feelings of worthlessness, with these peaking (although less impressively than vasomotor symptoms) just prior to the menopause.

Although the evidence for an increase in psychological symptoms around the menopause still remains somewhat inconsistent, there is fairly good evidence that such symptoms do in fact increase during the climacteric as a whole, neither confirming nor disproving a causal relationship to estrogen deficiency. Greene[81] suggests a vulnerability model to explain the increase in psychological symptoms reported during the climacteric, in which the physiological instability resulting from a decline in ovarian function, together with predisposing socioeconomic stresses and life events, results in the precipitation or exacerbation of psychological symptoms.

An increase in neurotic disorders has long been attributed to physiological changes occurring in women at the climacteric. Hallström and Samuelsson[86] carried out detailed psychiatric examinations on a representative sample of 899 middle-aged women in Gothenburg on 2 occasions with an interval of 6 years. Those women whose menopause started early (ages 30 to 39) had a higher prevalence of neurotic disorders than others and tended to have suffered from neurotic disorder even before the climacteric. On the whole, however, menopause was not identified as a precipitating factor for neurotic disorder.

5. Treatment

Treatment approaches to climacteric symptoms which need to be considered are (1) the degree of distress suffered, (2) that the decline in gonadal function is itself normal, and (3) the potential side effects of any therapy.

Clonidine has been found to be effective, despite a considerable placebo response.[87] However, the main approach to treatment has been estrogen replacement therapy, either orally, percutaneously, transvaginally, or by subcutaneous implant. Estradiol given by mouth is converted to estrone in the gut and largely inactivated by the liver.[83] Despite this, the majority of women gain significant relief of symptoms with oral estrogens. Greene[81] reviewed 15 controlled studies of the effect of estrogen replacement on climacteric symptoms. Despite differences in design, estrogen preparations, sample size, type of patients, and the time period for evaluating effects, all the studies showed that estrogen therapy reduced both vasomotor symptoms and insomnia, and, in all but three studies, a significant superiority of active drug over placebo was demonstrated. Apart from the study by Campbell and Whitehead[88] discussed below, the findings are less consistent regarding psychological symptoms, most studies identifying only one or two such symptoms for which estrogen therapy was significantly superior to placebo.

The two largest double-blind, placebo-controlled, cross-over studies published to date[88,89] both demonstrated a significant improvement in psychological symptoms with estrogen therapy over placebo, although in the Dennerstein et al.[89] study this was mainly due to a decrease in the anxiety items on the Hamilton Depression Rating Scale. Although both studies showed an improvement in a variety of symptoms, they used contrasting methods to assess psychological change.[90] Campbell and Whitehead[88] reported that certain symptoms (e.g., irritability) were only relieved secondary to abolition of hot flushes, although other symptoms (e.g., poor memory, anxiety, worry about age, and about self) improved significantly in non-flushing patients as well. They concluded that this provided further evidence of a role for estrogens in the production of psychological symptoms. These improvements were demonstrated in the 4-month trial, which included women with more severe symptoms; though in their 12-month trial, the only psychological symptoms to improve significantly more in the estrogen-treated group were insomnia and poor memory. The authors attribute this difference to the patients in this trial having less severe symptoms. Greene[81] suggests that another possible explanation may be that psychological symptoms initially improve secondary to relief of hot flushes, but subsequently return independently.

If oral medication is unsuccessful, or in cases of premature menopause, where estrogen therapy may be required for several years, subcutaneous implants may be preferred. Studd et al.[91] have shown that estradiol implants combined with subcutaneous testosterone improve psychosexual problems, such as loss of libido and anorgasmia, as well as core climacteric symptoms. Although estrogen therapy is of greatest benefit in relieving the vasomotor effects of the climacteric, it may also be of benefit in relieving psychological and sexual symptomatology, although the exact nature of the relationship between hormonal changes and mental symptoms remains unclear.

III. PREGNANCY

Neurotic disorders of normal pregnancy and the condition of pseudocyesis will be reviewed here.

A. Ante-Natal Anxiety and Depression

Pregnancy has long been regarded as a time of psychological well-being, and the epidemiological evidence is convincing that psychiatric hospitalization in pregnancy is rare.[92] Relatively minor neurotic disorders, and depression in particular, have been found, however, to be almost as prevalent antenatally as they are postnatally.[93,94]

1. Definition and Measurmenet

The lack of consistent diagnostic criteria across studies of antenatal anxiety and depression has rendered comparisons difficult. Early studies have also been flawed by a lack of standardization of interview techniques, inadequate measures of severity, and the absence of comparison groups. More recent studies[93,94] have utilized Research Diagnostic Criteria (RDC),[95] modified to specify an onset of anxiety and depression within the pregnancy, and standardized interview schedules[96,97] which include scales of known validity and reliability, measuring severity of several neurotic symptoms. The great majority of RDC "cases" fulfilled criteria for major, minor, or intermittent depression.

Pregnancy has an important pathoplastic influence on the clinical presentation of these neurotic states. Anxiety states in pregnancy usually reflect the mother's concern about the impending delivery and the physical well-being of the child.[98] In depressed patients, preoccupation and guilt over a previous termination and fears that the baby will be born damaged as a form of retribution, have been identified.[99]

2. Prevalence

Recent studies have estimated the prevalence of depression in pregnancy to be from 9 to 13% (compared to postpartum prevalences of 12 to 16%).[93,94,100] Kumar and Robson[93] found that the onset of depression was most likely to occur in the first trimester, while O'Hara's[94] estimate of 9% was obtained during a second trimester assessment. Watson et al.[100] reported 6% of cases to occur in first trimester and an additional 7% in the third trimester. They suggest that retrospective accounts by subjects whose first antenatal visit was in the second trimester may have resulted in an under reporting of depression earlier in the pregnancy.

These are consistent with the findings of an earlier study by Cox[101] of Ugandan women, in which 8% were found to be suffering from depression, and a further 8% from other neuroses. A control group of nonpregnant women was also included, and although there was an overall increase in depression and anxiety among pregnant women, this did not reach statistical significance. Nilsson and Almgren[102] also demonstrated an increase in depression and anxiety during pregnancy, but did not use a control group.

Dalton[103] and Cox et al.[98] have reported an increased prevalence for anxiety antenatally. However, Dalton, along with others,[104,105] has suggested an association between such antenatal anxiety and postnatal depression. This association has not been confirmed by more recent studies using reliable diagnostic criteria.

3. Etiology

Much research into antenatal depression has concentrated on psychosocial rather than biological factors. Studies of psychosocial causes of depression in pregnancy have produced conflicting results. Kumar and Robson[93] and O'Hara[94] have both confirmed earlier reports of a lack of relationship between ante- and postnatal depression (thus suggesting different etiologies); yet these two studies identified different psychosocial factors as possible causes for depression in pregnancy. Kumar and Robson[93] identified a number of factors that they found to be significantly associated with depression in primigravidae: high neuroticism and psychoticism scores, a history of past psychiatric problems, marital conflict, previous induced abortion, thoughts about having an abortion this time, and anxieties about the fetus. O'Hara,[94] only half of whose sample were primigravidae, found no difference in the incidence of stressful life events, level of marital satisfaction, or personal and family history of depression, between depressed and nondepressed subjects. Although a past history of induced abortion was not specifically mentioned, no difference in the obstetric histories of the two groups was noted. O'Hara postulated that somatic discomfort may have been a contributing factor to depression in pregnancy.

A number of biological variables have been measured at different stages of pregnancy, but generally only as part of the collection of baseline data for studies primarily concerned with postpartum disorders. An early study[106] found a negative correlation between levels of urinary noradrenaline excretion and scores on depression ratings in late pregnancy. The authors hypothesize that this reduction in urinary noradrenaline excretion may be interpreted as indicating an increased biological susceptibility to affective disorder in pregnancy, but point out that this assumes that urinary noradrenaline excretion reflects brain noradrenaline turnover, and that the depressive symptoms they measured are a legitimate model for true depressive illness.

Nott et al.[107] measured mood change and plasma levels of estrogen, progesterone, luteinizing hormone (LH), and follicle stimulating hormone (FSH) antenatally on three occasions, but their study focused primarily on postpartum mood change, and they did not provide detailed analysis of the relationship between antenatal mood and hormonal levels. Bonham Carter et al.[108] measured urinary tyramine output after a loading dose of oral tyramine in a group of 74 women in late pregnancy. Those with the lowest output had a higher lifetime incidence of depressive illness. None, however, was suffering from depression at the time. Unfortunately, the incidence of episodes of depression earlier in these pregnancies is not mentioned. The authors postulate that decreased excretion of tyramine is associated with vulnerability to depressive illness whether puerperal or nonpuerperal.

Ballinger et al.[109] measured a number of biochemical and behavioral parameters on 3 occasions during pregnancy (at 10 to 16, 32, and 38 weeks) and on 3 occasions after delivery. Relationships between mood and biochemical variables are presented only in terms of the comparison between those who showed a "positive mood change" after delivery and those who did not. The "positive mood change" group had lower hostility, depression, and anxiety ratings, as well as significantly lower plasma cyclic AMP levels, in pregnancy. The relationship between biochemical variables and concurrent mood was not reported, however.

These studies have demonstrated a number of biochemical changes in pregnancy related mainly to postpartum mood disturbance. The findings indicate the need for further studies of biochemical measures during pregnancy and their relationship to concurrent mood disturbance, and in particular for studies during early pregnancy, when abnormal mood change most frequently occurs.

4. Treatment

There have been no controlled studies of any treatment modality in the depressions of pregnancy. There is a legitimate reluctance to use drug treatment of any kind in pregnancy, and depression in pregnancy is often thought to be self-limiting and to need no more than psychotherapeutic intervention. The lack of consistent findings of a psychosocial etiology for these depressions, as well as recent findings suggestive of associated hormonal and biochemical abnormalities, implies that physical treatment may not, however, always be inappropriate. Tricyclic antidepressants (TCA) have, indeed, been shown to be effective in neurotic depressions without endogenomorphic features.[110] Most TCAs are relatively free of teratogenic effects,[111] and the avoidance of their use in pregnancy should be relative rather than absolute.

B. Pseudocyesis

Pseudocyesis is a distinctive condition that was observed by Hippocrates and is recognized to affect all races, all nations, and all strata of society.[112]

1. Definition and Prevalence

Pseudocyesis is a condition in which the physiological and psychological concomitants of pregnancy are present in the absence of the true gravid state.[113] Physiological changes range from disruption of normal menstruation, changes in abdominal size, quickening, and gastrointestinal symptoms through to enlarged breasts and lactation, cervical changes, and uterine enlargement. Bivin and Klinger[112] collected data on 444 cases of pseudocyesis drawn from 20 countries over the preceding 200 years. The age range was from 5 to 79 years, with the majority aged between 20 and 39 years. Most were married, and 41% had previously given birth. The pseudocyesis lasted for 9 months in 42% of cases. Recovery was occasionally spontaneous, but more often it was preceded by labor pains.

Accurate estimates of the prevalence of pseudocyesis have been difficult. In 1916-17, approximately 12 cases a year were seen at one New York Hospital.[112] In the late 1940s, Schopbach et al.[114] estimated the incidence to be 1 in 250 maternity admissions at a Philadelphia hospital. More recently, Brenner[115] estimated an incidence of 1 in 200 among new Black South African antenatal hospital patients, yet Cohen[113] suggests that the condition is rapidly disappearing. He postulates that many cases previously regarded as pseudocyesis are now diagnosed as suffering from the Galactorrhoea-Amenorrhoea-Hyperprolactinaemia Syndrome (GAHS). Both syndromes have a number of physical symptoms in common, although patients with GAHS do not usually believe themselves to be pregnant. Cohen argues for continued use of the term pseudocyesis for a small group of patients, who he considers to represent variants of monosymptomatic hypochondriasis or Munchausen's Syndrome.

2. Etiology

Classically, the origin of pseudocyesis has been attributed to fantasies about pregnancy which then lead to alterations in physiology.[116] However, the marked physiological changes that occur in pseudocyesis have led a number of researchers to look for endocrine abnormalities.

Clinical investigations of ovulatory status have obtained conflicting results, with active corpora lutea being found at laparoscopy in three cases quoted by Cohen[113] but not in two more recently reported cases.[117] PRL has been the subject of investigation in both animal and human studies. Coppola et al.[118] demonstrated in rats that by specifically depleting brain stores of noradrenaline with reserpine or α-methyldopa, they could inhibit PRL inhibitory factor release, resulting in both increased PRL levels and associated lactation, and in increased LH levels and pseudopregnancy. High levels of PRL have also been demonstrated in human subjects with pseudocyesis, although conflicting findings of LH and FSH levels have been recorded.[117,119]

A more recent investigation focused on the relationship between psychiatric and neuroendocrine abnormalities in two patients.[120] Both had abnormal growth hormone secretory patterns, demonstrated by a lack of sleep-associated peaks and the absence of a response to L-dopa. Estrogen and progesterone levels were raised, although PRL levels were normal. One patient, who met DSM III criteria[121] for major depressive disorder, had abnormally elevated LH levels. The authors conclude that no single neuroendocrine abnormality is common to all patients with pseudocyesis. It seems clear, however, that biological as well as psychodynamic variables need to be taken into account in further research into the etiology of this bizarre condition.

3. Treatment

Early treatments included emetics, purgatives, baths, curettage, massages, tonics, and even surgical procedures.[122] Earlier this century, psychotherapy became the treatment of choice, often supplemented by showing the patient pelvic x-rays or by parenteral administration of testosterone and diethylstilboestrol to induce menstruation.[123] Psychotropic drugs are seldom indicated,[99] and our present understanding of the biological basis for pseudocyesis is insufficient to suggest generally applicable physical treatment approaches.

IV. POSTPARTUM PERIOD

The maternity blues and postnatal depression will be reviewed in this section.

A. Maternity Blues

The maternity blues is such a commonly occurring and transient conditon that many regard it as a normal phenomenon following childbirth. It provides a convenient focus for examining the biological basis of mood disorder, since it is frequent, easily measured, occurs in the context of the major physiological changes of the early postpartum period, and appears to have a remarkable lack of psychological, social, and obstetric correlates.

1. Definition

The term "maternity blues" refers to a transitory syndrome of mood disturbance usually confined to the first 2 weeks postpartum. There is a wide variation in the choice of symptoms to be included in any definition of maternity blues, as well as in the criteria for onset, severity, and duration. The most frequently encountered symptoms include depression, tearfulness, irritability, elation, anxiety, confusion, forgetfulness, headache, and insomnia, but there is still no consensus about which symptoms should or should not be included.[2]

Pitt[124] considered the maternity blues to be a "trivial fleeting phenomenon" very different from postnatal depression, though Hopkins et al.[125] consider that the distinction between the maternity blues and clinical depression needs further corroboration. There is also a lack of agreement about the onset of symptoms and their pattern and frequency. Many early studies assessed women on only one occasion. Later studies[126-129] have identified peak symptoms on different days. Stein[130] made daily ratings during the first postpartum week in 37 normal women and found some symptoms (e.g., exhaustion, anorexia, and poor concentration) to be more frequent on day one, whereas other symptoms (e.g., depression, irritability, restlessness, and dreaming) have their peak between the fourth and sixth day.

Although this picture of distressing symptoms is what is classically regarded as the maternity blues, elation also frequently occurs. The peak incidence of elation is, however, on the first postpartum day; it occurs at this time in up to 80% of women, and is present in up to 40% of subjects on day four.[131-134] Much postpartum elation may be regarded as an understandable psychological reaction to having a new baby, but it can persist and be of an inappropriate degree. Robin[127] described a subgroup of women who were overexcited and garrulous when others were experiencing the maternity blues, whereas Ballinger et al.[132] estimated 80% of women in their study to have persistently excessive degrees of happiness and high spirits. No studies have described such postpartum elation occurring to an extent sufficient to be labeled as hypomanic.

2. Prevalence and Measurement

The incidence of the maternity blues in most series is 50 to 70%, although the range reported is from 15 to 80%.[2] Similar figures have been obtained in home- and hospital-based studies.[129] Studies in different countries[129,135,136] have also estimated episodes of minor psychological disturbance (particularly depression and lability of mood) as occurring in 50 to 70% of women after delivery.

Many earlier studies are flawed not only by the lack of suitable diagnostic criteria and measures of severity, but also by inadequate or unrepresentative patient numbers and by retrospective or infrequent assessments. These problems have been addressed to some extent by more recent studies which rated symptoms on a daily basis throughout the early puerperium and used instruments specifically designed for assessing the maternity blues. These have recently been reviewed in detail by Kennerley and Gath.[137] Interestingly, the four well-known scales[124,129,130,138] which were devised on the basis of clinical observations to measure the same syndrome have only three symptoms in common: depression (or sadness), anxiety (or nervousness), and crying (or weeping, tears).[137] A new 28-item rating scale developed by Kennerley and Gath[137] in Oxford has been shown to be reliable and valid, easy to administer, and acceptable to mothers. Cluster analysis revealed seven symptom clusters, with the most common containing seven items: tearful, tired, anxious, over-emotional, up and down in mood, low-spirited and muddled thinking. This cluster has been termed the "Primary Blues" by the authors and, in contrast to other rating scales, did not include "depressed", which appeared in another, much less frequent cluster. The development of this new instrument is to be welcomed as an essential tool for further studies of the relationship between postpartum mood disturbance and biological variables.

3. Etiology

An association between the maternity blues and mood disturbance in late pregnancy has consistently been reported.[107,130,135,139,140] and there is some evidence to suggest that a past history of neurotic disorder may be associated with more severe cases. An association with PMS has also been reported.[107,132] In general, however, studies attempting to find psychological, social or obstetric factors that might play a causal role in the production of the maternity blues have reported essentially negative results.[2]

The maternity blues syndrome has been postulated as a "model" for understanding the biochemical changes underlying mood disorder. Elucidation of its biochemical basis may be a gateway to an understanding of the etiology of postpartum depression and perhaps of affective disorders in general. A wide range of biological factors have been proposed and investigated.

a. Hormones
i. Gonadal Hormones

The very high plasma levels of estrogen, progesterone, and luteinizing hormone in late pregnancy are rapidly cleared from the circulation within a few days of delivery. The temporal relationship of these dramatic changes in hormonal levels to the maternity blues, as well as the known association between mood disturbance and the hormonal changes of the menstrual cycle, raise the possibility of a causal link.

Nott et al.[107] identified 4 weak correlations between hormonal levels and blues symptoms: (1) predelivery estrogen levels tended to be higher in those who were more irritable; (2) the greater the progesterone drop, the more likely were subjects to rate themselves as depressed within 10 days; but (3) the less likely they were to report sleep disturbance; and (4) the lower the estrogen levels postpartum, the more sleep disturbance was reported. These 4 correlations out of a possible total of 30 may, however, have occurred by chance. The authors also reported that mood disturbance was common in the 6 weeks prior to delivery, and that those who reported more distress during this time reported more immediately afterwards. Predelivery estrogen levels tended to be higher in those who were more irritable. The study is weakened by the small number of subjects and large number of variables examined.

More recently, however, in a pilot study of 40 primiparous mothers, the 5 who experienced most severe maternity blues were matched wtih 5 who were symptom free.[140] On the days of symptoms, saliva concentrations of estradiol and progesterone were significantly higher in the maternity blues group. The authors did not collect antenatal samples and were somewhat reluctant to suggest that there is a less precipitous fall in these hormones in maternity blues sufferers postpartum.

ii. Prolactin

George et al.,[141] in a study of 38 women with no past history of mental illness, investigated the relationship between serum PRL and the maternity blues. Subjects were interviewed on days 2, 4, and 6 after giving birth, using the Present State Examination (PSE);[142] and serum PRL levels were measured on the first day before breast feeding and at the time of each interview. Despite the relatively high threshold for symptom scoring on the PSE, the study demonstrated statistically significant correlations between serum PRL levels and anxiety, tension, and worries over the 6 days.

iii. Cortisol

Serum cortisol levels have received much attention and have been linked to a number of mental disorders, particularly depression.[143] In addition, the similarity of the maternity blues to the mental symptoms associated with Cushing's Disease had led researchers to postulate an association between cortisol levels and the maternity blues. As with gonadal hormones, cortisol rises during pregnancy, although most of this increase is bound to transcortin, the cortisol-binding globulin. Diurnal variation is diminished as in depressive illness, and the level rises during labor but falls rapidly soon after delivery.[2]

In a longitudinal study, Handley and colleagues[139] have reported an association between elevated plasma cortisol at 38 weeks of pregnancy and the subsequent development of relatively severe maternity blues. Their finding in an earlier study of a link between raised serum cortisol and elation was, however, not confirmed.[126] The dexamethasone suppression test (DST) has been examined as a possible marker for the maternity blues and postnatal depression, and a high nonsuppression rate has been reported,[144] but the immediate post-partum period is associated with false positive nonsuppression in 82% of subjects, inde-pendent of concurrent or subsequent psychiatric symptomatology.[144,145]

iv. Thyroid Function

Biochemical hypothyroidism, not usually associated with myxedema, commonly occurs in the postpartum period, may last for several months, and can be explained only partially in terms of the changes in specific binding proteins.[146] Regular screening has been advo-cated.[147] Hamilton[148] has reported encouraging responses to thyroxine, and particularly to triiodothyronine, in postnatal depression.

b. Biochemical

i. Body Weight

Weight loss in the postpartum period does not usually begin until the third or fourth day and is fairly rapid thereafter, with losses of 0.5 to 1.5 kg/day. Stein et al.[134] found that in those with an abrupt onset of mood disturbance, there was weight loss of 0.5 kg per 24 hr on the day of the mood swing, which continued for the next two days, weight having been relatively static prior to the onset of the mood swing. This finding was confirmed in a later study.[149] Similar changes in body weight have been reported in functional psychoses, in particular rapid-cycling manic depressive illness. A disturbance of hypothalamic functions as part of the mental illness has been suggested as a possible explanation.[150]

ii. Fluid and Electrolytes

Attempts to link fluid and electrolyte changes in the early postpartum period with mood disturbance have met with little success. Stein noted a rise in urinary sodium excretion which coincided with both mood disturbance and weight loss.[2] In addition, calcium was lower than in controls and fell during the first week postpartum. Riley[151] found significantly higher plasma calcium levels in patients with the maternity blues than in those without the syndrome, although they were still within the normal range, and mental disturbances due to hypercal-cemia are usually associated with considerably elevated calcium levels.

Stein measured urinary vasopressin daily from the second to the seventh day postpartum, but was unable to demonstrate any correlation to mental changes,[130] and has cast some doubt as to the extent to which urinary vasopressin is an accurate reflection of plasma vasopressin or posterior pituitary function, suggesting that further exploration of the renin-angiotensin system may prove fruitful, particularly as angiotensin II prevents the reuptake of noradren-aline in the brain.[149]

iii. Cyclic AMP

Cyclic AMP (adenosine 3' 5' monophosphate) excretion has been reported to be increased in mania, and decreased in depression.[152] Lower levels have been found in depression, with restoration of normal levels following treatment.[153,154] Ballinger et al.[132] have reported that the change in excretion of cyclic AMP following delivery was related more to mood change than to mood state at any particular time. This was particularly evident in those who experienced elation between days 2 and 4 following delivery. A subsequent study by Ballinger et al.[109] showed that in "positive mood change" subjects, urinary cyclic AMP, plasma cyclic AMP, whole blood cell cyclic AMP, and ATP (adenosine 3' 5' triphosphate), hematocrit, and urinary 11-OHCS (hydroxycortisol steroids) following delivery were different from those observed in the rest of the group, and comparable with the biochemical changes described during upswings in mood in rapid-cycling manic depressive subjects.

iv. Monoamines

Treadway et al.[106] were unable to demonstrate any significant differences in levels of urinary excretion of noradrenaline and depressive scores in postpartum or control subjects. The possibility of abnormalities of tryptophan metabolism in maternity blues have been more extensively investigated. Stein et al.[128] and Handley et al.,[126] in studies of relatively small groups of subjects, both found a significant association between a lowered free plasma tryptophan and mood disturbance. In a subsequent larger study of 71 subjects, Handley et al.[139] found free plasma tryptophan to be reduced in subjects with maternity blues, but only at the time of year when free tryptophan was normally high. They also reported that an absence of the normal rise in total tryptophan in late pregnancy and day 1 postpartum was significantly associated with severe maternity blues, and with depression at 6 months postpartum. The lack of a tryptophan peak in "maternity blues" cases has been confirmed recently by Gard et al.[155] in a study which restricted itself to the months when maternity blues is most prevalent, although they were unable to confirm an association with the later development of depression. They concluded that alterations in tryptophan metabolism are unlikely to be causative, citing the study by Harris[156] in which L-tryptophan supplementation failed to prevent the incidence or severity of maternity blues, and suggested that tryptophan levels are more likely to be an epiphenomenon of some underlying predisposing mechanism.

v. Monoamine Oxidase (MAO)

The menstrual cycle influences platelet MAO, with the highest activity occurring just prior to ovulation and the lowest in the post-ovulatory period.[2] George and Wilson[157] investigated platelet MAO activity in relation to the maternity blues in 38 women and found a significant association between a rise in platelet MAO activity and depression and, to a lesser degree, loss of concentration and obsessionalism. They stated that, as platelets have a life of about 10 days, the correlations may reflect both pre- and postpartum events, and concluded that the changes in mood and platelet MAO may be related to underlying but unspecified hormonal changes.

vi. Platelet Receptor Binding

Studies of platelet α_2-noradrenergic receptors (adrenoceptors) in depression have produced conflicting results.[158] Metz et al.[159] investigated platelet α_2-adrenoceptor binding characteristics in 28 healthy subjects during the peripartum period. The fall in circulating estrogen and progesterone levels after childbirth was paralleled by a fall in the number of α_2-adrenoceptors. In subjects who subsequently developed the maternity blues, platelet α_2-adrenoceptor binding capacity remained significantly higher than in either subjects who did not or in a control group of normally menstruating women. The authors concluded that platelet α_2-adrenoceptors may be affected by the circulating levels of endogenous estrogen and progesterone and that a delayed or diminished fall in platelet α_2-adrenoceptor binding capacity after childbirth may be associated with the development of the maternity blues. The authors have recently replicated the findings of this study and have extended assessment to 6 months postpartum, but were unable to demonstrate any differences in platelet α_2-adrenoreceptor binding between depressed and nondepressed subjects at this stage.[160]

Katona et al.[161] examined platelet ^{3}H-imipramine binding, which is thought to be functionally related to serotonin uptake, in a cross-sectional study of 70 women in pregnancy and the early postpartum period, and in 23 nonpregnant women of childbearing age. Concurrent mood was measured using self-administered questionnaires.[130,162] No significant differences in the number of binding sites between groups were found, and there was no convincing relationship between mood and binding capacity within or across groups, but the binding affinity was significantly low at 5 to 7 days postpartum. A second study, following 100 subjects prospectively through pregnancy and postpartum demonstrated a fall in ^{3}H-imipramine binding capacity in the early postpartum period, though this fall was unrelated to mood disturbance.[160]

vii. β Endorphins

Newnham et al.[163,164] noted that plasma concentrations of β endorphin are raised during pregnancy and labor and fall rapidly after delivery. They postulated that the sudden withdrawal of these endogenous opioids might be responsible for the maternity blues. They found a negative correlation between subjects' estimates of their pain in labor and the β endorphin concentrations postpartum, suggesting an analgesic role for β endorphins in labor. They also found a positive correlation between β endorphin concentrations at delivery and attitude to pregnancy at 36 weeks, and a negative correlation between the maternity blues score and the β endorphin concentration at 36 weeks. However, the maternity blues score did not correlate either with the β endorphin concentration at delivery or at 24 hr postpartum. The results neither confirm nor refute the opioid withdrawal hypothesis.

The relationship between early postpartum mood and absolute levels or rates of change in plasma estrogen and progesterone is less clear than clinical observations would suggest. Maternity blues, like PMS, shows associations with a number of biochemical variables, which themselves may be partly mediated by alterations in gonadal steroid levels. The tantalizing prospect remains that further elucidation of these relationships may contribute to our understanding of affective disorders.

4. Treatment

The benign and transient nature of most cases of maternity blues means that simple reassurance is all that is necessary. More severe cases should be viewed with caution and followed carefully as these women have an increased risk of developing postnatal depression which may require more specific treatment measures.

B. Postnatal Depression

While the main focus of research in the maternity blues has been biological rather than psychosocial, the reverse has been the case for postnatal depression. In particular, psychosocial research in postnatal depression has been stimulated by the clearly demonstrated association between stressful life events and psychiatric illness.[165] Recently, however, there has been a revival of interest in investigating the relationship of biological variables to postnatal depression.

1. Definition and Measurement

The use of poorly defined or inadequate criteria for assessing and diagnosing postnatal depression, and even for defining the duration of the puerperium, raises serious doubts about the findings of a number of earlier studies. Tod[104] failed to specify the criteria he used for diagnosing postnatal depression, while Dalton's[103] criteria were simply that subjects either be seeking psychiatric outpatient help, or receiving drug treatment from a general practitioner. Pitt[166] attempted to overcome some of these problems by laying down the following criteria for diaganosing postnatal depression in his study: (1) subjects should describe depressive symptoms, which (2) have developed since delivery, (3) are unusual in their experience and, to some extent, disabling, and (4) have persisted for more than 2 weeks.

Pitt termed postnatal depression "atypical", as it tended to be mild with neurotic rather than endogenous features predominating, and suicidal ideation relatively rare. The study did not include a nonpostpartum depressed control group, and nonstandardized interview ratings of depression were used. Apart from the finding by other workers[167] of the relative infrequency of suicidal ideation, there is no firm evidence to suggest major differences, either in the nature or severity of symptoms, between postnatal and nonpostnatal depression. Recent studies using the RDC and standardized interview schedules[96,97] have shown postnatal depression to fulfil criteria for major or minor depression.[93,94]

2. Prevalence

Pitt[166] estimated the incidence of postnatal depression to be 11%. This has been confirmed by many investigators,[99] although estimates vary considerably from 2.9[104] to 30%.[168] Such differences may relate to the size of samples studied and the methods used for rating and diagnosing postnatal depression.[167] Watson et al.[100] reported the prevalence of psychiatric disorder (the majority being affective) to be 16% at 6 weeks postpartum, rising to 26% at one year. The onset of depression preceded childbirth in one third of subjects. It is clear that most cases of postnatal depression in the community go undetected. In the light of this, Cox[169] has recently validated a brief self-report rating scale, designed specifically to identify cases of postnatal depression, which can be administered by health visitors and thus detect cases who would not otherwise come to professional attention.

3. Etiology

A consistent association with marital conflict has been reported.[93,95,98,100] With this exception, investigation of psychosocial variables has again resulted in conflicting findings. Kumar and Robson[93] and Cox et al.[98] did not find an increased incidence of a past history of depression, although O'Hara[94] and Watson et al.[100] found that depressed postpartum subjects reported significantly more previous episodes of depression than their nondepressed counterparts. O'Hara also reported that depressed subjects were more likely to have at least one depressed first-degree relative.

The identification of psychosocial vulnerability factors or precipitants in no way precludes the etiological significance of biological variables. Much biological research has, however, been limited to the early postpartum period. This may be due in part to inherent difficulties in obtaining a large enough sample for repeated measures. The majority of postnatal depression "cases" have their onset after discharge from the hospital and are managed in the community by general practitioners, if they come to medical attention at all.

Early observations[103,166] of a relationship between menstrual difficulties (PMS and dysmenorrhea) and postnatal depression suggested the possibility of a biological basis for the latter disorder. This association has not, however, been replicated.[94]

The evidence of a link between hormonal levels and postnatal depression is inconclusive.[107] Gard et al.[155] found no association between postnatal depression and cortisol, estrogen, progesterone, or tryptophan levels, although they reported an altered pattern of decline in nonesterified fatty acids in their depression cases. Harris et al.[170,171] recently examined salivary gonadal hormone levels (which reflect free rather than total plasma levels) of 150 mothers at 6 weeks after delivery. Overall, depression was associated with higher levels of progesterone. However, further analysis of the data revealed a significant negative correlation between depression (as measured on the Beck Scale) and progesterone in breast feeders, but a significant positive correlation in bottle feeders.

In general, biochemical and hormonal studies, particularly those examining estrogen and progesterone levels or their rates of change, have been inconclusive. The sensitivity of brain receptors for these hormones may, however, be more relevant. Preliminary evidence[172,173] shows that estrogen receptor sensitivity (as measured by the stimulation of secretion of estrogen-sensitive neurophysin) was lower in subjects assessed on the fourth postpartum day than in nonpregnant controls. This change in research focus toward receptor sensitivity parallels that of recent developments in biological research in affective disorders.[158]

4. Treatment

The lack of consensus on the management of postnatal depression is a reflection both of the inconsistent psychosocial evidence regarding its etiology and treatment and of the paucity of biological investigations. Riley[174] reported significantly reduced depressive symptomatology in postpartum subjects given pyridoxine in a double-blind placebo-controlled trial of 94 randomly selected recently delivered women. This reduction was particularly marked in subjects with a past history of PMS, who overall were more likely to have depressive symptoms in both the early postpartum period and at one year. Unfortunately, this study did not define criteria for PMS nor did it use a standardized interview, rendering the results difficult to interpret. To our knowledge, however, there are no controlled trials of any other physical treatment in postnatal depression.

In contrast, two recent studies have examined the effect of minimal psychosocial intervention on the incidence of postnatal depression and have produced some encouraging results. Holden et al.[175] randomly allocated 50 mothers to 2 treatment groups at approximately 3 months postpartum. One group received standard postnatal care, whereas the other, in addition, received 8 half-hour individual counseling sessions of a nondirective supportive nature from a health visitor. At the end of the 8 sessions, the prevalence of depression in the intervention group, as assessed by the standardized psychiatric interview[96] and according to RDC[97] for major and minor depression, was significantly lower. Similar results were obtained by Leverton and Elliott[176,177] in a study of 99 first- and second-time pregnant women, identified as either socially or psychologically vulnerable from a total of 220 subjects. Of the vulnerable group, 48 were randomly assigned to an intervention package, which included an invitation to attend a monthly support group (from the third month of pregnancy to 6 months postpartum) as well as early contact with their health visitor. This was in addition to routine antenatal care. The prevalence of depression (borderline or definite cases), as assessed using the PSE and the Bedford College Criteria,[178] was significantly lower in the intervention group at 3 months postpartum. This was particularly marked in first-time mothers, who had been able to attend the support groups more regularly.

For the present, the management of postnatal depression must be determined by clinical judgment with relatively little scientific backing. Although social and psychological support for the depressed woman and her family forms part of any overall management plan, the role of physical treatment should not be dismissed. Pitt[179] suggests that because of the "atypical" nature of postnatal depressions, they may be expected to respond particularly well to monoamine oxidase inhibitors (MAOIs). There is, however, little evidence to support either the "atypical" nature of postnatal depressive symptoms or the therapeutic advantages of MAOIs in this patient group. The decision whether to give antidepressants and the choice between them, must therefore follow similar principles to those in force outside the context of childbirth. Breast feeding may complicate the issue to some extent. Lithium, in particular, may be ingested in sufficient quantity from breast milk to affect the infant adversely.[180]

V. CONCLUSION

Despite the widely held assumption that psychiatric symptoms associated with childbirth and the menstrual cycle have a prominent psychosocial etiology, there is considerable evidence for an important biological component. Paradoxically, the strongest such evidence comes from the evaluation of biological treatment interventions rather than from studies focusing directly on etiology. In particular, studies of hormone levels in these conditions have been uniformly inconclusive, yet hormonal treatments are often clearly effective. Improvements in research methodology offer the prospect not only of rapidly increasing understanding of the biological basis of these conditions, but through them, perhaps also of the neuroses and affective disorders in general.

REFERENCES

1. **Steiner, M., Haskett, R. F., and Carroll, B. J.,** Premenstrual tension syndrome: The development of research diagnostic criteria and new rating scales, *Acta Psychiatr. Scand.,* 62, 177, 1980.
2. **Stein, G.,** The Maternity Blues, in *Motherhood and Mental Illness,* Kumar, R. and Brockington, I. F., Eds., Academic Press, London, 1982.
3. **Frank, R. T.,** The hormonal causes of premenstrual tension, *Arch. Neurol. Psychiatry,* 26, 1053, 1931.
4. **Kramp, J. L.,** Studies of the premenstrual syndrome in relation to psychiatry, *Acta Psychiatr. Scand.,* (suppl. 203), 261, 1968.
5. **Sutherland, H. and Stewart, I.,** A critical analysis of premenstrual syndrome, *Lancet,* 1, 1180, 1965.
6. **Dalton, K.,** *Premenstrual Syndrome and Progesterone Therapy,* 2nd ed., Heinemann, London, 1984, chap. 1.
7. **Moos, R. N.,** The development of a menstrual distress questionnaire, *Psychosom. Med.,* 30, 853, 1968.
8. **Dennerstein, L. and Burrows, C. D.,** Affect and the menstrual cycle, *J. Affect. Disorders,* 1, 77, 1979.
9. **Slade, P.,** Premenstrual emotional changes in normal women: fact or fiction, *J. Psychosom. Res.,* 28, 1, 1984.
10. **Clare, A. W.,** Psychiatric and social aspects of premenstrual complaint, *Psychol. Med. Mon.,* (suppl. 4), 1983.
11. **Magos, A. and Studd, J.,** The premenstrual syndrome, in *Progress in Obstetrics and Gynaecology,* vol. 4, Studd, J. W. W., Ed., Churchill Livingstone, Edinburgh, 1984, chap. 24.
12. **Katz, M.,** *The Premenstrual Syndrome,* Postgraduate Centre Series, Update Publications, Guildford, Surrey, 1984.
13. **Haskett, R. F. and Abplanalp, J. M.,** Premenstrual Tension Syndrome: Diagnostic Criteria and Selection of Research Subjects, *Psychiatry Res.,* 9, 125, 1983.
14. **d'Orban, P. T. and Dalton, J.,** Violent crime and the menstrual cycle, *Psychol. Med.,* 10, 353, 1980.
15. **Dalton, K.,** Menstruation and Crime, *Br. Med. J.,* 2, 1752, 1961.
16. **Dalton, K.,** Cyclical criminal acts in premenstrual syndrome, *Lancet,* 2, 1070, 1980.
17. **Halbreich, U., Endicott, J., Schachts, S., and Nee, J.,** The diversity of premenstrual changes as reflected in the Premenstrual Assessment Form, *Acta Psychiatr. Scand.,* 65, 46, 1982.
18. **Bancroft, J. and Bäckstrom, T.,** Premenstrual Syndrome, *Clin. Endocrin.,* 22, 313, 1985.
19. **Andersch, B.,** Epidemological hormone and water balance studies on premenstrual tension, Thesis, University of Gotheburg, quoted in Bancroft, J. and Bäckstrom, T., *Premenstrual Syndrome, Clin. Endocrinol.,* 22, 313, 1985.
20. **Janiger, O., Riffenburgh, R., and Kersh, R.,** Crosscultural study of premenstrual symptoms, *Psychosomatics,* 13, 226, 1972.
21. **Stout, A. L., Grady, T. A., Steege, J. F., Blazer, D. G., George, L. K., and Melville, M. L.,** Premenstrual Symptoms in Black and White Community Samples, *Am. J. Psychiatry,* 143, 1436, 1986.
22. **Clare, A. W. and Wiggins, R. D.,** The construction of a modified version of the Menstrual Distress Questionnaire for use in general practice populations, in *Emotions and Reproduction,* Vol. 20A, Larenza, L. and Zichella, L., Eds., Academic Press, London, 1979, 177.
23. **Sanders, D.,** Hormones and Behaviour during the Menstrual Cycle, PhD. Thesis, University of Edinburgh, 1981.
24. **Rubinow, D. R., Roy-Byrne, P., Hoban, M. C., Gold, P. N., and Post, R. M.,** Prospective Assessment of Menstrually Related Mood Disorders, *Am. J. Psychiatry,* 141, 684, 1984.
25. **Logue, C. M. and Moos, R. H.,** Perimenstrual Symptoms: Prevalence and Risk Factors, *Psychosom. Med.,* 48, 388, 1986.
26. **Endicott, J., Halbreich, U., Schacht, H. T., and Nee, J.,** Premenstrual changes and affective disorders, *Psychosom. Med.,* 43, 519, 1981.

27. **Sanders, D., Warner, P., Bäckstrom, T., and Bancroft, J.,** Mood, Sexuality, Hormones and the Menstrual Cycle: 1. Changes in Mood and Physical State: Description of Subjects and Method, *Psychosom. Med.,* 45, 487, 1983.

28. **Ruble, D. N.,** Premenstrual symptoms: a reintepretation, *Science,* 197, 291, 1977.

29. **Bäckstrom, C. T., Boyle, H. J., and Baird, D. T.,** Persistence of symptoms of premenstrual tension in hysterectomised women, *Br. J. Obstetr. Gynaecol.,* 88, 530, 1981.

30. **Bäckstrom, T. and Carstensen, H.,** Estrogen and progesterone in plasma in relation to premenstrual tension, *J. Steroid Biochem.,* 5, 257, 1974.

31. **Bäckstrom, T., Wide, L., Soderga, R., and Carstensen, H.,** FSH, LH, TeBG-capacity, estrogen and progesterone in women with premenstrual tension during the luteal phase, *J. Steroid Biochem.,* 7, 473, 1976.

32. **Munday, M. R., Brush, M. G., and Taylor, R. W.,** Correlations between progesterone, oestradiol, and aldosterone levels in the premenstrual syndrome, *Clin. Endocrinol.* (Oxford), 14, 1, 1981.

33. **Abraham, G. E., Elsner, C. W., and Lucas, L. A.,** Hormonal and behavioural changes during the menstrual cycle, *Serologica,* 3, 33, 1978.

34. **Andersen, A. N., Larsen, J. F., Steerup, O. R., Svendstrup, B., and Nielsen, J.,** Effect of bromocriptine on the premenstrual syndrome: a double-blind clinical trial, *Br. J. Obstetr. Gynaecol.,* 84, 370, 1977.

35. **Andersch, B., Abrahamson, L., Wendestam, C., Ohman, R., and Hahn, L.,** Hormone profile in premenstrual tension: effect of bromocriptine and diuretics, *Clin. Endocrinol.,* (Oxford), 11, 657, 1979.

36. **Taylor, J. W.,** Plasma progesterone, oestradiol 17 beta and premenstrual symptoms, *Acta Psychiatr. Scand.,* 60, 76, 1979.

37. **Bäckstrom, T. and Mattsson, B.,** Correlations of symptoms in premenstrual tension to oestrogen and progesterone concentrations in blood plasma, *Neuropsychobiology,* 1, 80, 1975.

38. **Rubinow, D. R. and Roy-Byrne, P.,** Premenstrual Syndromes: Overview from a Methodological Perspective, *Am. J. Psychiatry,* 141, 163, 1984.

39. **O'Brien, P. M. S., Selby, C., and Symonds, E. M.,** Progesterone, fluid and electrolytes in premenstrual syndrome, *Br. Med. J.,* 2, 1161, 1980.

40. **Halbreich, U., Assael, M., Ben-David, M., and Bornstein, R.,** Serum prolactin in women with premenstrual syndrome, *Lancet,* 2, 654, 1979.

41. **Munday, M.,** Hormone levels in severe premenstrual tension, *Curr. Med. Res.,* 4 (suppl. 4), 1977.

42. **Brush, M. G.,** Endocrine and other biochemical factors in the aetiology of the premenstrual syndrome, *Curr. Med. Res.,* 6 (suppl. 19), 1979.

43. **Dalton, K.,** Progesterone, fluid and electrolytes in premenstrual syndrome, *Br. Med. J.,* 2, 61, 1980.

44. **Tam, W. Y. K., Chan, M-Y, and Lee, P. H. K.,** The Menstrual Cycle and Platelet 5-HT Uptake, *Psychosom. Med.,* 47, 352, 1985.

45. **Jakubowicz, D. L., Godard, E., and Dewhurst, J.,** The Treatment of Premenstrual Tension with mefenamic acid: analysis of prostaglandin concentration, *Br. J. Obstetr. Gynaecol.,* 91, 78, 1984.

46. **Bäckstrom, T., Sanders, D., Leask, R., Davidson, D., Warner, P., and Bancroft, J.,** Mood, sexuality, hormones and the menstrual cycle II. Hormone levels and their relationship to the premenstrual syndrome, *Psychosom. Med.,* 45, 503, 1983.

47. **Reid, R. C. and Yen, S. S. C.,** Premenstrual Syndrome, *Am. J. Obstetr. Gynaecol.,* 139, 85, 1981.

48. **Magos, A. L., Brincat, M., and Studd, J. W. W.,** Treatment of the premenstrual syndrome by subcutaneous oestradiol implants and cyclical oral norethisterone: placebo controlled study, *Br. Med. J.,* 292, 1629, 1986.

49. **Sampson, G. A.,** Premenstrual syndrome: a double blind controlled trial of progesterone and placebo, *Br. J. Psychiatry,* 135, 209, 1979.

50. **Haspels, A. A.,** A double-blind, placebo-controlled, multicentre study of the efficacy of dydrogesterone (Duphaston), in *The Premenstrual Syndrome,* Van Keep, P. A. and Utian, W. H., Eds., MTP Press, Lancaster, England, 1981.

51. **Muse, K. N., Cetel, N. S., Futterman, L. A., and Yen, S. S. C.,** The Premenstrual Syndrome — Effects of "Medical Ovariectomy", *N. Engl. J. Med.,* 311, 1345, 1984.

52. **Dennerstein, L., Spencer-Gardner, C., and Burrows, G. D.,** Mood and the Menstrual Cycle, *J. Psychiatr. Res.,* 18, 1, 1984.

53. **Steiner, M., Haskett, R. F., Osmun, J. N., et al.,** Treatment of premenstrual tension with lithium carbonate, *Acta Psychiatr. Scand.,* 61, 96, 1980.

54. **Singer, K., Cheng, R., and Schou, M.,** A controlled evaluation of lithium in the premenstrual tension syndrome, *Br. J. Psychiatry,* 124, 50, 1974.

55. **Mattsson, B. and von Schoultz, B.,** A comparison between lithium, placebo, and a diuretic in premenstrual tension, *Acta Psychiatr. Scand.,* (suppl. 255), 75, 1974.

56. **Stokes, T. and Mendels, J.,** Pyridoxine and premenstrual tension, *Lancet,* 1, 1177, 1972.

57. **Abraham, G. E. and Hargrove, J. T.,** Effect of vitamin B6 on premenstrual symptomatology in women with premenstrual tension syndrome: a double-blind crossover study, *Infertility,* 3, 155, 1980.

58. **Wood, C. and Jakubowicz, D.,** The treatment of premenstrual symptoms with mefenamic acid, *Br. J. Obstetr. Gynaecol.,* 87, 627, 1980.

59. **Lamb, E. J.,** Clinical Features of Primary Dysmenorrhoea, in *Dysmenorrhoea,* Dawood, M. Y., Ed., Williams and Wilkins, Baltimore, 1981, chap. 6, 107.

60. **Andersch, B. and Milsom, I.,** An epidemiologic study of young women with dysmenorrhoea, *Am. J. Obstetr. Gynaecol.,* 144, 655, 1982.

61. **Owen, P. R.,** Prostaglandin synthetase inhibitors in the treatment of primary dysmenorrhoea, *Am. J. Obstetr. Gynaecol.,* 148, 96, 1984.

62. **Ylikorkala, O. and Dawood, M. Y.,** New concepts in dysmenorrhoea, *Am. J. Obstetr. Gynaecol.,* 130, 833, 1978.

63. **Pilowsky, I.,** Pain and illness behaviour: assessment and management, in *Pain,* Melzack, R. and Wall, P. D., Eds., Churchill Livingstone, Edinburgh, 1984.

64. **Coppen, A.,** The prevalence of menstrual disorders in psychiatric patients, *Br. Med. J.,* 111, 155, 1965.

65. **Spero, J. R.,** A study of the relation between selected functional disorders and interpersonal conflict, *Diss. Abstr. Int.,* 29, 2905, 1969.

66. **Bloom, L. J., Shelton, J. L., and Michaels, A. C.,** Dysmenorrhoea and personality, *J. Personality Assessment,* 42, 272, 1978.

67. **Cox, D. J. and Santirocco, B. B. G.,** Psychological and Behavioural Factors in Dysmenorrhoea, in *Dysmenorrhoea,* Dawood, M. Y., Ed., Williams and Wilkins, Baltimore, 1981, 75.

68. **Shapiro, S. S.,** Flurbiprofen for the Treatment of Primary Dysmenorrhoea, *Am. J. Med.,* 80, 71, 1986.

69. **Lundstrom, V.,** The myometrial response of intrauterine administration of PGF2 α and PGE2 in dysmenorrheic women, Acta *Obstetric. Gynaecol. Scand.,* 56, 167, 1977.

70. **Lumsden, M. A., Keky, R. W., and Baird, D. T.,** Primary Dysmenorrhoea: the importance of both prostaglandins E2 and F2, *Br. J. Obstetr. Gynaecol.,* 90, 1135, 1983.

71. **Lumsden, M. A.,** Dysmenorrhoea, in *Progress in Obstetrics and Gynaecology,* Vol. 5, Studd, J. W. W., Ed., Churchill Livingstone, Edinburgh, 1985, chap. 18.

72. **Maathius, J. B. and Keky, R. W.,** Concentrations of prostaglandins F2 and E2 in the endometrium throughout the human menstrual cycle, after the administration of clomiphene or an oestrogen-progestogen pill and in early pregnancy, *J. Endocrinol.,* 77, 362, 1978.

73. **Akerlund, M., Anderssen, K., and Ingemarssen, I.,** Effects of terlontaline on myometrial activity, uterine blood flow and lower abdominal pains in women with primary dysmenorrhoea, *Br. J. Obstetr. Gynaecol.,* 83, 673, 1976.

74. **Akerlund, M., Stromberg, P., and Forsling, M. L.,** Primary dysmenorrhoea and vasopressin, *Br. J. Obstetr. Gynaecol.,* 86, 484, 1979.

75. **Kay, C.,** *Oral Contraceptives and Health,* Pitman Medical, London, 1974.

76. **Vessey, M., Doll, R., and Peto, R.,** A long term follow up study of women using different methods of contraception: an interim report, *J. Biosoc. Sci.,* 8, 373, 1976.

77. **Kremser, E., and Mitchen, G. M.,** Treatment of primary dysmenorrhoea with combined type oral contraceptive — a double-blind study, *J. Am. College Hlth. Assoc.,* 19, 195, 1971.

78. **Van Keep, P. A., Greenblatt, R. B., and Albeaux-Fernet, M.,** *Consensus on Menopause Research,* MTP Press, Lancaster, U.K., 1976.

79. **McKinlay, S. and Jeffreys, M.,** An investigation of age at menopause, *J. Biosoc. Science,* 4, 161, 1972.

80. **Jaszmann, L., Van Lith, N., and Zaat, J.,** The perimenopausal symptoms: the statistical analysis of a survey, *Med. Gynaecol. Sociol.,* 4, 268, 1969.

81. **Greene, J. G.,** *The Social and Psychological Origins of the Climacteric Syndrome,* Gower, Aldershot, U.K., 1984.

82. **Greene, J. G. and Cooke, D. J.,** Life stress and symptoms at the climacteric, *Br. J. Psychiatry,* 136, 486, 1980.

83. **Whitehead, M. I.,** The Climacteric, in *Progress in Obstetrics and Gynaecology,* Studd, J. W. W., Ed., Churchill Livingstone, Edinburgh, 1985, chap. 21.

84. **Utian, W. H.,** *Menopause in Modern Perspective: A Guide to Clinical Practice,* Appleton Century Crofts, New York, 1980.

85. **Bungay, G., Vessey, M., and McPherson, C.,** Study of symptoms in middle life with special reference to the menopause, *Br. Med. J.,* 281, 181, 1980.

86. **Hallstrom, T. and Samuelsson, S.,** Mental health in the Climacteric, *Acta Obstetr. Gynaecol. Scand.* (suppl.), 130, 13, 1985.

87. **Clayden, J. R., Bell, J. W., and Pollard, P.,** Menopausal flushing: double-blind trial of a non-hormonal medication, *Br. Med. J.,* 1, 409, 1974.

88. **Campbell, S. and Whitehead, M. I.,** Oestrogen therapy and the menopausal syndrome, *Clin. Obstetr. Gynaecol.,* 4, 31, 1977.

89. **Dennerstein, L., Burrows, G. D., Hyman, G., and Wood, C.,** Menopausal hot flushes — double blind comparison of ethinyloestradiol and norgestal, *Br. J. Obstetr. Gynaecol.,* 85, 85, 1978.

90. **Dennerstein, L., Burrows, G. D., Hyman, G., and Sharpe, K.,** Hormone Therapy and affect, *Maturitas,* 1, 247, 1979.

91. **Studd, J. W. W., Collins, W. P., Chakravati, S., Newton, J. R., Oram, D. H., and Parsons, A.,** Oestradiol and testosterone implants in the treatment of psychosexual problems in postmenopausal women, *Br. J. Obstetr. Gynaecol.,* 84, 314, 1977.

92. **Paffenbarger, R. S.,** Epidemiological aspects of parapartum mental illness, *Br. J. Prevent. Soc. Med.,* 18, 189, 1964.

93. **Kumar, R. and Robson, K. M.,** A Prospective Study of Emotional Disorders in Childbearing Women, *Br. J. Psychiatry,* 144, 35, 1984.

94. **O'Hara, M. W.,** Social support, life events, and depression during pregnancy and the puerperium, *Arch. Gen. Psychiatry,* 43, 569, 1986.

95. **Spitzer, R. L., Endicott, J., and Robins, E.,** Research Diagnostic Criteria: Rationale and reliability, *Arch. Gen. Psychiatry,* 35, 773, 1978.

96. **Goldberg, D. P., Cooper, B., Eastwood, M. R., Kedward, H. B., and Shepherd, M.,** A standardized psychiatric interview for use in community surveys, *Br. J. Prevent. Soc. Med.,* 24, 18, 1970.

97. **Endicott, J. and Spitzer, R. L.,** A diagnostic interview: The Schedule for Affective Disorders and Schizophrenia, *Arch. Gen. Psychiatry,* 35, 837, 1978.

98. **Cox, J. L., Connor, Y., and Kendell, R. E.,** Prospective Study of the Psychiatric Disorders of Childbirth, *Br. J. Psychiatry,* 140, 111, 1982.

99. **Kumar, R.,** Neurotic Disorders in Childbearing Women, in *Motherhood and Mental Illness,* Kumar, R. and Brockington, I. F., Eds., Academic Press, London, 1982.

100. **Watson, J. P., Elliott, S. A., Rugg, A. J., and Brough, D. I.,** Psychiatric Disorder in Pregnancy and the First Postnatal Year, *Br. J. Psychiatry,* 144, 453, 1984.

101. **Cox, J. L.,** Psychiatric morbidity: a controlled study of 263 semi-rural Ugandan women, *Br. J. Psychiatry,* 134, 401, 1979.

102. **Nilsson, A., and Almgren, O.,** Paranatal emotional adjustment. A prospective investigation of 165 women, *Acta Psychiatr. Scand. Suppl.,* 220, 65, 1970.

103. **Dalton, K.,** Prospective study into puerperal depression, *Br. J. Psychiatry,* 118, 689, 1971.

104. **Tod, E. D. M.,** Puerperal depression, a prospective epidemiological study, *Lancet,* 2, 264, 1964.

105. **Meares, R., Grimwade, J., and Wood, C.,** A possible relationship between anxiety in pregnancy and puerperal depression, *J. Psychosom. Res.,* 20, 605, 1976.

106. **Treadway, C. R., Kane, F. J., Jarrehi-Zadeh, A., and Lipton, M. A.,** A psychoneuroendocrine study of pregnancy and the puerperium, *Am. J. Psychiatry,* 125, 1380, 1969.

107. **Nott, P. N., Franklin, M., Armitage, C., and Gelder, M. G.,** Hormonal Changes and Mood in the Puerperium, *Br. J. Psychiatry,* 128, 379, 1976.

108. **Bonham Carter, S. M., Reveley, M. A., Sandler, M., Dewhurst, J., Little, B. C., Hayworth, J., and Priest, R. G.,** Decreased urinary output of conjugated tyramine is associated with lifetime vulnerability to depressive illness, *Psychiatry Res.,* 3, 13, 1980.

109. **Ballinger, C. B., Kay, D. S. G., Naylor, G. J., and Smith, A. H. N.,** Some biochemical findings during pregnancy and after delivery in relation to mood change, *Psychol. Med.,* 12, 549, 1982.

110. **Johnstone, E. C., Owens, D. G., Frith, C. D., McPherson, K., Dowie, C., Riley, G., and Gold, A.,** Neurotic illness and its response to anxiolytic and antidepressant treatment, *Psychol. Med.,* 10, 321, 1980.

111. **Blackwell, B.,** Adverse effects of antidepressant drugs: Part I: Monoamine Oxidase inhibitors and tricyclics, *Drugs,* 21, 201, 1981.

112. **Bivin, G. and Klinger, M.,** *Pseudocyesis,* Principia Press, Bloomington, Ind., 1937.

113. **Cohen, L. M.,** A Current Perspective of Pseudocyesis, *Am. J. Psychiatry,* 139, 1140, 1982.

114. **Schopbach, R. R., Fried, P. H., and Rakoff, A. E.,** Psuedocyesis, a psychosomatic disorder, *Psychosom. Med.,* 14, 129, 1952.

115. **Brenner, B. N.,** Pseudocyesis in blacks, *S. African Med. J.,* 50, 1757, 1976.

116. **Fried, P. H., Rakoff, A. E., and Schopbach, R. R.,** Pseudocyesis, a psychosomatic study in gynaecology, *JAMA,* 145, 1329, 1951.

117. **Zarate, A., Canales, E. S., and Soria, J.,** Gonadotropin and prolactin secretion in human pseudocyesis, *Ann. Endocrinol.,* (Paris), 35, 445, 1974.

118. **Coppola, J. A., Leonardi, R. G., and Lippman, W.,** Induction of pseudopregnancy in rats by depletors of endogenous catecholamines, *Endocrinology,* 77, 485, 1965.

119. **Yen, S. C. C., Rebar, R. W., and Quesenberry, W.,** Pituitary function in pseudocyesis, *Br. J. Clin. Pract.,* 43, 132, 1976.

120. **Starkman, M. N., Marshall, J. C., La Ferla, J., and Kelch, R. P.,** Pseudocyesis: psychologic and Neuroendocrine Interrelationships, *Psychosom. Med.,* 47, 46, 1985.

121. **American Psychiatric Association,** *Diagnostic and Statistical Manual of Mental Disorders III,* 1980.

122. **Rutherford, R. N.,** Pseudocyesis, *N. Engl. J. Med.,* 224, 639, 1941.

123. **Murray, J. L. and Abraham, G. E.**, Pseudocyesis, a review, *Obstetr. Gynaecol.*, 51, 627, 1978.
124. **Pitt, B.**, Maternity Blues, *Br. J. Psychiatry*, 122, 431, 1973.
125. **Hopkins, J., Marcus, M., and Campbell, S. B.**, Postpartum Depression: A Critical Review, *Psychol. Bull.*, 95, 498, 1984.
126. **Handley, S. L., Dunn, T., Baker, J. M., Cockshott, C., and Gould, S.**, Mood changes in the puerperium plasma tryptophan and cortisol, *Br. Med. J.*, 2, 18, 1977.
127. **Robin, A. M.**, Psychological changes associated with childbirth, *Psychiatry Q.*, 36, 129, 1962.
128. **Stein, G. S., Milton, F., Bebbington, P., Wood, K., and Coppen, A.**, Relationship between mood disturbance and free plasma tryptophan in postpartum women, *Br. Med. J.*, 2, 457, 1976.
129. **Yalom, I., Lunde, D., Moos, R., and Hamburg, D.**, Postpartum blues syndrome, *Arch. Gen. Psychiatry*, 18, 16, 1968.
130. **Stein, G. S.**, The pattern of mental change and body weight change in the first postpartum week, *J. Psychosom. Res.*, 24, 165, 1980.
131. **Klaus, M. H., Trause, M. A., and Kennel, J. H.**, Does human maternal behaviour after delivery show a characteristic pattern?, *CIBA Foundation Symposium*, 33, 69, 1975.
132. **Ballinger, C. B., Buckley, D. E., Naylor, G. J., and Stansfield, D. A.**, Emotional disturbance following childbirth: clinical findings and urinary excretion of cyclic AMP (adenosine 3' 5' cyclic monophosphate), *Psychol. Med.*, 9, 293, 1979.
133. **MacFarlane, A.**, *The Psychology of Childbirth*, Fontana Open Books, London, 1979.
134. **Stein, G. S., Marsh, A., and Morton, J.**, Mood, weight and urinary electrolytes in the first postpartum week, *J. Psychosom. Res.*, 25, 395, 1981.
135. **Harris, B.**, Maternity Blues, *Br. J. Psychiatry*, 136, 520, 1980.
136. **Davidson, J. R. T.**, Post-partum Mood Change in Jamaican Women: a Description and Discussion on its Significance, *Br. J. Psychiatry*, 121, 659, 1972.
137. **Kennerley, H. and Gath, D.**, Maternity Blues Reassessed, *Psychiatric Dev.*, 1, 1, 1986.
138. **Kendell, R. E., McGuire, R. J., Connor, Y., and Cox, J. L.**, Mood changes in the first three weeks after childbirth, *J. Affect. Disorders*, 3, 317, 1981.
139. **Handley, S. L., Dunn, T. L., Waldron, G., and Baker, J. M.**, Tryptophan, Cortisol and Puerperal Mood, *Br. J. Psychiatry*, 136, 498, 1980.
140. **Feksi, A., Harris, B., Walker, R. F., Riad-Fahmy, D., and Newcombe, R. G.**, 'Maternity Blues' and hormone levels in saliva, *J. Affect. Disorders*, 6, 357, 1984.
141. **George, A. J., Copeland, J. R. M., and Wilson, K. C. M.**, Prolactin Secretion and the postpartum blues syndrome, *Br. J. Pharmacol.*, 70, 102P, 1980.
142. **Wing, J. K., Cooper, J., and Sartorius, N.**, *The Measurement and Classification of Psychiatric Symptoms*, Cambridge University Press, Cambridge, 1974.
143. **Abou-Saleh, M. T.**, Dexamethasone Suppression Tests in Psychiatry: Is there a place for an Integrated Hypothesis?, *Psychiatric Dev.*, 3, 275, 1985.
144. **Singh, B., Gilhotra, M., Smith, R., Brinsmead, M., Lewin, T., and Hall, C.**, Post-partum Psychoses and the Dexamethasone Suppression Test, *J. Affect. Disorders*, 11, 173, 1986.
145. **Greenwood, J. and Parker, G.**, The Dexamethasone Suppression Test in the Puerperium, *Austral. N. Zealand J. Psychiatry*, 18, 282, 1984.
146. Editorial, Thyroid disease and pregnancy, *Br. Med. J.*, 2, 977, 1975.
147. **How, J. and Bewsher, P. D.**, Thyroid disease in pregnancy, *Br. Med. J.*, 2, 1568, 1978.
148. **Hamilton, J. A.**, *Postpartum Psychiatric Problems*, C. V. Mosby, St. Louis, 1962, chap. 2.
149. **Stein, G., Morton, J., Marsh, A., Hartshorn, J., Ebeling, J., and Desaga, U.**, Vasopressin and Mood During the Puerperium, *Biol. Psychiatry*, 19, 1711, 1984.
150. **Crammer, J. L.**, Disturbance of Water and Sodium in a Manic Depressive Illness, *Br. J. Psychiatry*, 149, 337, 1986.
151. **Riley, D. M.**, A study of serum calcium in relation to puerperal mental illness, in *Emotion(s) and Reproduction*, Carenza, L. and Zichella, L., Academic Press, London, 1979.
152. **Abdulla, Y. H. and Hamadah, K.**, Adenosine monophosphate in depression and mania, *Lancet*, 1, 378, 1970.
153. **Naylor, G. J., Stansfield, D. A., White, S. F., and Hutchinson, F.**, Urinary excretion of Adenosine 3' 5' Cyclic Monophosphate in depressive illness, *Br. J. Psychiatry*, 125, 275, 1974.
154. **Sinanen, E., Keating, A. M. B., Beckett, P. G. S., and Clayton Cove, W.**, Urinary Cyclic AMP in endogenous and neurotic depression, *Br. J. Psychiatry*, 126, 49, 1975.
155. **Gard, P. R., Handley, S. L., Parsons, A. D., and Waldron, G.**, A multivariate investigation of postpartum mood disturbance, *Br. J. Psychiatry*, 149, 567, 1986.
156. **Harris, B.**, Prospective Trial of L-tryptophan in Maternity Blues, *Br. J. Psychiatry*, 137, 233, 1980.
157. **George, A. J. and Wilson, K. C. M.**, Monoamine Oxidase Activity and The Puerperal Blues Syndrome, *J. Psychosom. Res.*, 25, 409, 1981.

158. **Katona, C. L. E., Theodorou, A. E., Davies, S. C., Yamaguchi, Y., Tunnicliffe, C. A., Hale, A. S., Horton, R. W., Kelly, J. S., and Paykel, E. S.,** Platelet binding and neuroendocrine responses in depression, in *The Biology of Depression*, Deakin, J. F. W., Ed., Gaskell, London, 1986, chap. 7.

159. **Metz, A., Stump, K., Cowen, P. J., Elliott, J. M., Gelder, M. G., and Grahame-Smith, D. G.,** Changes in Platelet α_2-Adrenoceptor Binding Postpartum: Possible Relation to Maternity Blues, *Lancet*, 1, 495, 1983.

160. **Best, N. R.,** personal communication, 1987.

161. **Katona, C. L. E., Theodorou, A. E., Missouris, C. G., Bourke, M. P., Horton, R. N., Moncrieff, D., Paykel, E. S., and Kelly, J. S.,** Platelet ^{3}H-Imipramine Binding in Pregnancy and the Puerperium, *Psychiatry Res.*, 14, 33, 1985.

162. **Cox, J. L., Connor, Y. M., Henderson, I., McGuire, R. J., and Kendell, R. E.,** Prospective study of the psychiatric disorders of childbirth by self-report questionnaire, *J. Affect. Disorders*, 5, 1, 1983.

163. **Newnham, J. P., Tomlin, S., Ratter, S. J., Bourne, G. L., and Rees, L. H.,** Endogenous opioids in pregnancy, *Br. J. Obstet. Gynaecol.*, 90, 535, 1983.

164. **Newnham, J. P., Dennett, P. H., Ferron, S. A., Tomlin, S., Legg, C., Bourne, G. L., and Rees, L. H.,** A Study of the relationship between circulating endorphin-like immunoreactivity and postpartum "blues", *Clin. Endocrinol.*, (Oxford), 20, 169, 1984.

165. **Brown, G. and Harris, T.,** *Social Origins of Depression*, Tavistock, London, 1978.

166. **Pitt, B.,** Atypical depression following childbirth, *Br. J. Psychiatry*, 14, 1325, 1986.

167. **Hopkins, J., Marcus, M., and Campbell, S. B.,** Postpartum Depression: A Critical Review, *Psychol. Bull.*, 95, 498, 1984.

168. **Gordon, R. E., Kapostins, E. E., and Gordon, K. K.,** Factors in postpartum emotional adjustment, *Obstetr. Gynaecol.*, 25, 158, 1965.

169. **Cox, J. L.,** *Postnatal Depression: A Guide for Health Professionals*, Churchill Livingstone, Edinburgh, 1986.

170. **Harris, B., Fungh, H., Walker, R., McGregor, A., and Richards, C.,** Postnatal Depression, Saliva, Cortisol and Progesterone and Serum Binding Globulins, *Abstr. Marcé Soc. Conf.*, Nottingham, August, 1986.

171. **Harris, B.,** personal communication, 1987.

172. **Bearn, J., Fairhall, K., and Checkley, S.,** A New marker of Estrogen Receptor Sensitivity with Potential Application to Post Partum Depression, *Abstr. Marcé Soc. Conf.*, Nottingham, August 1986.

173. **Checkley, S.,** personal communication, 1987.

174. **Riley, D.,** unpublished data, 1987.

175. **Holden, J. M., Sagovsky, R., and Cox, J. L.,** personal communication, 1987.

176. **Leverton, T. J. and Elliott, S. A.,** Can a psychosocial intervention prevent postnatal depression?, *Abstr. Marcé Society Conf.*, Nottingham, August 1986.

177. **Leverton, T. J.,** personal communication, 1987.

178. **Finlay-Jones, R., Brown, G. W., Duncan-Jones, P., Harris, T., Murphy, E., and Prudo, R.,** Depression and anxiety in the Community: replicating the diagnosis of a case, *Psychol. Med.*, 10, 445, 1980.

179. **Pitt, B.,** Depression and Childbirth, in *Handbook of Affective Disorders*, Paykel, E. S., Ed., Churchill Livingstone, Edinburgh, 1982.

180. **Loudon, J. B.,** Prescribing in Pregnancy: Psychotropic drugs, *Br. Med. J.*, 294, 167, 1987.

INDEX

Printed and bound by CPI Group (UK) Ltd, Croydon, CR0 4YY

22/10/2024

01777633-0007